# COMPUTATION OF UNSTEADY INTERNAL FLOWS

## Fundamental Methods with Case Studies

# COMPUTATION OF UNSTEADY INTERNAL FLOWS

Fundamental Methods with Case Studies

Paul G. Tucker

KLUWER ACADEMIC PUBLISHERS

**Distributors for North, Central and South America:**
Kluwer Academic Publishers
101 Philip Drive
Assinippi Park
Norwell, Massachusetts 02061 USA
Telephone (781) 871-6600
Fax (781) 681-9045
E-Mail <kluwer@wkap.com>

**Distributors for all other countries:**
Kluwer Academic Publishers Group
Distribution Centre
Post Office Box 322
3300 AH Dordrecht, THE NETHERLANDS
Telephone 31 78 6392 392
Fax 31 78 6546 474
E-Mail <services@wkap.nl>

Electronic Services <http://www.wkap.nl>

**Library of Congress Cataloging-in-Publication Data**

Tucker, P.G.
Computation of unsteady internal flows: fundamental methods with case studies/ P.G. Tucker.
p.cm.
Includes bibliographical references and index.
ISBN 0-7923-7371-5 (alk. paper)
1. Unsteady flow (Fluid dynamics)—Mathematical models. 2. Unsteady flow (Fluid dynamics)—Mathematical models—Case studies. I. Title.

TA357.5U57 T82 2001
620.1'064—dc21

2001025719



*Printed on acid-free paper.*

Printed in the United States of America.

To Gwen, Daisy and Naomi

# CONTENTS

PREFACE ........ xiii
NOMENCLATURE ........ xv

**1 INTRODUCTION ........ 1**
1-1 The Importance of Modelling Unsteady Flows ........ 1
1-2 The Nature and Causes of Unsteadiness ........ 3
1-3 Flow Equations, their Mathematical Nature and Physical Implications 4
1-4 Summary of Book Content ........ 8

**2 OVERVIEW OF ELEMENTARY TEMPORAL DISCRETIZATIONS ........ 11**
2-1 Introduction ........ 11
2-2 Profile Assumptions for Variables ........ 12
*Dependent Variable Changes with Time* ........ 12
*Spatial Variation of the Time Derivative* ........ 13
2-3 The Generic Discretized Equation ........ 14
2-4 Two-level Schemes ........ 17
*Basic Discretization Approach* ........ 17
*Simple Accurate Crank-Nicolson Scheme Linearization Example* ... 18
*Δ Form Based on Differential Equation* ........ 20
*Δ Form Based in Discretized Equation* ........ 20
*Stability and Accuracy of Two Level Schemes* ........ 21
*Two Level Exponential and Modified Crank-Nicolson Schemes* ...... 22
*General Explicit Schemes* ........ 24
*Lax-Wendroff Scheme* ........ 25
2-5 Three-level Schemes ........ 25
*Multipoint Methods* ........ 25
*Leapfrog* ........ 26
*DuFort-Frankel* ........ 26
2-6 Predictor-Corrector Methods ........ 27
2-7 Splitting Methods ........ 29
2-8 Time Marching and Stiffness ........ 30
*Time Marching* ........ 30
*Stiffness* ........ 32
2-9 Characterizing Stability and Accuracy of Schemes ........ 32
2-10 Boundary Conditions and Solution Initialization ........ 33
*Boundary Conditions* ........ 33
*Solution Initialization* ........ 33
2-11 Integration of Particle/Droplet Transport Equations ........ 33
2-12 Two-level Scheme Exemplar ........ 36
*Flow Structure* ........ 36
2-13 Conclusions ........ 41

**3 TEMPORAL AND SPATIAL DISCRETIZATION RELATIONSHIPS** ..... **45**
3-1 Introduction ..... 45
3-2 The Substantial Derivative ..... 46
*Making Flows More Steady* ..... 48
3-3 Convective Term Treatments ..... 50
*General Convective Schemes* ..... 50
*Unsteady Convective Schemes* ..... 54
*Current Convective Schemes* ..... 57
3-4 Dissipation, Dispersion ..... 58
*The Modified Equation* ..... 58
*Dissipation and Dispersion* ..... 59
*Grid Related Effects* ..... 61
*Stability Analysis* ..... 62
*Aliasing* ..... 63
3-5 Moving Boundaries ..... 63
*General Modelling Techniques* ..... 64
*Moving Boundaries with Body Fitted Grids* ..... 65
*Space Conservation Laws* ..... 67
*Inertial Coordinate Systems with Moving Boundaries* ..... 70
*Free Surface Modelling* ..... 71
*Free Surface Tracking Techniques* ..... 72
3-6 Time Adaptive Solutions ..... 74
*Spatial Grid Adaptation with Time* ..... 75
3-7 Conclusions ..... 77

**4 SOLVER ALGORITHMS AND GENERAL SOLUTION PROCEDURES** ..... **79**
4-1 Introduction ..... 79
4-2 Simultaneous Equation Solvers ..... 80
*Diagonal Dominance Issues* ..... 82
*TDMA Solver* ..... 85
*Convergence Criteria* ..... 85
*The Multilevel Algorithm* ..... 86
*Specific Multilevel Convergence Temporal Applications* ..... 90
*The Newton Method* ..... 91
4-3 Parallel Processing ..... 92
*Parallel Processing with Multilevel Convergence* ..... 92
*Domain Decompositions in Time* ..... 92
4-4 Evaluation of the Pressure Field ..... 94
*Streamfunction Vorticity Method* ..... 94
*Basic SIMPLE Type Methods - Including PISO* ..... 95
*SIMPLE2 and SIMPLE** ..... 98
*AVPI Scheme* ..... 99
*Consistent 'Time-Step' variant of SIMPLE* ..... 100

*Fractional Step Methods* ... 101
*Artificial Compressibility Methods* ... 103
*Penalty Function Method* ... 103
*Implicit Continuous-fluid Eulerian (ICE) Method* ... 104
*Fully Coupled Solvers* ... 105
*Reduced Pressures* ... 107
*Pressure/Velocity Locations* ... 107
*Pressure Subcycling* ... 109
4-5 Review of Commercial Programs and their Unsteady Flow Capabilities ... 110
*Codes Considered* ... 110
*General Code Features* ... 110
*Capabilities Specific to Unsteady Flows* ... 113
4-6 Features of Codes used for Case Studies ... 113
4-7 Conclusions ... 115

**5 SOLUTION ADAPTED TIME-STEPS ... 117**
5-1 Introduction ... 117
5-2 Numerical Methods ... 120
*Crude Time-Step Adaptation with no Direct Error Estimate* ... 120
*Time-Step Adaptation with Altered Scheme Estimate* ... 121
*Time-Step Adaptation with Altered Step Error Estimate* ... 122
*Time-Step Adaptation with Altered Order Error Estimate* ... 123
*Evaluating a Representative Solution Error* ... 124
*Relating Solution Error to New Time Steps* ... 126
*Solution Correction* ... 127
*Specific Implementation Details for Schemes.(I) – (IV)* ... 128
*Novel Sub-cycling with Adaptive Time Stepping* ... 130
*Novel Forcing Function Element Adaptation* ... 132
*Diagonal Dominance* ... 135
5-3 Discussion of Results ... 135
*Case (A), Transient Wall Driven Square Box Flow* ... 136
*Case (B), Transient Buoyancy Driven Flow between Cylinders* ... 141
*Case (C), Transient Wall Driven Free Disc Flow* ... 144
5-4 Conclusions ... 149

**6 TURBULENCE MODELLING ... 151**
6-1 Introduction ... 151
6-2 VLES/URANS Modelling ... 152
6-3 LES Modelling ... 155
6-4 Zonal LES/DES ... 156
6-5 URANS Models used in Case Studies ... 157
*Turbulence Averaging* ... 158
*URANS Equations* ... 161
*Turbulence and Scalar Transport Equations* ... 162
*Mixing Length Models* ... 163

*l-v Models* ..... 164
*k-l Models* ..... 164
*k-ε Model* ..... 165
*Zonal Models* ..... 166
*Non-linear eddy viscosity models* ..... 166
*Stagnation Modification* ..... 167
*Calculation of Normal Wall Distances* ..... 168
*Particle Trajectories in Turbulent Flows* ..... 169
6-6 Conclusions ..... 172

**7 CYCLIC ANNULAR CAVITY FLOWS ..... 173**
7-1 Introduction ..... 173
7-2 Numerical Details ..... 174
*Substantial Derivative* ..... 175
*Solution Initialization* ..... 176
*Grid Structure and Numerical Parameter Settings* ..... 176
*Convergence Criteria* ..... 176
7-3 Discussion of Results ..... 177
*Experimental Conditions and Fluid Properties* ..... 177
*Comparisons with Experimental Data* ..... 178
7-4 Conclusions ..... 182

**8 CYCLIC AERO ENGINE MOTIVATED CAVITY FLOWS ..... 183**
8-1 Introduction ..... 183
8-2 Numerical Details ..... 185
*Solution Initialization Boundary Conditions* ..... 186
*Convergence Criteria* ..... 186
*Numerical Parameter Settings* ..... 187
8-3 Lower Reynolds Number Results ..... 187
*Flow Structure Features* ..... 187
*Comparison between Numerical Predictions and Experimental Flow Visualization* ..... 189
*The Flow Drift* ..... 189
*The Substantial Derivative* ..... 190
*Performances of Different Numerical Schemes* ..... 192
8-4 Higher Reynolds Number Results ..... 193
*Computation of Bubble Vortex Breakdown* ..... 193
*URANS/VLES Predictions* ..... 195
8-5 Adaptive Time Stepping ..... 197
8-6 Conclusions ..... 198

**9 CYCLIC MOVING BOUNDARY FLOWS ..... 201**
9-1 Introduction ..... 201
9-2 Numerical Details ..... 203
*Boundary Conditions* ..... 203
*Dynamic Cavitation Modelling* ..... 203

*General Case Details* ... 203
*The Substantial Derivative* ... 204
9-3 Discussion of Results for Journal Bearing ... 205
*Analytical Isothermal Case* ... 205
*Experimental Isothermal Comparison* ... 205
9-4 Discussion of Results for Magnetic Bearing ... 208
9-5 Conclusions ... 210

**10 CYCLIC COMPLEX SYSTEM FLOW ... 211**
10-1 Introduction ... 211
10-2 Numerical Details ... 212
*Wall Distance Algorithm* ... 213
*Boundary Conditions and Solution Initialization* ... 213
*Structure of Grids* ... 214
*General Case Details* ... 215
10-3 Discussion of Results for Fluid Flow ... 216
*Spatial Velocity Variations* ... 216
*Temporal Velocity Variations* ... 219
*Spatial Turbulence Intensity Variations* ... 220
10-4 Adaptive-Time Stepping ... 223
10-5 Discussion of Results for Scalar and Particle Transport ... 224
*Scalar Transport* ... 224
*Particle Transport* ... 226
10-6 Conclusions ... 227

**11 TRANSIENT COMPLEX SYSTEM FLOW ... 229**
11-1 Introduction ... 229
11-2 Numerical Details ... 230
*Boundary Conditions and Solution Initialization* ... 230
11-3 Discussion of Results ... 231
*Steady Flow Comparisons* ... 231
*Transient Flow Comparisons* ... 232
*Adaptive Time Stepping* ... 234
11-4 Conclusions ... 237

**12 OVERALL CONCLUSIONS WITH FUTURE PERSPECTIVES... 239**

**APPENDICES**
**A DISCRETIZATION OF THE GOVERNING EQUATIONS AND MULTILEVEL CONVERGENCE OPERATORS ... 249**
A-1 General Equation Discretization ... 249
*Coefficients and Source Terms for Pressure Correction Equation* ... 255
A-2 Treatment of Variables at the Axis of Rotation ... 256
A-3 Discretization of Earth's Acceleration Due to Gravity ... 257
A-4 The Multilevel Restriction and Prolongation Operators ... 258
*Restriction Operators* ... 258

*Prolongation Operators* ........ 260

**B DERIVATION OF CONDIF SCHEME FOR VARIABLE DENSITY FLUID** ........ **261**

**C MULTILEVEL CONVERGENCE PERFORMANCE EXEMPLARS** ........ **265**
C-1 Driven Box Flow ........ 265
C-2 Grid Embedding ........ 267
C-3 Cluttered Geometries ........ 267
C-4 Grid Aspect Ratio ........ 271
C-5 Heated Rotating Cavity Flow ........ 273
C-6 Conclusions ........ 273

**D FURTHER CHAPTER 10 CASE DETAILS** ........ **275**
*Boundary Conditions and Solution Initialization* ........ 275
*Fluid-Journal Boundary (bc 1)* ........ 275
*Fluid-Groove Interface (bc 3)* ........ 276
*Fluid at Bearing Sides (bc 4)* ........ 276
*Shaft External to Bearing Width (bc 5)* ........ 276
*Bearing Outer Surface and Sides (bc 6)* ........ 276
*Dynamic Cavitation Modelling* ........ 276

**E FLO_CAV PROGRAM USER INFORMATION** ........ **279**
E-1 FLO_CAV Program ........ 279
E-2 Description of Program Structure and Subroutines ........ 279
E-3 Program Input ........ 283
E-4 Boundary Conditions ........ 295
E-5 Post Processing ........ 296

**F FLO_BEARING PROGRAM USER INFORMATION** ........ **299**
F-1 FLO_BEARING Program ........ 299
F-2 Description of Program Structure and Subroutines ........ 299
F-3 Program Input ........ 304
F-4 Boundary Conditions ........ 320
F-5 Input Files ........ 320

**G FLO_ELEC PROGRAM USER INFORMATION** ........ **321**
G-1 FLO_ELEC Program ........ 321
G-2 Description of Program Structure and Subroutines ........ 321
G-3 Program Input ........ 324
G-4 General Details ........ 331

REFERENCES ........ 333
INDEX ........ 349

# PREFACE

Most Engineering flows are unsteady but due to computational expense, this aspect is often ignored. However, with the ever increasing memory (unsteady simulations require more core memory and, if time histories are required, disc space) and processing speed of computers and their decreasing cost, the use of unsteady simulations is rising dramatically. This book presents specialized unsteady flow modelling algorithms, their traits, practical tips relating to their use and brief Case Studies. Algorithms, as used for steady flows, are also described along with their performances (and sometimes extensions) when modelling unsteady flows. Both traditional algorithms and future trends are discussed. Algorithms originating from various fields are presented including those intended for geophysical applications.

The focus of the present text is viscous internal and Newtonian flows. Initial chapters mostly review previous work and present numerical methods. However, these chapters are connected by common buoyancy driven flow exemplar prediction. This is intended to more immediately illuminate points made at a preliminary elementary level. The initial chapters are followed by brief Case Studies generally considering problems of greater complexity and practical significance. The Case Studies chosen reflect the author's recent interests explored at various research groups. These include the Universities of Sussex (Rolls Royce Technology Centre), Bath, Dundee (AT&T Research Centre) and Warwick (Fluid Dynamics Research Centre). The former work relates to rotating turbomachinery flows and the latter electromechanical systems. The Case Studies' presentation is harmonized as much as possible so that concepts can be readily connected and algorithm performances compared. Simple to understand and of a training nature source code and set-up files are included, enabling the reader to reproduce, extend some Case Studies and further explore algorithm performances.

The author has been Chairman of the National Association for Finite Element Methods and Standards (NAFEMS) CFD committee for several years and it is hoped the NAFEMS ethos of using CFD safely has been brought to the present text. A key part of using CFD safely involves understanding the nature of underlying algorithms. The discussion of these, in the initial chapters, is relatively non-mathematical. Any mathematical derivations, it is hoped, are used in a fashion that illuminates understanding of the physical implications of different numerical schemes. Physically intuitive mathematical concepts are used. Although the material is given in an accessible manner the presentation is nonetheless rigorous and thorough.

It is hoped that in reading this book an in-depth understanding of unsteady flow modelling and algorithms will be gained. This will enable selection of suitable algorithms and approaches for particular fields of application. The reader will then be able to begin implementation of algorithms through viewing of original references. Alternatively, the understanding of the behaviour of algorithms gained will allow practitioners to use them more safely in existing (possibly commercial) codes enabling meaningful results to be produced more economically.

In preparing this text, every effort has been made to remove errors of a typographical nature. However, if any errors are discovered, the author would be grateful for feedback. Also, it is the nature of larger codes that, when implemented on different platforms, minor problems can arise and information of this nature would be gratefully received.

The current unsteady flow modelling capabilities of various commercial codes are reviewed. There is now a wide range of quality commercial CFD software available. This makes it impossible to include details of all codes. Also, their capabilities are rapidly changing. It is hoped that no offence is caused by the omission of some commercial code types and information that will inevitably become dated. Bearing in mind the importance of such valuable data this risk was considered worthwhile.

I would like to express my gratitude to past Research Assistants who have helped run some of the simulations contained in this text. Especial thanks are due to Drs D. Adair and Z. Pan. Also, I would like to express thanks to my friends Drs D. Rayner and C. Vaughan for the hours of enjoyable discussions we have had on CFD. Finally, especial thanks are due to Dr Susan Holmes for her help preparing this manuscript.

Paul Tucker
Fluid Dynamics Research Centre
The University of Warwick, UK
Email: P.G.Tucker@warwick.ac.uk

# NOMENCLATURE

| | |
|---|---|
| $a$ | Inner radius |
| $A$ | Main grid point convection-diffusion coefficient, global representation of spatial disctretization or Fourier series coefficient |
| $A_\mu$ | Turbulence model constant |
| $[A]$ | Coefficients matrix for discretized equations |
| $b$ | Outer radius |
| $c_v$ | Specific heat capacity at constant volume |
| $C$ | Courant number ($u\Delta t/\Delta x$), convection coefficient (mass flow per unit length) or passive scalar concentration |
| $C_\mu$, $C_{\mu 1}$, $C_{\varepsilon 1}$, $C_{\varepsilon 2}$ | Turbulence model constants |
| $C_0$, $C_1$, $C_2$, $C_3$,... | Arbitrary constants; also functions |
| $c_p$ | Specific heat capacity at constant pressure |
| $d$ | Coefficient of pressure gradient term |
| $D$ | Time step to diffusion time scale ratio or transport equation diffusion conductance |
| $E = \Omega^2 r^2 /2\ Cp\ \Delta T$ | Eckert number |
| $E_1$,$E_2$,$E_3$ | Velocity components associated with the axis |
| $\boldsymbol{F}$ | Particle force |
| $g$ | Acceleration due to earth's gravity |
| $G=s/b$ | Cavity gap ratio |
| $G_D$ | Grid distortion term |
| $Gr$ | Grashof number |
| $h$ | Local height used to define transformed variable $\eta$ |
| $HOT$ | Higher Order Terms in the Taylor series |
| $i,j,k$ | Array or grid point location identifiers |
| $J$ | Jacobian or total flux (convection + diffusion) |
| $k$ | Thermal conductivity; turbulent kinetic energy or temporal weighting function component |
| $l$ | Turbulence length scale |
| $L$ | Distance scale evaluated from solution of a Poisson equation or characteristic length |
| $m$ | Number of grid levels; also wave number or particle mass |
| $\dot{m}$ | Mass flowrate |
| $M$ | Capacitance or mass operator |
| $n$ | General coordinate direction, |
| $Nu = qL/k\Delta T$ | Nusselt number |
| $p$ | Static pressure |
| $Pe = C/D$ | Peclet number (ratio of the strengths of convection and diffusion) |
| $Pr = \mu c_p/k$ | Prandtl number |
| $q$ | Heat flux |

| | |
|---|---|
| $r, \theta, z$ | Cylindrical polar spatial coordinates |
| $rms_\phi$ | Normalised root mean square change |
| $R$ | Gas constant |
| $\mathbf{R}$ | Local curvature parameter |
| $Re, Re_z, Re_\phi$ | General, axial and rotational Reynolds numbers |
| $Ri$ | Richardson number |
| $Ro = W/\Omega a$ | Rossby number |
| $[R]$ | Residual or defect array |
| $s$ | Axial distance between discs |
| $S_{1,\phi}$ | Primary source term in discrete equation |
| $S_{2,\phi}$ | Secondary source term in discrete equation (which augments $S_{1,\phi}$ when positive and $A_{P,\phi}$ when negative) |
| $Sc = \nu/\Gamma$ | Schmidt number |
| $S_{ij}$ | Mean strain rate tensor |
| $\overset{o}{S}_{ij}$ | Oldroyd derivative |
| $[S]$ | Discretized source term matrix |
| $t$ | Time |
| $T$ | Temperature |
| $Ta = 4\Omega^2 L^4/\nu^2$ | Taylor number |
| $T_i$ | Turbulence intensity |
| $T_i'$ | Total turbulence intensity |
| $u, v, w$ | Instantaneous velocity components |
| $U, V, W$ | Average velocity components |
| $U_o$ | characteristic velocity scale |
| $W_f$ | Weighting function |
| $x, y, z$ | Spatial coordinates |
| $X, Y, Z$ | Dimensionless coordinates |
| $y^+$ | Dimensionless wall distance |
| $\alpha_\phi, \alpha_0, \alpha_1$ | relaxation parameters |
| $\gamma = c_p/c_v$ | Ratio of specific heats |
| $\Gamma$ | Diffusion coefficient |
| $\delta$ | Boundary layer thickness |
| $\Delta t$ | Time-step length |
| $\varepsilon$ | Turbulence dissipation rate or specified error tolerance |
| $\eta$ | Parameter that defines time levels in discretized equations or transformed spatial variable |
| $\lambda$ | Temporal discretization control parameter or viscosity coefficient $(-2\mu/3)$ |
| $\mu$ | Laminar dynamic viscosity |
| $\mu_t$ | Turbulent viscosity |
| $\xi, \eta, \zeta$ | General transformed coordinates |
| $\rho$ | Fluid density |
| $\sigma$ | Diffusion Prandtl number or turbulence fluctuation scale |

| | |
|---|---|
| $\tau$ | Particle relaxation time or transformed temporal coordinate |
| $\phi$ | General variable |
| $\Phi$ | Flow discriminant |
| $\psi$ | Streamfunction |
| $\omega$ | Angular velocity or vorticity |
| $\Omega$ | Vorticity parameter |
| $\mathit{\Omega}$ | Angular velocity |

**Subscripts**

| | |
|---|---|
| $a$ | Pertaining to actual value |
| $ave$ | Average value |
| $f$ | Unsteadiness associated with vacillation |
| $fp$ | Relating to a particular moving fluid particle |
| $g$ | Pertaining to grid |
| $g0$ | Relating to a fixed location |
| $g1$ | Unsteadiness that can be reduced by grid movement |
| $i, j, k$ | Array subscripts pertaining to the axial, radial and tangential directions respectively |
| $ipl, iml$ | i+1 and i-1 respectively |
| $jpl, jml$ | j+1 and j-1 respectively |
| $kpl, kml$ | k+1 and k-1 respectively |
| $m$ | Grid level |
| $min, max$ | Minimum and maximum values |
| $n,e,s,w,f,b$ | Control volume faces neighbouring a central main grid point P |
| $N,E,S,W,F,B$ | Main grid points neighbouring a central main grid point P |
| $NB$ | General neighbour grid point |
| $P$ | Central grid point under consideration |
| $Q_1, q_2, q_3, q_4$ | Pertaining to control volume face |
| $Q_1, Q_2, Q_3, Q_4,...$ | Pertaining to main grid point |
| $ref$ | Reference value |
| $S$ | Pertaining to the surface |
| $SCL$ | Pertaining to quantity obeying the Space Conservation Law |
| $u, v, w$ | Pertaining to listed velocity components |
| $w$ | Pertaining to wall |
| $x, y, z$ | Pertaining to the x, y and z directions, respectively |
| $z, r, \theta$ | Pertaining to the axial, radial and tangential directions respectively |

**Superscripts**

| | |
|---|---|
| $'$ | Correction to approximate value |
| $*$ | Approximate value |
| $+$ | Dimensionless variable |
| $a$ | Approximate value |
| $c$ | Correction to approximate value |

| | |
|---|---|
| *f* | Value originating from a finer mesh |
| *m* | Fine mesh |
| *new* | Pertaining to new value |
| *o* | Value that originates from fine mesh |
| *old* | Pertaining to old value |
| *older* | Pertaining to older value |

**Special symbols**

| | |
|---|---|
| [ I A, B I ] | Maximum value of A and B |
| I ] A, B [ I | Minimum value of A and B |
| $\langle\ \rangle$ | Phase averaged variable |

# CHAPTER 1

# INTRODUCTION

## 1-1 Importance of Modelling Unsteady Flows

Virtually all fluid flows are unsteady. Often, this unsteadiness can be viewed as relating to: stochastic components associated with smaller turbulence scales, larger scale more orderly unsteadiness or a combination of the two. In many cases, especially for complex industrial systems, the larger scale, and lower frequency more orderly components are significant. They can cause relatively large mechanical structure load variations and, if the dominant unsteadiness flow component corresponds to the resonant frequency of the structure, serious damage can occur. A classic example of this is the Ticoma narrows bridge disaster caused by a special kind of von Karman vortex street shedding. Jenssen (1998) gives computations of unsteady bridge aerodynamic loadings. Unsteady flow modelling is also important for understanding drag and aerodynamic phenomena associated with ships (Miyata and Nishimura (1985)) and wings (Barakos and Drikakis (1999), Tatineni and Zhong (2000)). An aspect related to the latter point is the modelling of unsteady vortex wakes from aircrafts (see Steger and Kutler (1977)). These can pose a significant threat to smaller craft in the same airspace. Unsteady flow calculations are further important to public safety through, for example, weather forecasting and prediction of the dispersion of general pollutants. With this respect, real time guidance is the ideal to be aimed for. Hence, partly motivated by this aspect, there is much research into the development of fast, accurate algorithms for modelling unsteady flows. The successful application of a real time dispersion model to the Chernobyl reactor disaster is described by Pudykiewicz (1990). Further environmental, safety related work, is the modelling of flow from ruptured dams (see Harlow and Welch (1965) for a simple related exemplar) and the study of unsteady fire induced flow behaviour (see Mullen, Than and Savilonis (1993)). The prediction of unsteady flows is also important with respect to improving health by assisting with medical research. Its use to further the understanding of pulstatile blood flow through stenosis (constrictions in arteries) and passage bifurcations are described by Deville, Dheur and Vanderschuren (1992), Banerjee, Cho and Back (1993), and Tasciyan, Banerjee and Cho (1993).

Unsteady flow modelling is also important in many areas of general design. For example, the modelling of unsteady engine throttle valve and in-cylinder flows are discussed by Alsemgeest *et al.* (2000), and Gharakhani and Ghoniem (1999), respectively. Unsteadiness can cause significant heat transfer increases (generally the penalty, as with turbulence, is increased pumping losses). Studies, to different degrees, motivated by the idea of using unsteadiness to increase heat transfer in electronics are reported by Ghaddar *et al.* (1986a) (for two-dimensional flow in channels with rectangular blocks using a spectral element method), Ghaddar *et al.* (1986b), Patera and Mikic (1986), Nigen and Amon (1994), Amon (1995) and Nigen and Amon (1995). The latter extend the work to include realistic electronic component thermal modelling. A Case Study on the modelling of unsteady turbulent isothermal flow in a complex electronic system is presented in Chapter 11. The modelling of unsteadiness can also help explain a special failure mechanism for hydrodynamic journal bearings (see de Jongh and Morton (1995)). A Case Study relating to this is discussed in Chapter 10. The modelling of unsteady flow behaviour is also important in aiding the understanding of cooling flows in modern gas turbine aero engines (see Long, Morse and Tucker (1997)). Case Studies relating to this are given in Chapters 8 and 9.

Clearly, the modelling of flow unsteadiness can be a key part of a design analysis. Generally, if an unsteady flow is treated as steady, solution convergence cannot be achieved. Then, frequently, the predicted flow structure and associated parameters may not even be qualitatively correct and be of danger when used as part of a design analysis. The increasing speed and storage of computers along with numerical algorithm improvements is making the simulation of three-dimensional unsteady flows increasingly feasible. This point is partly illustrated by Flood (1985) who presents a graph of weather forecasting error as a function of forecast period for predictions made in 1974 and 1984. This graph is reproduced in *Figure 1-1*. As can be seen, the later model is significantly more accurate and most of this improvement is attributed to better numerical models and also the ability to use finer grids arising from improvements in the power of computers.

Despite these computing power and algorithm improvements, there is evidence to suggest that still, in a significant number of simulations unsteadiness is ignored through inadequate convergence (modulations in solution variables being ignored) or damped through excessively dissipative discretization schemes or models for turbulence.

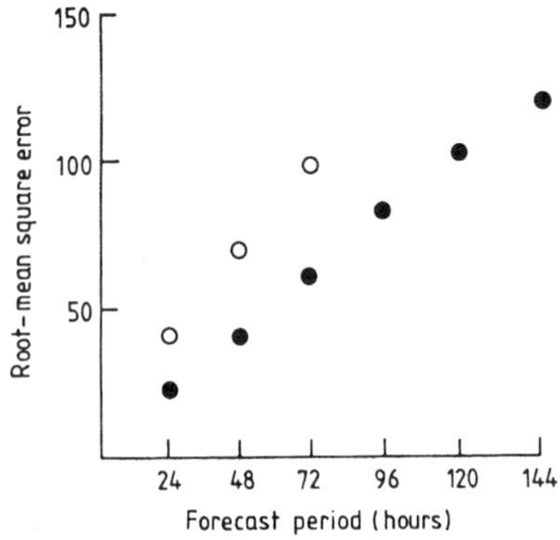


*Figure 1-1* Weather forecasting error as a function of forecast period (o 1974 predictions, • 1984 predictions)

**1-2 The Nature and Causes of Unsteadiness**

Two broad, but key, unsteady flow classifications are:

(I) *Cyclic, Periodic or Oscillatory* flows – these being flows that repeat in time and

(II) *Transient flows* – these being flows that reach a steady state.

A combination of the above is also possible, i.e. a sudden transient leading to a *Cyclic* flow. Alternatively, intermittent forcing can give rise to a series of transient flows which in a sense can be viewed as periodic. An example of this would be the opening and closing of a shutter. Often unsteadiness is a function of the solution frame of reference. For example, the use of inertial coordinate systems, possibly with sliding grids, can give rise to a flow that is steady or quasi-steady relative to the new frame of reference. Such concepts are addressed further in Chapter 3. Also, several Case Studies illustrating this point are presented (see Chapters 7 and 8). Certain problems can also be made steady by variable transformations rendering solutions that are similar in time. An analytical example of this is Thiriot's (1940) solution for flow over a suddenly started or stopped disc. Brebbia and Wrobel (1980) give a boundary element numerical method example relating to the solution of Laplace's equation. These coordinate system transformation points are briefly addressed in Patankar (1980, p. 19).

Unsteadiness can be caused by:

- *Boundary or Component Movement* (i.e. mechanical forcing) such as piston motion in reciprocating combustion engines, the free surface movement arising from the towing of an object or rupturing of a membrane. Arterial flow, where the walls pulsate, is an important unsteady biomedical flow. The boundary/component movement can be tangential or normal to the domain boundaries. Examples relating to the latter and former are given in Chapters 5 and 10, respectively.

- *Instability* related influences such as: Tollmien-Schlichting in boundary layers; baroclinic as found in many geophysical flows (see Chapter 7); hydraulic jumps in free surface flows (where a fluid layer suddenly changes cross-section and hence velocity); shear layers (adjacent fluid layers with quite different velocities) and Rayleigh-Bernard (where less dense fluid lays below more dense – this mechanism plays a part in the Chapter 8 Case Study).

Rockwell and Naudascher (1979) propose that *Oscillatory/Cyclic* flows can be caused by shear layer instabilities with impingement. Evidently, feedback from the impingement region maintains the oscillations. For three-dimensional flows this feedback mechanism is complex. The possibility of oscillatory flow can increase with decreasing viscosity (see Brackenridge (1960)). This suggests that for turbulent, eddy viscosity based predictions oscillatory behaviour will be sensitive to effective viscosity levels.

Generally, when instability controls flow unsteadiness, below some critical parameter (say for example a critical Reynolds number $Re_{cr}$) steady flow will exist. However, for boundary conditions above the critical parameter disturbances are no longer damped and the steady flow *bifurcates* to an unsteady flow solution.

## 1-3 Flow Equations, their Mathematical Nature and Physical Implications

As outlined by Henkes (1990), from partial differential equation theory it is possible that an initial flow is *attracted* to the following solutions as time ($t$) tends to infinity:

(a) steady state attractor;
(b) periodic attractor;
(c) quasi-periodic attractor and
(d) strange attractor.

The latter three are unsteady flows. The periodic attractor corresponds to what is known as a *limit cycle* where the flow structure modulates between distinct states. For the strange attractor, solutions show no distinct peaks when plotted in terms of amplitudes against frequencies (a Fourier spectrum). Therefore, as with turbulence, the strange attractor has a broad-based spectrum. For attractors (b) and (c), Fourier spectrum show distinct peaks. Temporal traces for attractor (c) display the orderly nature of those for (b) along with the less orderly of (d). Importantly, for the same boundary conditions, depending on initial conditions, different solutions (attractors) can be gained. The stability of a solution, i.e. it propensity to stay on a particular *branch* is frequently studied using linear stability theory. With this, a solution's stability with respect to small disturbances is considered. Solutions can require relatively large disturbances (hence requiring a non-linear stability theory to study) to change branch and this is referred to as *hysteresis*. Solutions can change from one branch to another at what are called bifurcation points. Two relatively well known bifurcations are the *Hopf* and *pitchfork*. The former (of most interest here) relates to a steady solution branching to an unsteady periodic solution (i.e. it relates to attractor (b)). The pitchfork relates to a steady solution changing to another steady solution (i.e. an exchange between steady state attractors).

As is traditional, for completeness, the full flow governing equations are now presented. They are given in a form applicable to flows with both density ($\rho$), viscosity ($\mu$), thermal conductivity ($k$) and specific heat ($c_p$) variations. Generally Case Studies are considered where the governing equations are solved in strictly Cartesian or cylindrical coordinate systems. When using geometries amenable to mapping in such coordinate systems, structured solvers, that do not rely on relatively complex grid transformations or other approaches, can be used. This avoids much of the uncertainty on how the spatial discretization impinges on temporal accuracy (see Yao *et al.* (2000)). Here, the formal stagnation enthalpy equation is substituted by one in which the temperature ($T$) is the dependent variable. The Cartesian transport equations forms, for an ($x$, $y$, $z$) coordinate system, where the general variable $\phi = u$, $v$, $w$ or $T$ are as below

$$\frac{\partial}{\partial t}(\rho\phi) + \frac{\partial}{\partial x}(\rho u\phi) + \frac{\partial}{\partial y}(\rho v\phi) + \frac{\partial}{\partial z}(\rho w\phi)$$
$$= \frac{\partial}{\partial x}\left(\Gamma_{\phi x}\frac{\partial\phi}{\partial x}\right) + \frac{\partial}{\partial y}\left(\Gamma_{\phi y}\frac{\partial\phi}{\partial y}\right) + \frac{\partial}{\partial z}\left(\Gamma_{\phi z}\frac{\partial\phi}{\partial z}\right) \tag{1-3.1}$$

The variables $u$, $v$ and $w$ correspond to velocity components in the $x$, $y$ and $z$ directions, respectively. The diffusion coefficients in the above are

$$\Gamma_{ux} = 2\mu+\lambda \;,\; \Gamma_{uy} = \mu \;,\; \Gamma_{uz} = \mu$$

$$\Gamma_{vx} = \mu \;,\; \Gamma_{vy} = 2\mu+\lambda \;,\; \Gamma_{vz} = \mu$$

$$\Gamma_{wx} = \mu \;,\; \Gamma_{wy} = \mu \;,\; \Gamma_{wz} = 2\mu+\lambda \tag{1-3.2}$$

$$\Gamma_{Tx} = k/c_p \;,\; \Gamma_{Ty} = k/c_p \;,\; \Gamma_{Tz} = k/c_p$$

where $\lambda = -2\mu/3$. Corresponding source terms are given below

$$S_u = -\frac{\partial p}{\partial x} + \frac{\partial}{\partial x}\left(\lambda\frac{\partial v}{\partial y}\right) + \frac{\partial}{\partial y}\left(\mu\frac{\partial v}{\partial x}\right) + \frac{\partial}{\partial x}\left(\lambda\frac{\partial w}{\partial z}\right) + \frac{\partial}{\partial z}\left(\mu\frac{\partial w}{\partial x}\right)$$

$$S_v = -\frac{\partial p}{\partial y} + \frac{\partial}{\partial x}\left(\mu\frac{\partial u}{\partial y}\right) + \frac{\partial}{\partial y}\left(\lambda\frac{\partial u}{\partial x}\right) + \frac{\partial}{\partial y}\left(\lambda\frac{\partial w}{\partial z}\right) + \frac{\partial}{\partial z}\left(\mu\frac{\partial w}{\partial y}\right) \tag{1-3.3}$$

$$S_w = -\frac{\partial p}{\partial z} + \frac{\partial}{\partial x}\left(\mu\frac{\partial u}{\partial z}\right) + \frac{\partial}{\partial z}\left(\lambda\frac{\partial u}{\partial x}\right) + \frac{\partial}{\partial y}\left(\mu\frac{\partial v}{\partial z}\right) + \frac{\partial}{\partial z}\left(\lambda\frac{\partial v}{\partial y}\right)$$

For Cartesian coordinate system cases considered here frictional and compressive heating are negligible and $S_T = 0$. The continuity equation relating to the above is

$$\frac{\partial \rho}{\partial t} + \frac{\partial}{\partial x}(\rho u) + \frac{\partial}{\partial y}(\rho v) + \frac{\partial}{\partial z}(\rho w) = 0 \tag{1-3.4}$$

In a cylindrical polar coordinate system the above Cartesian equations require radial weighting terms and become

$$\frac{\partial}{\partial t}(\rho\phi) + \frac{1}{r}\frac{\partial}{\partial r}(\rho r u\phi) + \frac{1}{r}\frac{\partial}{\partial \theta}(\rho v\phi) + \frac{\partial}{\partial z}(\rho w\phi)$$
$$= \frac{1}{r}\frac{\partial}{\partial r}\left(r\Gamma_{\phi r}\frac{\partial \phi}{\partial r}\right) + \frac{1}{r}\frac{\partial}{\partial \theta}\left(\frac{1}{r}\Gamma_{\phi\theta}\frac{\partial \phi}{\partial \theta}\right) + \frac{\partial}{\partial z}\left(\Gamma_{\phi z}\frac{\partial \phi}{\partial z}\right) \tag{1-3.5}$$

where the velocity components $u$, $v$ and $w$ correspond to the radial $r$, tangential $\theta$ and axial $z$ directions respectively. The diffusion coefficients for the above are

$$\Gamma_{ur} = 2\mu + \lambda \ , \ \Gamma_{u\theta} = \mu \ , \ \Gamma_{uz} = \mu$$

$$\Gamma_{vr} = \mu \ , \ \Gamma_{v\theta} = 2\mu + \lambda \ , \ \Gamma_{vz} = \mu$$

$$\Gamma_{wr} = \mu \ , \ \Gamma_{w\theta} = \mu \ , \ \Gamma_{wz} = 2\mu + \lambda \tag{1-3.6}$$

$$\Gamma_{Tr} = k / c_p \ , \ \Gamma_{T\theta} = k / c_p \ , \ \Gamma_{Tz} = k / c_p$$

and the corresponding momentum source terms

$$S_u = -\frac{\partial p}{\partial r} - \frac{2\mu u}{r^2} + \frac{\rho v^2}{r} + u\frac{\partial}{\partial r}\left(\frac{\lambda}{r}\right) - \frac{2\mu}{r^2}\frac{\partial v}{\partial \theta}$$

$$+ \frac{\partial}{\partial r}\left(\frac{\lambda}{r}\frac{\partial v}{\partial \theta}\right) + \frac{\partial}{\partial \theta}\left(\mu\frac{\partial}{\partial r}\left(\frac{v}{r}\right)\right) + \frac{\partial}{\partial r}\left(\lambda\frac{\partial w}{\partial z}\right) + \frac{\partial}{\partial z}\left(\mu\frac{\partial w}{\partial r}\right) + \rho g_r$$

$$S_v = -\frac{1}{r}\frac{\partial p}{\partial \theta} - \frac{\mu v}{r^2} - \frac{\rho u v}{r} - \frac{v}{r}\frac{\partial \mu}{\partial r} + \frac{2}{r^2}\frac{\partial}{\partial \theta}(\mu u) \tag{1-3.7}$$

$$+ \frac{1}{r^2}\frac{\partial}{\partial r}\left(\mu r\frac{\partial u}{\partial \theta}\right) + \frac{1}{r^2}\frac{\partial}{\partial \theta}\left(\lambda\frac{\partial}{\partial r}(ru)\right) + \frac{1}{r}\frac{\partial}{\partial \theta}\left(\lambda\frac{\partial w}{\partial z}\right) + \frac{1}{r}\frac{\partial}{\partial z}\left(\mu\frac{\partial w}{\partial \theta}\right) + \rho g_\theta$$

$$S_w = -\frac{\partial p}{\partial z} + \frac{1}{r}\frac{\partial}{\partial r}\left(\mu r\frac{\partial u}{\partial z}\right) + \frac{1}{r}\frac{\partial}{\partial z}\left(\lambda\frac{\partial}{\partial r}(ru)\right) + \frac{1}{r}\frac{\partial}{\partial \theta}\left(\mu\frac{\partial v}{\partial z}\right) + \frac{1}{r}\frac{\partial}{\partial z}\left(\lambda\frac{\partial v}{\partial \theta}\right)$$

For the flows considered here, compressive heating is insignificant. However, sometimes it is necessary to consider frictional heating. Then

$$S_T = \frac{2\mu}{c_p}\left[\left(\frac{\partial u}{\partial r}\right)^2 + \left(\frac{1}{r}\frac{\partial v}{\partial \theta} + \frac{u}{r}\right)^2 + \left(\frac{\partial w}{\partial z}\right)^2\right] \tag{1-3.8}$$
$$+ \frac{\mu}{c_p}\left[\left(\frac{\partial u}{\partial z} + \frac{\partial w}{\partial r}\right)^2 + \left(\frac{\partial v}{\partial z} + \frac{1}{r}\frac{\partial w}{\partial \theta}\right)^2 + \left(\frac{1}{r}\frac{\partial u}{\partial \theta} + \frac{\partial v}{\partial r} + \frac{v}{r}\right)^2\right]$$
$$+ \frac{\lambda}{c_p}\left[\left(\frac{1}{r}\frac{\partial}{\partial r}(ru) + \frac{1}{r}\frac{\partial v}{\partial \theta} + \frac{\partial w}{\partial z}\right)^2\right]$$

The continuity equation in a cylindrical polar coordinate system is

$$\frac{\partial \rho}{\partial t} + \frac{1}{r}\frac{\partial}{\partial r}(\rho r u) + \frac{1}{r}\frac{\partial}{\partial \theta}(\rho v) + \frac{\partial}{\partial z}(\rho w) = 0 \tag{1-3.9}$$

For the above equations, if the fluid density is constant, all terms involving $\lambda$ vanish. If both the density and viscosity are constant, all terms multiplied by viscosity in $S_\phi$ become zero and all $\Gamma_\phi = \mu$.

The time derivative makes the equations formally *parabolic*. However, it is more helpful to consider the flows here as *elliptic* in space and *parabolic* in time. There are exceptions to this, and for one of the examples in Chapter 5, the flow is parabolic in both space and time. It will be noted that the time derivative of pressure does not appear in the governing equations. Pressure is assumed to change instantaneously.

## 1-4 Summary of Book Content

For unsteady flows, the discretization process for the above equations needs to incorporate assumptions of how variables will change with space, time and also how this time variation varies over each cell. Spatial variable change assumptions are discussed in Chapter 3. There are numerous temporal discretization assumptions available, many originating from considering the solution of ordinary differential equations. The most common schemes, which have found application in CFD (Computational Fluid Dynamics), will be described in Chapter 2. As noted earlier, for a significant number of engineering systems, by careful choice of the solution coordinate system the fluid flow structure can be made steady or steadier. In Chapter 3, through considering the substantial derivative, the relationship between coordinate system choice, unsteadiness levels, and solution accuracy and economy are explored. Also, modifications to convective term treatments, ensuring compatibility with respect to accuracy for unsteady flows are discussed. The close interaction between spatial and temporal discretization schemes and the effects of this on solution accuracy and economy are also addressed in Chapter 3. Furthermore, types of discretization error are formalized. Chapter 3 also gives an overview of moving boundary, free surface flow modelling and adaptive grid techniques.

Once the temporal derivatives in the transport equations have been discretized as in Chapter 2 and the spatial terms as outlined in Chapter 3, the discretized equations must be arranged in a form that will be both *stable* and accurate in the chosen solution framework. They must also be in a form amenable to being solved in an efficient manner that has compatibility with the nature of the fluid flow and the architecture of the computer being used. These aspects, along with pressure evaluation strategies, are dealt with in Chapter 4. This chapter finishes with a review of algorithms used in some relatively well known commercial CFD codes. Importantly, the chapter contents focus around commercially used algorithms. Chapter 5 gives an extensive discussion of standard solution adaptive time-step approaches along with some new novel methods. Chapter 6 presents turbulence models and reviews their performances when modelling unsteady flows. The remaining chapters give Case Studies that apply the algorithms and concepts in the previous chapters to some challenging practically related unsteady flow problems.

# CHAPTER 2

# OVERVIEW OF ELEMENTARY TEMPORAL DISCRETIZATIONS

## 2-1 Introduction

A major element of any CFD procedure is the solution of the convection and diffusion transport equations. These govern such things as momentum and energy conservation and also the transport of passive scalars and certain turbulence parameters. To a lesser extent, CFD predictions can involve the solution of Lagrangian particle transport equations. For unsteady flows (and also the particle transport equations) the temporal discretization process needs to incorporate assumptions of how variables will change with time and also how this time variation varies over each cell. Generally, the process involves two key initial stages. First, just the spatial terms are discretized (again profile assumptions are involved and these are discussed in Chapter 3). These terms are generally represented here by the global symbol $A$. The temporal derivative remains, giving a semi-discrete Ordinary Differential Equation ( $\partial\phi/\partial t = A$ ). This process of reducing the full governing equations to semi-discrete Ordinary Differential Equations (ODEs) is called the *Method of Lines*. The second stage involves discretization of the ODEs. There are numerous discretization assumptions and methods originating from general solution procedures for ODEs. The most common procedures, which have found application in CFD, will be described in this chapter. Also, such things as their stability, accuracy, computational economy and compatibility with different computer architectures will be discussed. Most attention will be focused on schemes widely used in industrial CFD. The effects on the performances of schemes arising from different discretization assumptions will be explored. Efficient discretization strategies will be presented. Also, the relationship between time and space step sizes and stability will be discussed. No formal stability proofs are given. However, the stability limits for all the schemes presented are stated along with general stability traits. Discussion on when and where different schemes are most appropriate along with pros and cons for different schemes and some practical tips are given.

The following generic approaches are presented: two-level, three-level, predictor-corrector and splitting. The use of time marching as a stable means of gaining steady state solutions is also dealt with. Boundary and initial conditions, and their impact on accuracy is discussed. Schemes for the integration of particle/droplet transport equations are then presented. Finally, an example is given, illustrating the performances of some of the two-level schemes described in this chapter.

## 2-2 Profile Assumptions for Variables

*Dependent Variable Changes with Time*

When discretizing the unsteady flow equations an assumption must be made of how dependent variables change with time. Common profile assumptions are given in *Figure 2-2.1*.

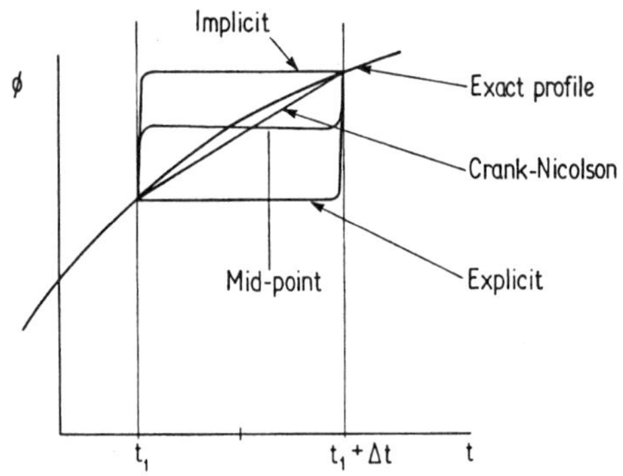


*Figure 2-2.1* Temporal profile assumptions

These can be represented mathematically as

$$\phi = W_f^{\,new}\phi^{new} + W_f^{\,old}\phi^{old} + W_f^{\,older}\phi^{older} \tag{2-2.1}$$

The weighting functions, $W_f$, give the different profiles and we have an:

**(I)** implicit scheme profile for $W_f^{old} = 0, W_f^{new} = 1, W_f^{older} = 0;$

**(II)** trapezoidal for $W_f^{old} = W_f^{new} = 0.5, W_f^{older} = 0;$

**(III)** explicit scheme profile for $W_f^{old} = 1, W_f^{new} = 0, W_f^{older} = 0$ and

**(IV)** the mid-point rule for $W_f^{new} = 0, W_f^{old} = 1$ and $W_f^{older} = 0$

For schemes **(I-III)**, $t_1 = t^{old}$ and $t_1 + \Delta t = t^{new}$ (giving what are called two-level methods, involving $\phi^{new}$ and $\phi^{old}$) and for the mid-point rule **(IV)** $t_1 = t^{older}$, $t_1 + \Delta t/2 = t^{old}$ and $t_1 + \Delta t = t^{new}$.

*Spatial Variation of the Time Derivative*

For consistency with other texts, to identify spatial grid point locations, the primarily geographical notation of Patankar (1980), shown in *Figure 2-2.2* is used. Moving from left to right, main nodal/grid points (where solved for variables are stored) are labelled $W$ (West), $P$ and $E$ (East), respectively. Cell faces are here considered to coincide with the $w$ and $e$ points corresponding to a spatial 'step' of $\Delta x$.

The time derivative, in the governing equations can either be assumed constant over a cell or to vary in some fashion *consistent* with underlying space discretization. With finite volume/difference methods, as used here, the former *lumped approach*, indicated for one-dimension in *Figure 2-2.3*, is adopted. For finite elements, the later *consistent formulation* is common (however, Gresho *et al.* (1984) use a lumped finite element formulation). With this, the time derivative is multiplied by what is called a mass (see Fletcher (1990)) or capacitance matrix (see Segerlind (1984)). Effectively, this spreads the time derivative over several nodes. For a one-dimensional uniform mesh case, involving linear shape functions/elements the capacitance or mass operators are such that $M_W = 1/6, M_P = 2/3, M_E = 1/6.$ This results in the piecewise-linear $\partial(\rho\phi)/\partial t$ profile indicated in *Figure 2-2.3*. The time derivative spreading

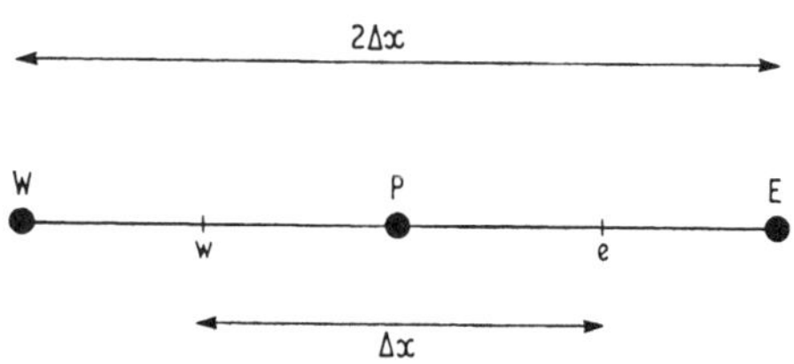


*Figure 2-2.2* Geographic spatial notation

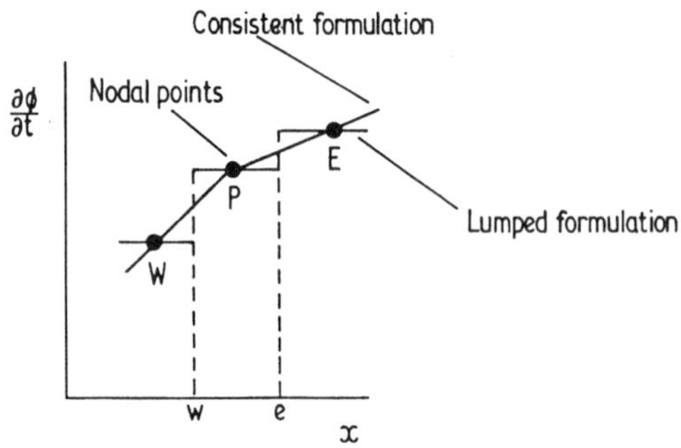


*Figure 2-2.3* Spatial variation of the time derivative

will introduce extra computational cost. However, there are accuracy benefits (see Fletcher (1997a)). It is possible to combine differing discretization techniques, such as having a finite element space discretization with finite difference temporal modelling (see Segerlind (1984)). It is worth pointing out here that what are known as *space-time coupled* finite-element discretizations are a further option. With these the finite element mesh extends into both the space and time domains (i.e. a one-dimensional unsteady flow is mapped using a two-dimensional space time mesh). For three-dimensional flows the mesh/problem size becomes excessively large. However, Bell and Surana (1994) present a procedure where the mesh in the time domain extends as the solution advances. In this way the computational cost is reduced. For some one-dimensional non-linear transport equation problems, even for relatively large time-steps, the method is shown to be accurate and can be considered to be unconditionally stable.

## 2-3 The Generic Discretized Equation

For finite volume, difference and element programs the discretized transport equations can be written with time derivatives released from the source and $A_{p\phi}$ terms, and expressed in two components, as shown below

$$(1+\eta)\frac{\Delta \rho_p \phi_p^{\ new}}{\Delta t} - \eta \frac{\Delta \rho_p \phi_p^{\ old}}{\Delta t} + A_{p\phi}\phi_p = \sum_{NB} A_{NB\phi}\phi_{NB} + S_{1\phi} + S_{2\phi}\phi_P \qquad (2\text{-}3.1)$$

The $\sum_{NB} A_{NB\phi}\phi_{NB}$ term represents the sum of the coefficients next to a node $P$. Unless otherwise stated, in this section, the fluid density $\rho$ is assumed constant and so

$$\frac{\Delta\left(\rho_P \phi_P{}^{new}\right)}{\Delta t} = \rho_P \frac{\left(\phi_P{}^{new} - \phi_P{}^{old}\right)}{\Delta t}, \qquad \frac{\Delta\left(\rho_P \phi_P{}^{old}\right)}{\Delta t} = \rho_P \frac{\left(\phi_P{}^{old} - \phi^{older}\right)}{\Delta t} \tag{2-3.2}$$

For the finite difference based, finite volume discretization of Patankar (1980) the typical form of the coefficients for Equation (2-3.1) are shown in Appendix A.

*Figure 2-3.1* Locations of discretized variables

With a three-dimensional, structured finite difference or volume grid in Cartesian, cylindrical or curvilinear co-ordinates there are generally six neighbouring coefficients. These involve main grid points labelled *W, E, N* (North), *S* (South), *F* (Front) and *B* (Back). Again, respective control volume faces are denoted using lower case letters. The application of the above notation to a cylindrical polar co-ordinate system is shown in *Figure 2-3.1*. The shaded area represents the control volume over which, for computations presented here, the time derivative $\partial(\rho\phi)/\partial t$ is assumed constant. The extension of the notation to Cartesian and curvilinear co-ordinates is evident.

| Scheme | $\eta$ | $\lambda$ | $k$ | Stability* | Order | Other Features |
|---|---|---|---|---|---|---|
| Explicit (Euler) | 0 | 0 | 0 | $0 \le C^2 \le 2D \le 1$ | 1 | - |
| Implicit (Backwards Euler) | 0 | 1 | 0 | Unconditionally Stable | 1 | - |
| Crank-Nicolson (one-step trapezoidal) | 0 | ½ | 0 | Unconditionally Stable | 2 | - |
| Leap frog | -½ | 0 | 0 | Unconditionally Unstable | 2 | - |
| Adams-Moulton | 0 | $5/12$ | $1/12$ | - | 3 | - |
| Galerkin | 0 | $3/5$ | 0 | Unconditionally Stable | 1 | - |
| Exponential | 0 | $\lambda = \frac{1}{1-e^{A_{P\phi}\Delta t}} - \frac{1}{A_{P\phi}\Delta t}$ | 0 | Unconditionally stable | 1 | - |
| Lax-Wendroff | 0 | 0 | 0 | $0 \le C^2 \le 2D^* \le 1$ | - | $\Gamma = \Gamma^*$ |
| DuFort-Frankel | 0 | 0 | 0 | $C < 1$, for accuracy $C^2 \le 1$ | 2 | $\phi_P^* = \frac{\phi_P^{new} + \phi_P^{older}}{2}$ |

*Table 2-3.1* Values of $\eta$, $\lambda$ and $k$ for commonly occurring time schemes (* See section 2-9 for definitions of $C$ and $D$)

Equation (2-3.1) offers the possibility of both two and three-level (involving $\phi^{new}, \phi^{old}, \phi^{older}$) schemes. To implement these, values of $\phi$ not involved in time derivatives $(\Delta(\rho_p \phi_p)/\Delta t)$ are replaced with Equation (2-2.1). The possible number of temporal discretizations available to CFD is vast. Some values of $\eta$, $\lambda$ and $k$, where

$$W_f^{new} = \lambda$$
$$W_f^{old} = 1 - \lambda - k \qquad (2\text{-}3.3)$$
$$W_f^{older} = k$$

for commonly occurring time schemes are given in *Table 2-3.1.* More values/schemes are given by Beam and Warming (1982). Importantly, three-level schemes (which use $\phi^{new}$, $\phi^{old}$ and $\phi^{older}$) require more (approximately 50%) storage, computational effort and attention to the starting solutions. However, computer storage is becoming cheaper and so the former point is becoming less of an issue. Despite this, for industrial computations, mostly two-level schemes are used and so initially these will be addressed. Then multi-point, predictor-corrector and splitting schemes will be discussed.

## 2-4 Two-level Schemes

*Basic Discretization Approach*

For the two-level schemes addressed here in equations (2-3.1) and (2-3.3) $k = \eta = 0$ and $0 \le \lambda \le 1$. Then, the equation resulting from substituting (2-2.1) (which for a two-level scheme can be considered as taking a weighted average of $\phi^{new}$ and $\phi^{old}$) into Equation (2-3.1) can be most usefully and compactly expressed as

$$\overline{A}_{P\phi}\phi_P^{new} = \sum_{NB} \overline{A}_{NB\phi}\,\phi_{NB}^{new} + \overline{S}_\phi \qquad (2\text{-}4.1)$$

Some coefficients of Equation (2-3.1) can be a function of $\phi$ themselves. This occurs in those originating from terms of the form $\partial(\rho u_i u_j)/\partial x_j$, where $i = j$ and hence the transport and transported velocity components are the same. The overbars in Equation (2-4.1) partly acknowledge this, and hence the possibility that some of the coefficients themselves can be discretized semi-implicitly (effectively an averaging process) using Equation (2-2.1). Essentially, the form of the coefficients in Equation (2-4.1) arising from the Equation (2-2.1) substitution are as below.

$$\overline{A}_{P\phi} = \frac{\rho_P^{old}}{\Delta t} + \lambda_P A_{P\phi} \qquad (2\text{-}4.2)$$

$$\overline{A}_{NB\phi} = \lambda_{NB} A_{NB\phi}$$

Assuming, for simplicity, $S_{2\phi} = 0$, the source term is

$$\overline{S}_\phi = S_{1\phi} + (1-\lambda_p)A_{p\phi}\phi_p^{old} + (1-\lambda_{NB})\sum_{NB} A_{NB\phi}\phi_p^{old} + \frac{\rho_p^{old}\phi_p^{old}}{\Delta t} \qquad (2\text{-}4.3)$$

If $S_{2\phi} \neq 0$, again, Equation (2-2.1) could be applied to the $\phi_P$ term pre-multiplied by $S_{2\phi}$. However, for predictions here the $S_{2\phi}$ term is treated in a fully implicit fashion with $W_f^{new} = 1$ and $W_f^{old} = W_f^{older} = 0$ in Equation (2-3.1). Thus, the equations have a mixed discretization. As already noted, coefficients which are a function of $\phi$ themselves can also be expressed semi-implicitly using Equation (2-2.1). They could then have the following form

$$A_{Q1,Q2} = \lambda_{Q1} A_{Q1,Q2}^{new} + (1-\lambda_{Q1}) A_{Q1,Q2}^{old} \qquad (2\text{-}4.4)$$

For cylindrical polar coordinates, as shown in *Figure 2-3.1*, the coefficient subscripts in Equation (2-4.4) would be

$Q1 = W,\ Q2 = w;\ Q1 = E,\ Q2 = w;\ Q1 = F,\ Q2 = v;$
$Q1 = B,\ Q2 = v;\ Q1 = N,\ Q2 = u;\ Q1 = S,\ Q2 = u.$

where the lower case letters refer to velocity components and capitals to grid point locations. For a Cartesian system the following can be written:

$Q1 = W,\ Q2 = u;\ Q1 = E,\ Q2 = u;\ Q1 = F,\ Q2 = w;$
$Q1 = B,\ Q2 = w;\ Q1 = N,\ Q2 = v;\ Q1 = S,\ Q2 = v.$

Equation (2-4.4) is inefficient, since coefficients need to be iteratively updated. Therefore, some form of linearization is preferable.

*Simple Accurate Crank-Nicolson Scheme Linearisation Example*
A linearisation for the problem coefficients, when using the Crank-Nicolson scheme, will now be demonstrated in simple finite difference framework. To illustrate the linearisation the following one-dimensional Euler type equation is sufficient

$$\frac{\partial \phi}{\partial t} + \frac{\partial(\phi\phi)}{\partial x} = 0 \qquad (2\text{-}4.5)$$

For the Crank-Nicolson scheme Equation (2-4.5) can be discretized with the following weighted average of the convective operators corresponding to $\lambda = 0.5$.

$$\frac{\partial \phi}{\partial t}+\frac{\partial}{\partial x}\left(\tfrac{1}{2}(\phi\phi)^{old}+\tfrac{1}{2}(\phi\phi)^{new}\right)=0 \tag{2-4.6}$$

For a one-dimensional *W-P-E* finite difference stencil (see *Figure 2-2.2*) Equation (2-4.6) can be expressed as

$$\frac{\phi_P^{new}-\phi_P^{old}}{\Delta t}+\frac{1}{2}\frac{(\phi_E\phi_E-\phi_W\phi_W)^{old}}{2\Delta x_P}+\frac{1}{2}\frac{(\phi_E\phi_E-\phi_W\phi_W)^{new}}{2\Delta x_P}=0 \tag{2-4.7}$$

Applying the Taylor series to the problem $(\phi\phi)$ products gives

$$(\phi\phi)_P^{new}=(\phi\phi)_P^{old}+\Delta t\left(\frac{\partial(\phi\phi)}{\partial t}\right)_P^{old}+\Delta t^2\left(\frac{\partial(\phi\phi)}{\partial t^2}\right)_P^{old} \tag{2-4.8}$$

Still maintaining second order accuracy, to be compatible with the Crank-Nicolson scheme, Equation (2-4.8) can be re-expressed as

$$(\phi\phi)_P^{new}=(\phi\phi)_P^{old}+\left(\frac{\partial\phi\phi}{\partial\phi}\right)_P^{old}\left(\phi_P^{new}-\phi_P^{old}\right)+O\left(\Delta t^2\right) \tag{2-4.9}$$

where $(\partial(\phi\phi)/\partial\phi)_P^{old}$ is called a Jacobian ($J$). Since the Jacobian $J=(\partial(\phi\phi)/\partial\phi)_P^{old}=2\phi_P^{old}$ Equation (2-4.9) can be simplified to

$$(\phi\phi)_P^{new}=2\left(\phi^{old}\phi^{new}\right)_P-(\phi\phi)_P^{old} \tag{2-4.10}$$

Substituting Equation (2-4.10) into (2-4.7) gives the following

$$\frac{\phi_P^{new}-\phi_P^{new}}{\Delta t}+\frac{1}{2\Delta x_p}\left[\left(\phi^{old}\phi^{new}\right)_E-\left(\phi^{old}\phi^{new}\right)_W\right]=0 \tag{2-4.11}$$

The linearised Equation (2-4.11) is second order accurate and clearly cheaper to solve. The method just described above relates to the work of Beam and Warming (1976) and an example of its application, which is widespread, is given by Steger and Kutler (1977). An alternative technique is described by Briley and McDonald (1975). More details relating to the Beam and Warming approach are given below.

*Δ Form Based on Differential Equation*
The differential Equation (2-4.5) itself can more generally expressed in the above linearized form as follows. Firstly, the equation is rewritten as

$$\frac{\partial \phi}{\partial t} + \lambda \frac{\partial F^{new}}{\partial x} + (1-\lambda)\frac{\partial F^{old}}{\partial x} = 0 \tag{2-4.12}$$

where here $F = \phi\phi$. Equation (2-4.12) can then be rearranged as

$$\frac{\partial \phi}{\partial t} + \lambda\left(\frac{\partial F^{new}}{\partial x} - \frac{\partial F^{old}}{\partial x}\right) = -\frac{\partial F^{old}}{\partial x} \tag{2-4.13}$$

Based on Equation (2-4.9) the following can be stated

$$F^{new} = F^{old} + J(\phi)^{old}\left(\phi^{new} - \phi^{old}\right) \tag{2-4.14}$$

From Equation (2-4.14) it can be written that

$$\frac{\partial F^{new}}{\partial x} = \frac{\partial F^{old}}{\partial x} + \frac{\partial}{\partial x}\left(J(\phi)^{old}\,\Delta\phi\right) \tag{2-4.15}$$

Substituting Equation (2-4.15) into (2-4.13) allows the equation below to be written

$$\frac{\partial \phi}{\partial t} + \lambda \frac{\partial}{\partial x}\left(J(\phi)^{old}\,\Delta\phi\right) = -\frac{\partial F^{old}}{\partial x} \tag{2-4.16}$$

Since $\Delta\phi$ appears as a dependent variable, the above equation is sometimes referred to as being in the Delta (*Δ*) form.

*Δ Form Based on Discretized Equation*
With $k = \eta = 0$ and the variable *A* defined as

$$A = \sum_{NB} A_{NB\phi}\phi_{NB} - A_P\phi_P + S_{1\phi} + S_{2\phi}\phi_P \tag{2-4.17}$$

Equation (2-3.1) simplifies to

$$\frac{\Delta\phi}{\Delta t} = A \tag{2-4.18}$$

or

$$\frac{\Delta\phi}{\Delta t} = \lambda A^{new} + (1-\lambda)A^{old} \tag{2-4.19}$$

Again based on Equation (2-4.9) the following can be written

$$A^{new} = A^{old} + J(\phi)^{old}\Delta\phi + O(\Delta t^2) \tag{2-4.20}$$

where

$$J(\phi)^{old} = \frac{\partial A}{\partial \phi} \tag{2-4.21}$$

Substitution of Equation (2-4.20) into (2-4.19) gives the following

$$\left[1 - \lambda\Delta t J(\phi)^{old}\right]\Delta\phi = \Delta t A^{old} \tag{2-4.22}$$

$\Delta$ form of the discretized equation. Many other approaches to address the linearization problem are possible (see Lacor (1999)). For example, Jameson (1991) introduces a derivative of pseudo time.

From the preceding discussion it can be seen that essentially the same scheme can be discretized in different ways. Therefore, for the same parameter settings different codes can have different accuracies and rates of convergence. Generally, with commercial codes, robustness is a key issue and this can be at the sacrifice of accuracy.

*Stability and Accuracy of Two-level Schemes*

For $0.5 \le \lambda \le 1$, two-level schemes are unconditionally stable. This does not mean that for $\lambda < 1$ solutions will be physically realistic but instead that errors will reduce with time. This can be appreciated by following Patankar and Baliga (1978), and considering the finite difference solution of the unsteady one-dimensional diffusion equation

$$\frac{\partial \phi}{\partial t} = \Gamma \frac{\partial^2 \phi}{\partial x^2} \tag{2-4.23}$$

on the uniform grid given in *Figure 2-2.2.* For this simple case, the coefficients in Equations (2-4.2) and (2-4.3) are

$$A_{E\phi} = A_{W\phi} = \frac{\Gamma}{(\Delta x)^2} \tag{2-4.24}$$

and

$$A_{P\phi} = A_{E\phi} + A_{W\phi} = \frac{2\Gamma}{(\Delta x)^2} \tag{2-4.25}$$

Considering a discretized spatial domain consisting of just 3 grid points with for $t \geq 0, \phi_W = \phi_E = 0$, and $t = 0$, $\phi_P > 0$; Equation (2-4.1) multiplied by $\Delta t$ reduces to

$$\frac{\phi_P^{new}}{\phi_P^{old}} = \frac{1 - A_{P\phi} \Delta t (1 - \lambda_P)}{1 + \lambda_P A_{P\phi} \Delta t} \tag{2-4.26}$$

For the scheme to behave physically sensibly $\phi_P^{new} / \phi_P^{old}$ should stay positive and tend to zero as $A_{P\phi}\Delta t$ becomes larger. Again following Patankar and Baliga, *Figure* 2-4.*1* shows the variation of $\phi_P^{new} / \phi_P^{old}$ with increasing $A_{P\phi}\Delta t$ for $\lambda_P = 0, 0.5, 0.6$ and $1$ corresponding to explicit, Crank-Nicolson, Galerkin and implicit schemes. Also, the line labelled 'exponential scheme' represents the exact solution. Clearly, as $\Delta t$ becomes large the Crank-Nicolson scheme will give physically unrealistic solutions.

*Two-level Exponential and Modified Crank-Nicolson Schemes*

The exact solution for the variation of $\phi_P^{new} / \phi_P^{old}$ is easily deduced by writing the discretized transport Equation (2-3.1) as

$$\frac{\partial \phi_P}{\partial t} = -A_{P\phi} \phi_P \tag{2-4.27}$$

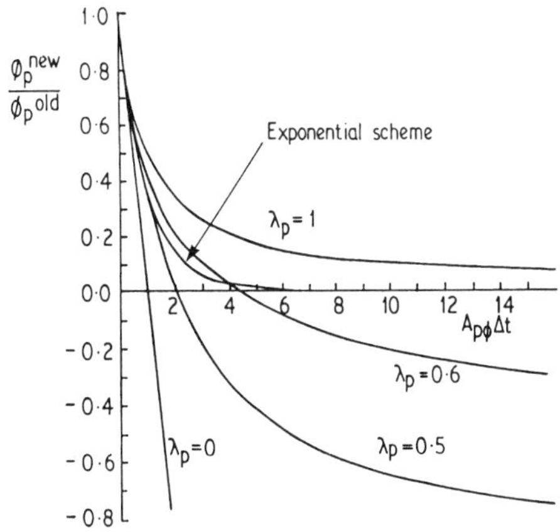


*Figure 2-4.1* Variation of $\phi_p^{new}/\phi_p^{old}$ with increasing $A_{p\phi}\Delta t$ for $\lambda = 0$, 0.5, 0.6 and 1 (Adapted from Patanker and Baliga (1978))

The above comes from Equation (2-3.1) with $\eta = S_{1\phi} = S_{2\phi} = 0$, $\Delta\left(\rho\phi_p^{new}\right)/\Delta t$ replaced with $\partial\phi_p / \partial t$ and the effective consequences of $\phi_W = \phi_E = 0$. Separating the variables and integrating gives

$$\frac{\phi_P^{new}}{\phi_P^{old}} = e^{-\left(A_{P\phi}\right)\Delta t} \tag{2-4.28}$$

Comparing Equations (2-4.26) and (2-4.27) enables the following to be written

$$\lambda_P = \frac{1}{1 - e^{-A_{P\phi}\Delta t}} - \frac{1}{A_{P\phi}\Delta t} \tag{2-4.29}$$

By simply replacing subscripts, similar expressions can be stated for the $\lambda_{NB}$'s. The use of Equation (2-4.29) in Equations (2-4.2) - (2-4.4) gives the exponential scheme of Patankar and Baliga (1978) where $\lambda$ is a function of both space and time.

For Equation (2-4.29), when $A_{p\phi}\Delta t$ is large, $\lambda_p$ tends to unity giving the fully implicit scheme. As $A_{p\phi}\Delta t$ becomes small, it tends to 0.5, corresponding to the Crank-Nicolson Scheme. In practice, exponentials are expensive to compute and are best approximated using power law expressions. From a practical viewpoint the exponential/power law scheme is first order accurate. Although robust and interesting its use is probably best avoided. Like its exponential and power law spatial equivalents (see Patankar (1980)), its benefits are essentially cosmetic - temporal solution wiggles being surpressed. Since these wiggles are a good indicator that $\Delta t$ is too large this is not desirable. A solution with damped wiggles is in no way superior to a solution with wiggles. Notably, Hujeirat and Rannacher (1998) (for predictions where $\Delta t$ is adapted based on solution errors) use the following modified form of the Crank-Nicolson scheme

$$\lambda = \frac{(1 + \alpha\Delta t)}{2} \tag{2-4.30}$$

The term $\alpha\Delta t$ (where $\alpha$ is a free positive parameter) is added to damp wiggles (which can affect the Crank-Nicolson scheme's long time-period integration properties). Since $\lambda = f(t)$, $\lambda$ is temporally varying and this approach has some similarities to the above exponential scheme where $\lambda$ is also a function of time. However, unlike the exponential scheme, the use of Equation (2-4.30) (which in the discretization process itself becomes multiplied by $\Delta t$) still maintains second order accuracy.

*General Explicit Schemes*

As noted earlier, schemes for $0.5 \leq \lambda \leq 1$ give unconditionally stable solutions. When $\lambda = 0$ (with $k = \eta = 0$) the explicit forward difference Euler scheme is gained. This first order scheme is stable for $C^2 \leq 2D \leq 1$ where $C = u\Delta t/\Delta x$ and $D = \Gamma\Delta t/\rho\Delta x^2$ (the parameters $C$ and $D$ will be discussed in more detail in Section 2-9). The restriction $2D \leq 1$ implies that when the mesh spacing is halved the time-step must be made four times smaller. This is a severe time-step restriction. Therefore, the explicit Euler scheme is only suitable for problems where high temporal resolution is required. For problems where $\phi$ is changing slowly with time, the time-step restriction makes the scheme inefficient and implicit schemes should be used. Generally, implicit schemes are more stable than explicit schemes.

Higher order explicit schemes can be implemented. However, generally the higher the scheme's order, the more severe the stability restriction. Where diffusion is high, stability restrictions are worsened (this can be seen be considering the parameter $D$). However, explicit methods do not require simultaneous equation solvers. Therefore, they lend themselves to vector and parallel processing. Consequently, the advent of parallel and vector machines has stimulated greater interest in explicit methods. For many industrial CFD applications, the stability of implicit schemes is attractive and for cases where relatively large time-steps can be used they are efficient. Hence, in the majority of general purpose industrial CFD codes implicit schemes are used. To finish this section on two-level schemes, for completeness, the explicit Lax-Wendroff scheme is now described.

*Lax-Wendroff Scheme*

The Lax-Wendroff scheme is only of practical use for inviscid compressible flows. It essentially involves producing a second order Taylor series approximation to the time derivative giving

$$\frac{\partial \phi_p}{\partial t} \simeq \frac{\phi_p^{new} - \phi_p^{old}}{\Delta t} - \left(\frac{\Delta t}{2!}\right) \frac{\partial^2 \phi}{\partial t^2} \qquad (2\text{-}4.31)$$

Since $\Delta t = \Delta x/u$, the latter term in the above can be expressed as $(0.5\,\Delta t)\,u^2$ $\partial^2\phi/\partial x^2$. This extra diffusion like term can be accounted for using a modified diffusion coefficient $\Gamma^* = \Gamma + (0.5\Delta t)u^2$ in the discretized transport equation coefficients ( $\Gamma$ is an already present coefficient). The Lax-Wendroff scheme can be implemented using Equation (2-3.1) with $\eta = \lambda = k = 0$. A key feature of the Lax-Wendroff scheme is that temporal influences are accounted for through spatial terms. This idea is explored further in Chapter 3.

## 2-5 Three-Level Schemes

*Multipoint Methods*

As the name suggests, multipoint methods use data stored at several (> 2) levels/points. Generally for industrial CFD applications, no more than 3-4 levels are implemented. For three points, in the present nomenclature these will be the 'new', 'old' and 'older' levels. The most common multipoint methods are the Adams types. Using polynomial temporal fits both explicit and implicit Adams type schemes can be constructed. The former are referred to as Adams-Bashforth and the latter Adams-Moulton methods. Adams methods are relatively easy to implement but clearly require extra storage. Also, at the start of computations, solutions at several levels must be available. This can present difficulties, since the quality of the initial solution can have a profound influence on the solution's success. The starting solutions can be produced using lower order two-level schemes. However, time-steps must be small enough to give solutions consistent with the accuracy of the multipoint scheme being used. Also, as computations progress in time, with multipoint methods non-physical solutions can be produced. Periodically stopping and restarting solutions can overcome this difficulty, but clearly this action reduces certainty regarding computational accuracy and can affect the algorithms efficiency. In conjunction with Equation (2-3.1) various Adams multipoint methods can be gained by selecting different values of $\eta, \lambda$ and $k$. Parameter values, for the implicit 3rd order Adams-Moulton method are given in *Table 2-3.1*. Values for other schemes are given by Beam and Warming (1982). Flow predictions showing the application of the Adams-Bashforth method are given by Kim and Moin (1985). The three-level multipoint leapfrog scheme will now be discussed.

*Leapfrog*

The leapfrog scheme approximates the time derivative using a second order central difference of the form

$$\frac{\partial \phi}{\partial t} = \frac{(\phi^{new} - \phi^{older})}{2\Delta t} \tag{2-5.1}$$

The time derivative leaps the spatial operators, which are approximated at the $t^{old}$ level, in an uncoupled fashion. The scheme corresponds to Equations (2-3.1) and (2-3.2) with $\eta = 1/2, \lambda = k = 0$. Strictly, for purely diffusion problems, leapfrog should be called the *Richardson method*, but this method is of no practical use, being unconditionally unstable. For pure convection problems when $C \leq 1$ Leapfrog is stable. However, the lack of coupling between the old and new/older levels can give rise to computational difficulties and averaging of solutions at the new and old levels can be required. Another problem, is that to start leapfrog, two solution levels are required. Furthermore, unmodified, the leapfrog scheme is unconditionally unstable for convection-diffusion transport equations. For industrial CFD its practical use is limited. However, surprisingly, with modifications the scheme is used for meteorological (see Hignett *et al.* (1985)) and oceanographic predictions.

*DuFort-Frankel*

The DuFort-Frankel is a three-level, multipoint, second order scheme. It addresses the leapfrog scheme's decoupling problem. This is achieved by defining the variable

$$\phi_p^* = \frac{(\phi_p^{new} + \phi_p^{older})}{2} \tag{2-5.2}$$

which is substituted for $\phi_p^{old}$ in the diffusion related spatial operators. Again, because it is a three-level scheme, starting solutions is awkward. For convection-diffusion equations, the scheme is stable for $C \leq 1$. However, for accuracy there is the severe restriction $C^2 << 1$. This arises from the substitution of $\phi_p^*$ and limits the schemes practical use.

**2-6 Predictor-corrector Methods**

A well-known member of the class of predictor-corrector methods are the Runge-Kutta schemes. These are a general technique for the solution of ordinary differential equations and have been used for some specialised CFD applications (see Jameson *et al.* (1981), Schmidt and Jameson (1982) and Satofuka (1986)). The explicit Euler method is in fact part of the Runge-Kutta family of schemes. It can be classified as a single stage first order explicit Runge-Kutta scheme. However, generally more than one stage is used. Both explicit and implicit formulations are possible, but, for implicit schemes, simultaneous equation sets must be solved at each stage, making them computationally expensive. Essentially, the Runge-Kutta technique can be considered as using a carefully selected blend of lower order methods to give a desired higher order of accuracy. The 4$^{th}$ order, 4-stage technique is popular and is used in particle integrations to be presented later. This technique will now be described. To give insight this is done in simple graphical sense.

Dividing Equation (2-3.1) by $\rho$, setting $\eta = 0$ and grouping terms involving $A_{p,\phi}$, $A_{NB,\phi}$, $S_{1\phi}$ and $S_{2\phi}$ into the single function $A$ (cf. Equation (2-4.18)) gives

$$\frac{\partial \phi}{\partial t} = A \qquad (2\text{-}6.1)$$

With the Runge-Kutta approach the numerical solution $\phi^{new}$ to Equation (2-6.1) can be viewed as a weighted average of four solutions $\phi_1$, $\phi_2$, $\phi_3$ and $\phi_4$ such that

$$\phi^{new} = \phi^{old} + (\frac{1}{6}\phi_1 + \frac{1}{3}\phi_2 + \frac{1}{3}\phi_3 + \frac{1}{6}\phi_4) \qquad (2\text{-}6.2)$$

The estimates $\phi_1$ to $\phi_4$ involve gradients taken at points (1) to (4) shown in *Figure 2-6.1* extrapolated through the ($t^{old}$, $\phi^{old}$) point. Evaluation of the solution estimates is summarised below:

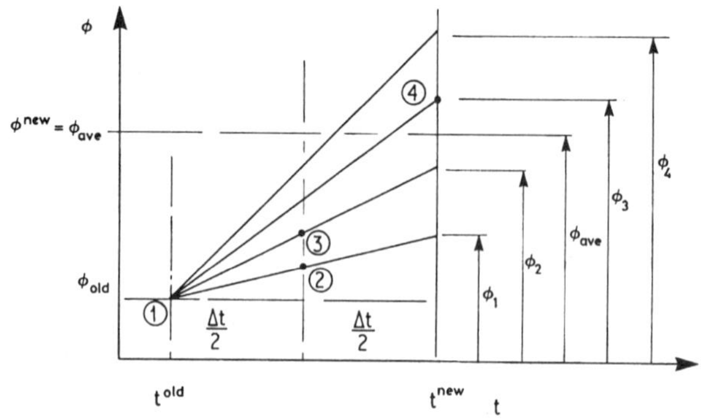


*Figure 2-6.1* Graphical interpretation of 4th order, 4 stage Runge-Kutta technique

(I) Gradient (1) estimate (forward difference based)

$$\phi_1 = \phi^{old} + \Delta t\ A\ (t^{old},\ \phi^{old}) \tag{2-6.3}$$

(II) Gradient (2) estimate (a centered estimate)

$$\phi_2 = \phi^{old} + \Delta t\, A\left( t^{old} + \frac{\Delta t}{2}, \frac{\phi_1 + \phi^{old}}{2} \right) \tag{2-6.4}$$

(III) Gradient (3) estimate (a centred estimate)

$$\phi_3 = \phi^{old} + \Delta t A\left( t^{old} + \frac{\Delta t}{2}, \frac{\phi_2 + \phi^{old}}{2} \right) \tag{2-6.5}$$

(IV) Gradient (4) estimate (backwards difference based)

$$\phi_4 = \phi^{old} + \Delta t\ A(\, t^{new}, \phi_3\, ) \tag{2-6.6}$$

As can be seen, repeated evaluations of the coefficients of the discretized equations are required. For each time-step integration, this will be the major expense. However, an advantage of the Runge-Kutta method is that high temporal accuracy is gained without the need, as for multipoint methods, for extra storage. Also, unlike multipoint schemes, there is no problem starting solutions. Importantly, for the same order (see Ferziger and Peric (1997)),

Runge-Kutta schemes are more stable and accurate (the constant multiplying the truncation error is smaller) than multipoint schemes.

## 2-7 Splitting Methods

A disadvantage of implicit methods is the expense of solving large simultaneous equation sets. For explicit methods this is not necessary, but instead there are greater stability restrictions. With splitting methods, implicit and explicit schemes are combined. In certain circumstances improvements in computational performance are made. The most widely known splitting method is the Alternating Direction Implicit (ADI) scheme of Peaceman and Rachford (1955). With this, in two-dimensions, one full time-step $\Delta t$, essentially consists of two $\Delta t/2$ half steps. For the first half time-step, spatial derivatives in say a *W-E* coordinate direction are treated implicitly while those in the *S-N* direction are treated explicitly. For the second half step, this procedure is reversed. For each half step, a simultaneous equation set needs to be solved for just one coordinate direction. This can be done efficiently using Gaussian elimination, which is also used one-dimensionally as part of the efficient TDMA algorithm (see Section 4-2). The two-dimensional ADI scheme is gained by considering Equation (2-3.1) with $\eta = 0$. After splitting the $\Sigma A_{NB}$ operator into $\Sigma A_{WE}$ and $\Sigma A_{NS}$ components the following can be written:

$$\rho\frac{(\phi_p^{old} - \phi_p^{older})}{\Delta t/2} + A_{p\phi}\phi_p^{old} = \sum\nolimits_{WE} A_{NB\phi}\phi_{NB}^{old} + \sum\nolimits_{NS} A_{NB\phi}\phi_{NB}^{older} + S_{1\phi} + S_{2\phi}\,\phi_p^{old} \qquad (2\text{-}7.1)$$

for the first half step, where $\phi_p^{old}$ is solved for, and

$$\rho\frac{(\phi_p^{new} - \phi_p^{old})}{\Delta t/2} + A_{p\phi}\phi_p^{new} = \sum\nolimits_{WE} A_{NB\phi}\phi_{NB}^{old} + \sum\nolimits_{NS} A_{NB\phi}\phi_{NB}^{new} + S_{1\phi} + S_{2\phi}\phi_p^{new} \qquad (2\text{-}7.2)$$

for the second step where $\phi_p^{new}$ is solved for. In two-dimensions ADI is unconditionally stable, second order accurate and economical. For three-dimensions, three $\Delta t/3$ steps are required and the scheme is still economical, but conditionally stable. From Equation (2-3.1), other splitting schemes can be derived. For example, with $\eta = k = 0$ and $\lambda \geq 0.5$, Douglas and Gunn, (1964) presents a second order splitting scheme which in three-dimensions is unconditionally stable. Three-level ($k \neq 0$) splitting schemes are also possible (see Fletcher (1997a)). The successful application of ADI to transient fluid flow problems is shown by Briley and Walls (1970), and Chergui and Daube (1989).

## 2-8 Time Marching and Stiffness

*Time Marching*

Unless there are obvious reasons, an effective strategy is to model all flows as unsteady. This is consistent with reality, where generally most flows are of this nature. Such an approach can avoid frequent fruitless attempts to gain converged solutions that, due to unsteadiness, cannot be supported by the governing equations. In an iterative sense, with under-relaxation, steady state solutions have similarities to unsteady computations. Therefore, the 'steady', under-relaxed solution, to an unsteady flow may exhibit temporal-like oscillations in residuals and dependent variables. This is a useful clue that a flow is in reality unsteady.

For stability reasons, steady flows are often modelled using unsteady solution procedures; convergence to the steady state being carried out via a series of semi-converged time-steps. This technique is sometimes called *time marching*. One advantage of time marching is that it allows the application of efficient time-dependent splitting approaches such as ADI and other efficient temporal methods, for steady flows. Also, it enables in the absence of under-relaxation, the use of certain iterative solution techniques which could not otherwise be applied to non-diagonally dominant equations (see Section 4-2). Obviously, when using time marching, for efficiency, it is desirable to make the time-steps as large as possible. However, generally, different regions for a particular flow will have different unsteadiness time scales. For example, see Fletcher (1997a), with flow over a backward facing step, convergence in the recirculation region is generally slower than elsewhere. Therefore, maximum efficiency is gained using spatially varying time-steps. Sedaghat, Ackroyd and Wood (1999), and Yee and Harten (1987) use the following flow structure independent spatially varying time-step ($\Delta t_P$) prescription

$$\Delta t_P = \frac{\alpha_{\Delta t}}{1 + \sqrt{1/V}} \tag{2-8.1}$$

where $\alpha_{\Delta t}$ is a constant, tuned for best convergence, and $V$ is the cell volume. For a semi-implicit method typically $\alpha_{\Delta t} = 0.3 - 0.5$. Examples of the application of varying/distorted time-step approaches when time marching are given by De Socio, Sparrow and Eckert (1973), and Mallinson and De Vahl Davis (1973). The former use a fully explicit time marching scheme with time-step selection based on stability limits. The latter use an ADI related approach with distorted time-steps and also different time-steps for different equations. As can be seen, there are a significant number of possibilities for time-step selection and use, and the performances of different approaches is most probably highly problem dependent.

Runge-Kutta schemes are also sometimes used for time marching. When this is the case they are best used in more stable modified forms that have good *smoothing* properties and hence compatibility with multilevel convergence acceleration (see Section 4-2). Time-marched, Runge-Kutta procedures essentially consist of a series of explicit forward difference Euler solutions with cyclic time-step values. A technique to improve stability and smoothing is to split the right-hand side of Equation (2-6.1) into terms associated with for example convection ($A_c$), diffusion ($A_d$) and false diffusion ($A_{fd}$) (see Section 3-4) and then take a blend of these terms from different stages of the solution. The following under-relaxation (see Section 4-2) blend type is presented by Lacor (1999)

$$A^{new} = A_c^{new} + \alpha( A_d^{new} + A_{fd}^{new} ) + ( 1 - \alpha )( A_d^{old} + A_{fd}^{old} ) \tag{2-8.2}$$

In the above, the superscripts *new* and *old* refer to values at current and previous stages and $\alpha$ is a stability tuned parameter. More general expressions, involving weightings at more than two stages and also involving weightings of $A_c$ are possible.

*Residual smoothing* is another stability enhancement approach allowing the use of larger time-steps (see Lerat (1979) and Jameson (1983)). With this, the following smoothed ( $\tilde{A}$ ) right hand side term (*Residual*) in Equation (2-6.1) is solved for

$$\tilde{A} = A + \varepsilon \nabla^2 \tilde{A} \tag{2-8.3}$$

where $\varepsilon$ is a free parameter. Uniform, two-dimensional grid, central difference based discretized equation coefficients for Equation (2-8.3) are as follows

$$\begin{aligned} A_{P\tilde{A}} &= 1 + 4\varepsilon \\ A_{NB.\tilde{A}} &= \varepsilon \\ S_{\tilde{A}} &= A \end{aligned} \tag{2-8.4}$$

The discretized equation defined through (2-8.4) must be solved using some form of efficient simultaneous equation solver. As a general rule of thumb, with residual smoothing the maximum allowable Courant number (see Equation (2-9.1)) or time-step can be increased by a factor of $\sqrt{1+4\varepsilon}$ . The typical practical value $\varepsilon = 0.75$ allows a factor of 3 increase in time-step or Courant number. Other variants on the above residual smoothing approach exist. For example, when forming (2-8.4), rather than using a central difference discretization (Section 3-3), upwind based methods (see Section 3-3) can be used (Zhu *et al.* (1993) and Blazek *et al.* (1991)).

*Stiffness*

The existence of significant spatial variations in time scales in the context of needing to efficiently integrate to a steady state (this could be time cyclic) has been briefly mentioned. Equations/systems that contain a significant range of time scales are called *Stiff*. Since, when using solution techniques which have time-step restrictions (due to for example stability reasons, the occurrence of oscillations, as with the Crank-Nicolson Scheme, or the need to accurately resolve the temporal history of a moving boundary) large time-steps cannot be used, integration of stiff equations to a steady state can be expensive. A practical journal bearing problem which is unusually stiff (a factor of 20,000 difference in key time scales) is considered in Chapter 9.

## 2-9 Characterising Stability and Accuracy of Schemes

The following dimensionless parameters, relating the time and space steps, are important when selecting time-step sizes for stable, accurate solutions

$$C = \frac{u\Delta t}{\Delta x} \qquad (2\text{-}9.1)$$

$$D = \frac{\Gamma\Delta t}{\rho(\Delta x)^2} \qquad (2\text{-}9.2)$$

The former is called the Courant number. In this, the ratio $\Delta x/u$ is the time required for a quantity or fluid particle to be convected through a distance $\Delta x$. Therefore, the Courant number can be viewed as a time-step to convection time scale ratio. The ratio $\rho(\Delta x)^2/\Gamma$, in the latter equation, is the time for a quantity to diffuse a distance $\Delta x$. Therefore, $D$ is a time-step to diffusion scale ratio. The restriction $C < 1$ (the Courant-Friedrichs-Lewy (CFL) condition) is frequently observed. This implies a fluid particle should not convect more than one grid cell per time-step.

## 2-10 Boundary Conditions and Solution Initialization

*Boundary Conditions*

As with the internal spatial discretization accuracy, it is important that the order of the boundary condition discretization is consistent with the temporal discretization. For example, the combination of a first order Neumann condition (see Fletcher (1997a)) with a higher order temporal discretization can reduce the temporal accuracy to first order. Also, with ADI, if a domain is moving with time, to maintain formal accuracy domain boundary positions and conditions should be updated at the end of half steps ($3^{rd}$ steps in three-dimensions) and not just full steps. Generally, the stability analyses for time schemes are based on Dirichlet boundary conditions. For Neumann conditions stability can be uncertain and oscillatory solutions may arise (see Mitchell and Griffiths (1980)).

*Solution Initialization*

Difficulties can arise if a boundary condition value, say $\phi_b$ for $t > 0$ is significantly different to the initial conditions $\phi_i$. When this is the case, as the boundary condition for the first step, it can be worthwhile for stability and accuracy reasons (see Fletcher (1997a)), to take an average of $\phi_b$ and $\phi_i$. For three-level schemes, values for $\phi^{old}$ and $\phi^{oldest}$ are required. To start the solution, a two-level scheme can be used. If time evolution is being studied, it is important that the accuracy of the initial two-level solution is consistent with the order of the three-level scheme. However, if the objective is to integrate to a cyclic steady state this latter issue is of no great importance. It may then be worth while carrying out initial time integrations with a lower order, faster to converge, temporal scheme with perhaps larger time-steps. The temporal accuracy can then be refined as the steady state is approached. This is a similar idea to overlaying a coarse spatial step solution on a fine grid to gain an initial guess. However caution is required, since if the order of the temporal scheme used for the initial guess is too low, the nature of the initial solution may be significantly different to the actual and hence the initial guess of no use. Also, if the final solution is dependent on the initial conditions it will be difficult to control the solution branch gained. To minimise the build up of machine round off error, through time integrations, double precision arithmetic is recommended.

## 2-11 Integration of Particle/Droplet Transport Equations

Frequently, as part of a solution the need to integrate Lagrangian particle trajectory equations arises. This can occur, to name but a few examples, when modelling the dispersion of pollutants, fuel droplets in combustors and particles in filtration processes. Also vortex methods (noted further in

Chapter 3) require the solution of such equations. In the former applications, particle trajectories are computed by solving both the trajectory

$$\frac{d\boldsymbol{x}}{dt}=\boldsymbol{U}_p \tag{2-11.1}$$

and momentum

$$m\frac{d\boldsymbol{U}_p}{dt}=\Sigma\boldsymbol{F} \tag{2-11.2}$$

equations above, where $m$ is the particle mass, $\mathbf{x}$ the particle position vector, $\mathbf{U}_p$ the particle velocity and $t$ time. $\Sigma$ $\mathbf{F}$ represents the sum of external forces acting on a particle.

Equation (2-11.1) can generally be expressed with sufficient accuracy in the following straightforward discretized form based on Equation (2-1.1)

$$\boldsymbol{x}^{new}=\boldsymbol{x}^{old}+\left[(1-\lambda)\boldsymbol{U}_P^{old}+\lambda\boldsymbol{U}_P^{new}\right]\Delta t \tag{2-11.3}$$

where, for example with $\lambda = 1/2$ a time-centred approximation is gained. However, an exponential time scheme similar in nature to that described in Section 2-4 can also be used. Incorporating Stokes drag, Equation (2-11.2) can be re-expressed as

$$\frac{d\boldsymbol{U}_P}{dt}=\frac{\boldsymbol{U}-\boldsymbol{U}_P}{\tau}+\frac{\Sigma\boldsymbol{F}}{m} \tag{2-11.4}$$

where $\boldsymbol{U}$ is the fluid velocity and if $\rho_P >> \rho$ ( $\rho_P$ is the particle density) the particle relaxation time (the time required for a suddenly released stationary particle to reach around 2/3[rds] the fluid's velocity) can be expressed as

$$\tau=\frac{\rho_P d^2}{18\mu} \tag{2-11.5}$$

The 4[th] order Runge-Kutta technique is popular for solving Equation (2-11.4) but has the restriction $\Delta t < 2\tau$ (see Barton (1996)). If it is assumed that the fluid velocity $\boldsymbol{U}$ remains constant through a time-step $\Delta t$ , Equation

(2-11.4) can be integrated to give the following (see Crowe *et al.* (1997)) stable discretized particle momentum equation

$$U_P^{new} = U_P^{old} e^{-\Delta t/\tau} + \left( U^{old} - \frac{\Sigma F}{m} \right)\left( 1 - e^{-\Delta t/\tau} \right) \qquad (2\text{-}11.6)$$

However, Equation (2-11.6), although popular, has the disadvantage that as $\tau$ tends to zero (with $\Sigma F$=0) it gives $U_P^{new} = U^{old}$, erroneously implying the particle velocity lags the fluid's by one time increment. This results from assuming $U$ is constant through the increment. To remedy this, Barton (1996) proposes, prior to the analytical integration of Equation (2-11.4), that the following Taylor series based fluid velocity variation assumption is made

$$U(t) = U^{old} + C_0 t + C_1 t^2 \qquad (2\text{-}11.7)$$

where

$$C_0 = \frac{U^{new} - U^{older}}{2\Delta t} \qquad (2\text{-}11.8)$$

and

$$C_1 = \frac{U^{new} - 2U^{old} - U^{older}}{2\Delta t^2} \qquad (2\text{-}11.9)$$

If desired, it can be assumed $C_1 = 0$, giving a less accurate linear fluid velocity variation assumption. When using equations (2-11.8) and (2-11.9) (which give rise to exponentially based temporal schemes) it is clear velocities need to be stored at three time levels. Also, $U^{new}$ is itself an unknown. To deal with this, Barton proposes a predictor corrector approach. With this, based on latest particle positions, $U^{new}$ is repeatedly updated. Evidently, generally only one correction/update is sufficient. When using Equation (2-11.7), as $\tau$ tends to zero, the sensible limit $U_P^{new} = U^{new}$ is now gained. For most practical engineering applications, with small enough time-steps Equation (2-11.6) is sufficient (see Morgan and Barton (2000)).

## 2-12 Two-level Scheme Exemplar

Performances of two-level schemes are now compared for a cylindrical geometry buoyancy driven flow. Lagrangian particle transport is also briefly considered (a more sophisticated example is given in Chapter 10). *Figure 2-12.1* shows a schematic of the geometry.

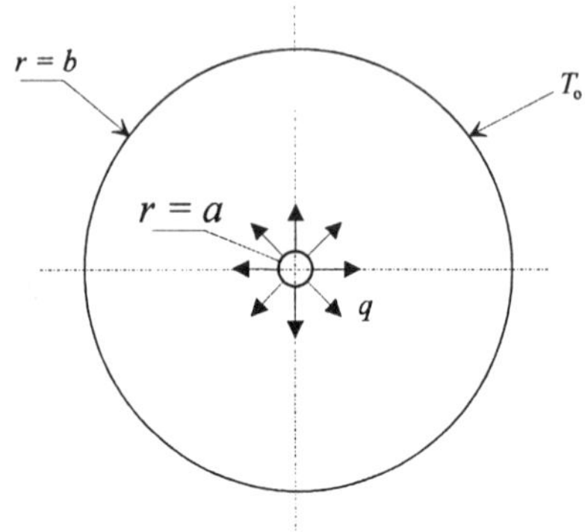


*Figure 2-12.1* Schematic of cylindrical geometry

The geometry consists of a suddenly heated (with $q$ = 156 W) inner cylinder of radius $a$ = 8mm inside an isothermal outer cylinder of radius $b$ = 95mm at a temperature $T_0$ =17.5°C. The geometry is intended to be infinitely long into the page. The fluid has a Prandtl number $Pr$ = 7 and a Grashof number $Gr = 9\times 10^7$. In the latter, the characteristic length is $b-a$ and temperature difference $\Delta T = aq/k$, where $k$ is the fluid's thermal conductivity. These conditions are chosen to match those used in flow visualization experiments by Castrejon and Spalding (1988). The earth's acceleration due to gravity is perpendicular to the inner cylinder axis. The conservative source term discretization of this body force is outlined in Appendix A (this is needed if use of the Boussinesq approximation is to be avoided).

*Flow Structure*

In the experiments, fluid below the horizontal centre line contains a dye-salt solution (this causes around a 10 s lag between the heat source being switched on and the plume growth). To make comparison with the flow visualisation, weightless particle tracks can be integrated through time. Since the particles are weightless the momentum Equation (2-11.2) reduces to $\boldsymbol{U}_p = \boldsymbol{U}$ and so only the Equation (2-11.1) for particle position need be solved. For the present geometry this is best carried out in polar coordinates. Following the experiments of Castrejon and Spalding, the lower half of the stationary annulus is seeded with imaginary particles (around 1 million are used, which, as will be seen later, gives the appearance of a continuous dyed fluid region). The variation of each particle's position, at grid point $P$, with respect to time is found from (the polar coordinate equivalent of Equation (2-11.1),

$$d\theta_P / dt = \Omega_P \tag{2-12.1}$$

where $\Omega_P$ is the particle's angular velocity and

$$dr_P / dt = u_P \tag{2-12.2}$$

where $u_P$ is radial velocity. These may be written in finite difference form using the 1[st] order forward Euler approach, as,

$$\theta_P = \theta^{old}{}_P + \Omega_P \Delta t \tag{2-12.3}$$

$$r_P = r^{old}{}_P + u_P \Delta t \tag{2-12.4}$$

corresponding to $\lambda = 0$ (see Equation (2-11.3)), where $\theta_P$ and $r_P$ are new positions of the particle at $t + \Delta t$, and $\Omega_P$ and $u_P$ are the current velocities evaluated at the old particle locations $\theta_P^{old}$, $r_P^{old}$. Particle velocities are obtained by direct linear interpolation from surrounding node points.

The following equation of state is used

$$\rho = C_0 - C_1 T \tag{2-12.5}$$

The constants $C_0$ and $C_1$ in the Equation (2-12.5) are 1066.4 kg/m$^3$ and 0.2347 kg/m$^3$K, respectively.

At time $t = 0$, the initial conditions for the variables $u$, $v$, $p$, $T$ and $\rho$ inside the cavity (i.e. $a \le r \le b$ and $0 < \theta \le 2\pi$ ) are:

$u = v = 0;$

$\rho = C_0 - C_1 T_0;$

$p = 1.0 \times 10^5 + \rho g r \sin\theta$ and

$T = T_0 = 17.5^{\circ}\mathrm{C}$

For $t > 0$ the shroud temperature does not change, i.e. for $r = b$ and $0 < \theta \leq 2\pi, T = T_0$. At $t > 0$, the inner cylinder ($r = a$ and $0 < \theta \leq 2\pi$) dissipates heat at a constant rate $q$. At the inner and outer (shroud) cylinders $\phi = 0$ where $\phi = u$ or $v$. All real variables are solved to 16 significant figures. The discretized equations are not under-relaxed (Section 4-2) – generally, as explained in Section 4-2, for unsteady flow this is not necessary. To increase the speed of convergence non-linear multilevel convergence acceleration is used with two grid levels (see Section 4-2).

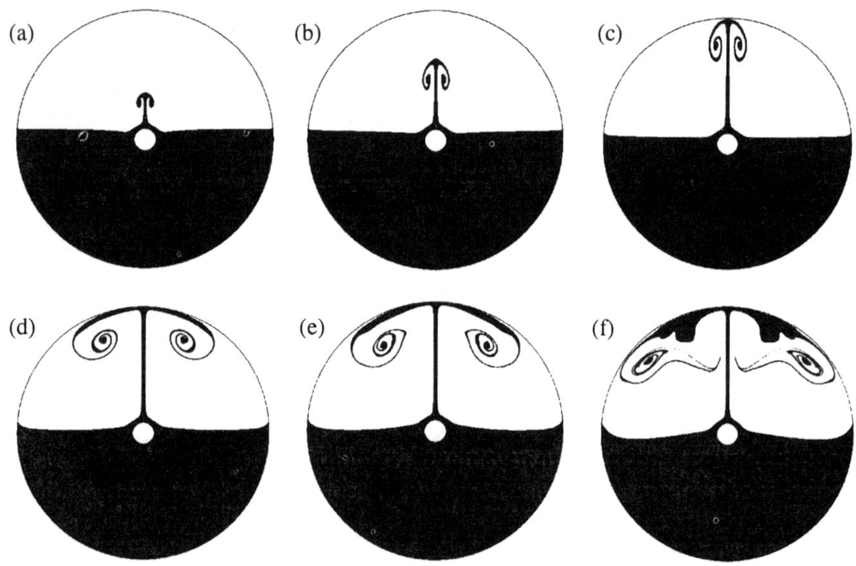

*Figure 2-12.2* Development with time of flow around the cylinder: (a) $t \approx 9.5$ s; (b) $t \approx 13.5$ s; (c) $t \approx 19.5$ s; (d) $t \approx 31.0$ s; (e) $t \approx 37$ s and (f) $t \approx 55$ s.

The computed temporal plume growth, induced by the heated cylinder, is shown in *Figure 2-12.2*. For these predictions, an essentially second order stabilised convective term discretization (CONDIF – see Section 3-3) is used with a 65 x 120 $(r, \theta)$ grid having a mild geometric refinement in the radial direction. Fixed time-steps of 0.05 s are used (predictions for this case, where time-steps are adjusted, based on temporal variable changes are presented in Chapter 5). Plots are gained by coupling the particle trajectory algorithm with the Navier-Stokes solver. Frames (a) to (e) show the flow at various times for $9 < t < 55$ s. Overall, there is approximate agreement with the experimental flow visualization. However, a key difference is that at around $t = 45$ s the experiments show the plume end filaments tend to break up, having a turbulent flow appearance. The coarse grids and time-steps (relative to the turbulence scales) used above cannot capture this. Based on the particle tracking information, *Figure 2-12.3* compares the predicted temporal variation in dimensionless ($X = r/b$) plume height with measurements (before the plume reaches the outer cylinder). The approximate 10 s lag in plume height, referred to earlier, is estimated by aligning this data, and so the rate of plume growth is the main thing to be compared in this plot. Considering possible experimental errors, agreement is encouraging.

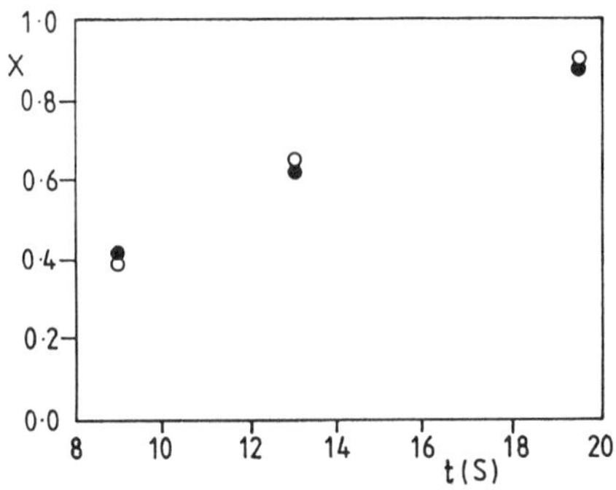


*Figure 2-12.3* Variation in dimensionless ($X = r/b$) plume height with time (• Measurements, o Current predictions)

*Figure 2-12.4* shows predicted vertical plane, radial variations of $u$, at $t = 15$ s for different schemes and time-steps. Each graph is for a different time scheme and in general show six curves. These correspond to time-steps of : $\Delta t =$ 15.0, 3.0, 1.0, 0.5, 0.25 and 0.125 s. Differences between schemes are best observed by comparing the spread of the different solution curves at around $r = 0.08$m (indicated using arrows). These predictions, unlike those for *Figure 2-12.2*, are certainly not grid independent. Since several computations are being made, as a computational economy measure, coarse grids are used. However, this constant error will not detract from the points to be made here.

Frame (a) shows the first order backwards Euler ($\lambda = 1$) results. As can be seen, for $\Delta t = 0.25$ s the solution is sensibly time-step independent. The SIMPLEC (Section (4-4)) and AVPI (Section (4-4)) schemes, discussed later, are used to give the pressure field. The latter has lower storage arising

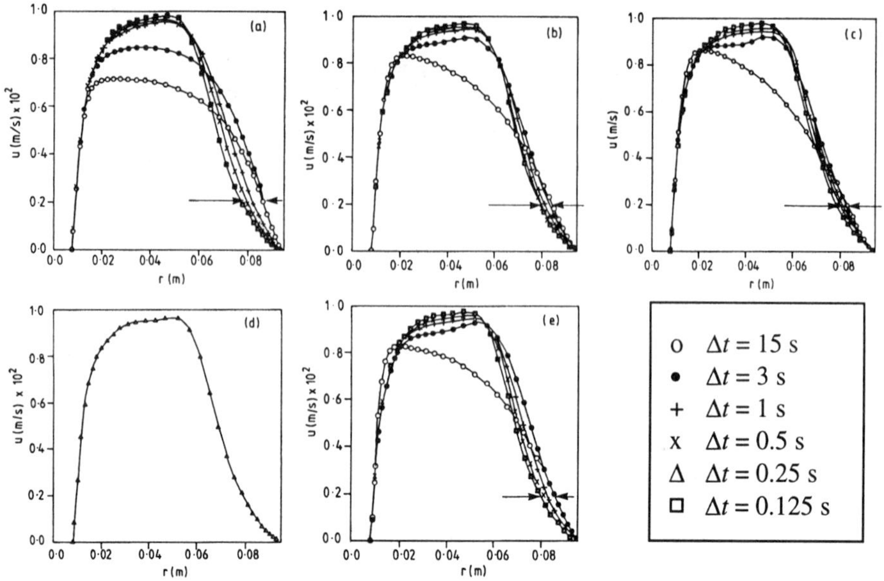


*Figure 2-12.4* Comparison between computed vertical plane $u$ variations at $t = 15$ s: (a) Fully implicit scheme; (b) Galerkin scheme; (c) Crank-Nicolson scheme; (d) Explicit scheme and (e) Crank-Nicolson scheme (fully implicit coefficients)

from a simplification facilitated by modelling the flow as unsteady. *Table 2-12.1* shows numbers of iterations to convergence. These are similar for the two pressure schemes.

| $\Delta t$ (s) | Number of Iterations | |
|---|---|---|
| | SIMPLEC | AVPI |
| 15.0 | 18 | * |
| 3.0 | 29 | 52 |
| 1.0 | 51 | 52 |
| 0.5 | 80 | 80 |
| 0.25 | 131 | 132 |

*Table 2-12.1* Number of iterations for convergence against time-step length for the implicit time scheme

Frame (b) gives the Galerkin scheme ($\lambda$ = 0.6) results. When compared to the implicit solutions, for $\Delta t$ = 15 s. and $\Delta t$ = 3 s., the Galerkin results (which are also formally first order) are more accurate. Frame (c) shows the performance of the Crank-Nicolson scheme ($\lambda$ = 0.5). As can just be seen, for larger time-steps, the method is slightly more (1-2%) accurate than the

Galerkin ($\lambda$ = 0.6) scheme. Frame *d* shows the performance of the explicit $1^{st}$ order forwards Euler scheme ($\lambda$ = 0). Only a solution obtained using $\Delta t$ = 0.25 s is shown. For step lengths larger than this, solutions diverged. Finally, frame (e) shows the performance of the Crank-Nicolson method with coefficients treated in a $1^{st}$ order fully-implicit fashion $\left(A_{NB,\phi} = f\left(\phi^{new}\right)\right)$.

Comparing frames (c) and (e) indicates that (as would be expected) implicit coefficients reduce accuracy. Except for the explicit method, numbers of iterations to convergence are within $\pm 8\%$ of those in *Table 2-12.1*. Therefore, the Crank-Nicolson scheme has the best performance - for larger time-steps, it gives the most accurate solutions without any significant increase in computing time. However, if time-steps are too large it can give physically unrealistic solutions. The Galerkin and backwards Euler are both formally $1^{st}$ order but clearly the former gives more accurate results.

## 2-13 Conclusions

For certain schemes, when discretizing the unsteady flow equations an assumption must be made of how dependent variables change with time. Some common profile assumptions can be expressed using weighting functions giving different significance's to dependent variable values at different time levels.

The time derivative, in the governing equations, can either be assumed constant over a cell (the *lumped* approach) or to vary in some fashion *consistent* with underlying space discretization. The *consistent* discretization will introduce extra computational cost. However, there are accuracy benefits. It is possible to combine differing discretization techniques, such as having a finite element space discretization with finite difference temporal modelling. Two and three-level schemes can be expressed in a generic form, different methods being gained by varying coefficients. The same non-discrete temporal scheme can be discretized in different ways. Therefore, for the same parameter settings and 'schemes' different codes can have different accuracies and convergence rates.

Three-level schemes require more storage, computational effort and attention to the starting solutions. For industrial computations, mostly two-level schemes are used. With Crank-Nicolson type methods, certain coefficients can be a function of the variable being solving for. Then the Beam and Warming linearization or something of this nature is desirable.

If time-steps are too large, solutions for unconditionally stable schemes may not be physically realistic. For stable, accurate solutions, consideration of Courant and time-step to diffusion scale ratio parameters is important. For the explicit Euler, the latter parameter gives a severe time-step restriction. Therefore, this scheme is only suitable for problems where high temporal resolution is required. For problems where variables are changing slowly, implicit schemes should be used. Generally speaking, implicit schemes are more stable. For many industrial problems this is attractive and so such schemes are popular in commercial codes. Generally, the higher the order of an explicit scheme, the more severe the stability restriction. However, explicit methods lend themselves to vector and parallel processing. For all temporal predictions, to minimise the build up of machine round off error through time integrations, double precision arithmetic is generally recommended.

The most common multipoint methods are the Adams types. The starting solutions can be produced using lower order two-level schemes. However, time-steps must be small enough to give solutions consistent with the order of the multipoint scheme. Also, as computations progress in time, with multipoint methods non-physical solutions can be produced and hence remedial action can be required. Also, with Leapfrog, the lack of coupling between time levels can make solution averaging necessary. The DuFort-Frankel method addresses this problem, but for accuracy, has a severe time-step restriction. A well-known member of the class of predictor-corrector methods are the Runge-Kutta schemes. These are more stable and accurate than multipoint schemes. Splitting methods combine implicit and explicit schemes. In this way significant improvements in computational performance can be produced. Although robust and interesting, the exponential temporal scheme is probably best avoided - its benefits are essentially cosmetic. However, when modelling particle transport using the Lagrangian method, for small enough time-steps an exponential approach is satisfactory. When using the Runge-Kutta technique to solve Lagrangian particle transport equations the time-step stability limit is based on the particle relaxation time.

Unless there are obvious reasons, an effective strategy is to model all flows as unsteady. Time marching allows stable (this is especially important when sophisticated turbulence models are to be implemented with higher order spatial discretization schemes) integrations to steady states and also the application of efficient splitting and other temporal schemes. Equations with differing time scales are called *stiff*. Implicit methods can be helpful when modelling such problems.

As with the internal spatial discretization accuracy, it is important that the order of the boundary condition discretization and also the order of the initialisation data is consistent with the temporal discretization.

Performances of numerical schemes were considered for a buoyancy driven flow. SIMPLEC and the storage saving AVPI pressure prediction scheme were contrasted. These showed similar convergence rates. Also, the two-level Galerkin, Crank-Nicolson, explicit 1$^{st}$ order forward Euler and 1$^{st}$ order backward Euler schemes were compared. Crank-Nicolson predictions with coefficients treated in a fully-implicit and trapezoidal fashion were also considered. As would be expected, the implicit coefficients reduced accuracy. The Crank-Nicolson scheme had the best performance. For accurate industrial computations it appears a good temporal scheme choice with both reasonable accuracy and robustness.

# CHAPTER 3

# TEMPORAL AND SPATIAL DISCRETIZATION RELATIONSHIPS

## 3-1 Introduction

For a significant number of engineering systems, by careful choice of the solution coordinate system the fluid flow structure can be made steady or steadier. The reduction of unsteadiness, may in some situations involve using several coordinate systems. This practice is common in turbomacinery design where sets of adjacent stationary and rotating blades are modelled in stationary and rotating coordinate system, respectively. Use of this approach, with a sliding grid interface, enables meaningful steady solutions to be produced which can aid design analysis. In this chapter, through considering the substantial derivative, the relationship between coordinate system choice, unsteadiness levels, and solution accuracy and economy are explored.

For certain temporal schemes it was shown in Chapter 2 that the time variation of dependent variables could be expressed using weighting functions associated with different points in time. Here, this idea is extended to the spatial domain. An overview of some popular convective term spatial discretizations (convective term treatments) is given. Modifications to convective term treatments, ensuring compatibility with respect to accuracy for unsteady flows are discussed. The close interaction between spatial and temporal discretization schemes and the effects of this on solution accuracy and economy is addressed. Types of discretization error are formalized. The physical manifestations of discretization errors in solutions when modelling unsteady flows are discussed.

An overview is given of moving boundary and free surface flow modelling techniques, and potential sources of solution error for these. Different methods for tracking free surfaces are introduced along with the pros and cons for each with respect to accuracy, computational effort and ease of implementation. The latter can impact on reliability. Finally, some adaptive solution strategies are discussed where, based on discretization error estimates, space steps are refined (time-step adaptation is discussed in Chapter 5).

### 3-2 The Substantial Derivative

When computing unsteady flows, concepts relating to the substantial derivative should be appreciated. For a Cartesian coordinate system this is defined below (strictly, for the substantial derivative, the vector ***U*** should be outside the dot product)

$$\left(\frac{\partial \phi}{\partial t}\right)_{fp} = \left(\frac{\partial \phi}{\partial t}\right)_{g0} + \nabla\cdot(\phi \mathbf{U}) = \left(\frac{\partial \phi}{\partial t}\right)_{g0} + u\frac{\partial \phi}{\partial x} + v\frac{\partial \phi}{\partial y} + w\frac{\partial \phi}{\partial z} \qquad (3\text{-}2.1)$$

The substantial derivative, $(\partial\phi/\partial t)_{fp}$, is more usually expressed as $D\phi/Dt$. However, here the subscript *fp* is used (as suggested in Anderson (1995)) to enforce its meaning, this being the time rate of change in $\phi$ when a particular fluid particle is locked onto and followed. The $(\partial\phi/\partial t)_{g0}$ term is the time rate of change at a fixed coordinate location. In a CFD context, this is the temporal change of $\phi$ relative to the grid. Generally, for flows classified as *unsteady* it is meant $(\partial\phi/\partial t)_{g0} \neq 0$ ($(\partial\phi/\partial t)_{fp} \neq 0$, even for steady flows).

For certain systems (classified as exhibiting unsteady flow), the fluid structure may stay exactly or much the same, but simply translate (at a velocity $\tilde{\mathbf{U}}$) or rotate (at an angular velocity $\tilde{\boldsymbol{\omega}}$) in space, relative to the coordinate system/grid. The mechanism for this translation or rotation could be fluid dynamic in origin or alternatively as the result of boundary movement. Examples of the former are a plane isolated shock wave propagating at say a velocity $\tilde{U}$ (see Fletcher (1997, p. 351)), a wave train moving around a cylindrical domain (see Hignett *et al.* (1985)) with an angular velocity, $\tilde{\omega}$, about the cavity rotation axis or vortex wakes propagating downstream from aircraft wings (see Steger and Kutler (1977)). Translational movement of a constant geometrical distortion at $\tilde{U} > 0$ with $\tilde{V} = \tilde{W} = 0$ (where, $\tilde{U}$, $\tilde{V}$ and $\tilde{W}$ are the upper horizontal surface *x*, *y* and

$z$ velocity components, respectively), as shown in the *Figure 3-2.1a* geometry (mapped by a curvilinear grid), is an example of the latter. For all the above examples, when viewed in a moving frame of reference, the flow can be made steady ($(\partial\phi/\partial t)_{g0} = 0$).

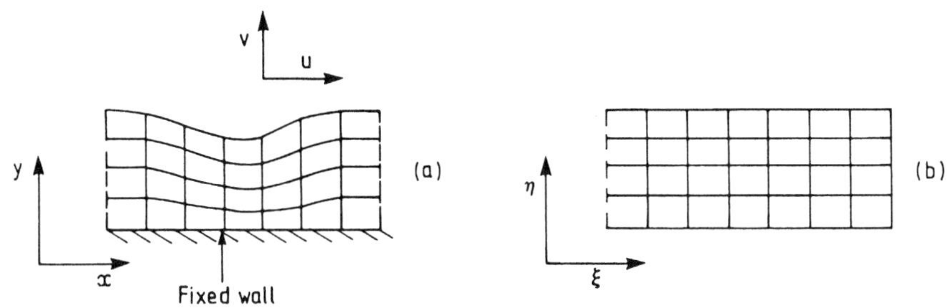


*Figure 3-2.1* Grids for a channel with a moving distorted top wall : (a) Physical plane and (b) transformed plane

Essentially, the substantial derivative is the left-hand side, as traditionally presented, of the Navier-Stokes equations. These are Galilean invariant (i.e. in a moving frame of reference their form does not change). Hence, when solved in a moving frame, the form of the substantial derivative also does not change. However, the convective velocities must be defined relative to the moving frame of reference. Therefore, for a coordinate system moving just horizontally at $U_g$, tracking the apparent translational source of flow unsteadiness in the examples given above (i.e. $U_g = \tilde{U} \neq 0$, $\tilde{V} = \tilde{W} = 0$), the Cartesian substantial derivative becomes

$$\left(\frac{\partial\phi}{\partial t}\right)_{fp} = \left(\frac{\partial\phi}{\partial t}\right)_{g0} + \left(u - U_g\right)\frac{\partial\phi}{\partial x} + v\frac{\partial\phi}{\partial y} + w\frac{\partial\phi}{\partial z} \qquad (3\text{-}2.2)$$

Assuming the transform has made the flow steady, $(\partial\phi/\partial t)_{g0} = 0$.

Incidentally, some equations related to CFD such as those for particle transport and the vortex method (see Khatir (2000)) can be solved in a completely Lagrangian frame which follows fluid points. The 'substantial derivative' then appears as

$$\left(\frac{\partial\phi}{\partial t}\right)_{fp} = \left(\frac{\partial\phi}{\partial t}\right)_{g0} \qquad (3\text{-}2.3)$$

This can be appreciated for a flow with $u = w = 0$. When $u = U_g$ (see Equation (3-2.2)) only the $(\partial\phi/\partial t)_{g0}$ term remains in Equation (3-2.1) and unsteady solution techniques must be used.

*Making Flows more Steady*

For most practical flows, it will not be possible to make the flow completely steady ($(\partial\phi/\partial t)_{g0} = 0$) by a simple coordinate system change. However, for some it is possible to make $(\partial\phi/\partial t)_{g0}$ smaller. For these it is helpful to break the local time derivative down into say two components

$$\left(\frac{\partial\phi}{\partial t}\right)_{g0} = \left(\frac{\partial\phi}{\partial t}\right)_{g1} + \left(\frac{\partial\phi}{\partial t}\right)_{f} \tag{3-2.4}$$

The term $(\partial\phi/\partial t)_{g1}$ is one that can be reduced by minimizing $|\mathbf{U}_g - \tilde{\mathbf{U}}|$ (the word minimize is used, since frequently $\tilde{\mathbf{U}}$ is an unknown to be itself estimated). Clearly for $\mathbf{U}_g = \tilde{\mathbf{U}}$, $(\partial\phi/\partial t)_{g1} = 0$. The latter $(\partial\phi/\partial t)_f$ term is the unsteadiness remaining after the coordinate system transformation. Clearly, in the interests of computational speed, minimizing $(\partial\phi/\partial t)_{g1}$ is worthwhile. Also, reducing the magnitude of temporal terms should bring down some temporal solution errors. Furthermore, moving coordinate systems can significantly lower convective term magnitudes. These terms are generally problematic, giving rise to false diffusion (see Section 3-4) and hence damping of temporal behaviour. Finally, convective term magnitude reductions can significantly increase numerical precision. Tucker (1997) gives examples of dramatic solution accuracy gains by the use of rotating coordinate systems.

In summary, the Cartesian substantial derivative, for a flow/system with $\tilde{V} = \tilde{W} = 0$ and $\tilde{U} \neq 0$, is considered here to be as follows

$$\left(\frac{\partial\phi}{\partial t}\right)_{fp} = \left(\frac{\partial\phi}{\partial t}\right)_{g1} + \left(\frac{\partial\phi}{\partial t}\right)_{f} + \left(u - U_g\right)\frac{\partial\phi}{\partial x} + v\frac{\partial\phi}{\partial y} + w\frac{\partial\phi}{\partial z} \tag{3-2.5}$$

By analogy with the above, for cylindrical rotating systems (see later) it is convenient to consider the following substantial derivative form

$$\left(\frac{\partial\phi}{\partial t}\right)_{fp} = \left(\frac{\partial\phi}{\partial t}\right)_{g1} + \left(\frac{\partial\phi}{\partial t}\right)_{f} + \left(v - \Omega_g r\right)\frac{1}{r}\frac{\partial\phi}{\partial\theta} + \frac{u}{r}\frac{\partial r\phi}{\partial r} + w\frac{\partial\phi}{\partial z} \qquad (3\text{-}2.6)$$

where $u$, $v$ and $w$ are the velocities in $r$, $\theta$ and $z$ directions respectively, and $\Omega_g$ is the grid/coordinate system angular velocity. Equations (3-2.5) and (3-2.6) can be expressed more generally as

$$\left(\frac{\partial\phi}{\partial t}\right)_{fp} = \left(\frac{\partial\phi}{\partial t}\right)_{g1} + \left(\frac{\partial\phi}{\partial t}\right)_{f} + \left(\mathbf{U} - \mathbf{U}_g\right).\nabla\phi + G_D \qquad (3\text{-}2.7)$$

where the $G_D$ term relates to grid distortion (Section 3-5), being applicable to moving boundary cases. For a three-dimensional transformed grid $(\xi,\eta,\zeta)$ system

$$G_D = \left(\frac{\partial\xi}{\partial t}\right)_{g1}\frac{\partial\phi}{\partial\xi} + \left(\frac{\partial\eta}{\partial t}\right)_{g1}\frac{\partial\phi}{\partial\eta} + \left(\frac{\partial\zeta}{\partial t}\right)_{g1}\frac{\partial\phi}{\partial\zeta} \qquad (3\text{-}2.8)$$

As implied earlier, the coordinate system can also be chosen to minimize $G_D$. For the moving distortion shown in *Figure 3-2.1*, for $U_g = \tilde{U}$ the grid will not change with time. Practical examples of making flows steady, quasi-steady or removing a dominant unsteadiness component by careful coordinate choice are shown in Chapters 7 to 9.

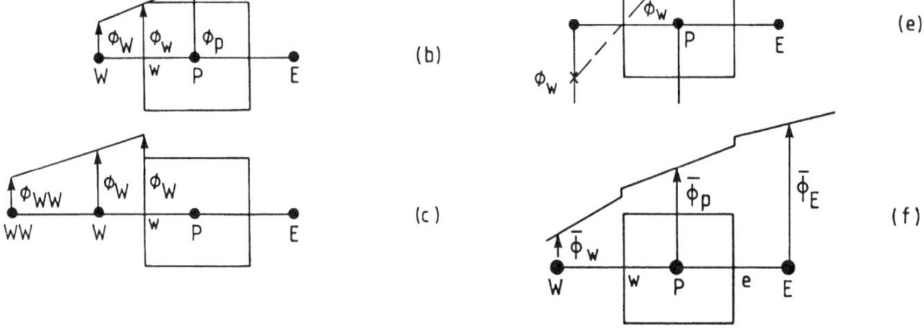


*Figure 3-3.1* Schematics of typical uniform Cartesian grid '*w*' control volume face interpolations

## 3-3 Convective Term Treatments

*General Convective Schemes*

As part of the convective term discretization, dependent variables are interpolated from grid points to control volume faces. *Figure 3-3.1* gives schematics of typical uniform Cartesian grid '*w*' control volume face interpolations. In a similar fashion to Equation (2-2.1), this interpolation can frequently be expressed as

$$\phi_w = W_{fWW}\phi_{WW} + W_{fW}\phi_W + W_{fP}\phi_P \tag{3-3.1}$$

where $W_f$ is, this time, a spatial weighting function. *Table 3-3.1,* summarizes uniform grid Equation (3-3.1) weighting functions for some *Figure 3-3.1* schemes. The order of the differential in omitted leading terms (*HOT* – Higher Order Terms) in the discretizations is also tabulated. The substitution of Equation (2-2.1) into (3-3.1) establishes the time-level of right hand side $\phi$ values in (3-3.1).

| Scheme | $W_{fWW}$ | $W_{fW}$ | $W_{fP}$ | *HOT* |
|---|---|---|---|---|
| First order upwind | 0 | 1 | 0 | $\partial^2\phi/\partial x^2$ |
| Central differencing | 0 | 1/2 | 1/2 | $\partial^3\phi/\partial x^3$ |
| Second order upwind | -1/2 | 3/2 | 0 | $\partial^3\phi/\partial x^3$ |
| Quadratic interpolation | -1/8 | 6/8 | 3/8 | $\partial^4\phi/\partial x^4$ |
| MUSCL | $(k$-1)/4 | (2-$k$)/2 | (1+$k$)/4 | NA |

*Table 3-3.1* Uniform grid Equation (3-3.1) weighting functions

Frame (a), of *Figure 3-3.1*, shows the interpolation for inaccurate first order upwinding (in a sense, this is the spatial equivalent of the 1st order Euler time differencing scheme). The scheme simply utilizes one upstream grid node. Second order central differencing (CD) (Frame (b)) involves an equal weighting of upstream and downstream node point values (this is the spatial equivalent of the Crank-Nicolson scheme). Without modification, the CD scheme can, under certain conditions (higher Reynolds numbers), give rise to undesirable negative coefficients (resulting from a loss in diagonal dominance of the governing equations, as discussed in Section 4-2) and, like Crank-Nicolson solutions, oscillations. In spatial differencing schemes these oscillations are sometimes referred to as *wiggles* (see Leonard (1979)) or *non-monotonic* behaviour. The CD scheme is also unstable. However, if the equations are solved time-dependently stability can be achieved (see Mallinson and de Vahl Davis (1973), Mansour and Morel (1983), Kim and Moin (1985), Adair and Tucker (1998) and Tucker and Long (1996)). Second order upwinding (see Vanka (1987)) uses two upstream grid points. The third order Quadratic Upstream Interpolation (QUICK) scheme of Leonard (1979) (Frame (d)) involves a quadratic fit to two upstream and one downstream point. When used with ADI (Section 2-7), the involvement of the $\phi_{WW}$ point detracts from QUICK's efficiency

Frame (e), shows skew upwinding (see Hassan, Rice and Kim (1983)) where the directions of local streamlines are used to indicate the position of the upwind value applied at the control volume face. Finally, frame (f), shows the Monotone Upstream Centered Scheme for Conservation Laws (MUSCL) of Van Leer (1977,1979). Unlike the others described above, it uses control volume average values, such that, for example $\phi_P = (1/\Delta x)\int_w^e \phi(x)dx$. Control volume face values are based on extrapolations involving gradients at the central grid point 'P'. Essentially, MUSCL is built about the first order upwind scheme, $k$ giving more refined schemes of different orders. For example $k$=-1, 1 and ½ correspond to the second order upwind, CD and QUICK schemes, respectively. *Figure 3-3.2* shows time-dependent, one-dimensional, non-dissipative transport equation (see Equation (3-4.1)) MUSCL solution profiles for a convected variable, $\phi$, in a uniform flow of velocity $u$. The flow is moving from left to right. Frame (a), gives the initial, 'to be modelled', distribution at $t$ = 0 (full line) and the MUSCL approximated profile (dashed line). Frame (b), shows the previous $t$ = 0 MUSCL profile, shifted a distance $u\Delta t$ (the dashed line). The full line gives the actual, revised MUSCL distribution at $t+\Delta t$. As can be seen, the solution profile looks like a series of time varying discontinuous lines.

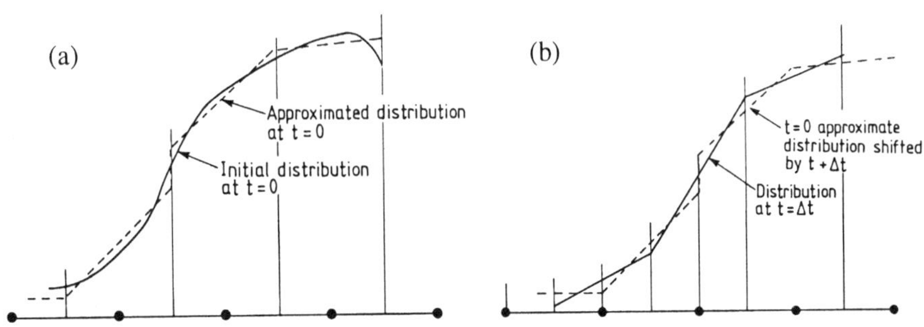


*Figure 3-3.2* Time-dependent, one-dimensional non-dissipative transport equation MUSCL solution profiles: (a) profiles at $t$ = 0 and (b) profiles at $t+\Delta t$

Analytical flow solution based interpolations/techniques have also been tried (see Spalding (1972) and Chen, Naseri-Neshat and Ho (1981)) along with compact spline based methods (see for example Nasser and Leschziner (1985)). Compact spline techniques offer the potential of high accuracy without involving more grid points than used in simple schemes such as CD. This is advantageous for ADI type methods. Lele (1992) outlines the compact difference scheme approach that has found more widespread use especially in LES and DNS work (see Section 6-3).

As with higher order temporal schemes, in highly convecting flows, with sharp velocity gradients, higher order spatial interpolation schemes can result in local oscillations (as noted earlier for the CD scheme). Therefore, these spatial schemes are sometimes bounded in some way (that is, limits are set on the value of $\phi$ predicted at each control volume face relative to neighbouring values). Bounding algorithms are sometimes called *Flux Limiters*. Many of these algorithms are in essence quite crude, however, others are at the cutting edge of applied mathematics. So-called blending schemes, where a mixture of the control volume interface value predicted by, for example, quadratic interpolation and upwinding, when local oscillations are detected, have been used by Zhu and Leschziner (1988). Various composite schemes have been proposed; one, for example, by Leonard (1989), switches between exponential function upwinding or linear extrapolation depending on the curvature of local variables. This has similarities to the much older and still popular HYBRID scheme of Spalding (1972), which for cell Reynolds ( $\rho u \Delta x/\mu$ ) or Peclet ( $\rho\phi \Delta x/\Gamma$, where $\Gamma$ is a diffusion coefficient) numbers less than two, switches to the CD scheme, as an alternative first order upwinding.

Essentially the above stabilization techniques supress oscillations by introducing *numerical* or *false diffusion* (effectively increasing diffusion coefficients and thus the smoothing properties of the modelled equation). An interesting variant of these methods is the CONDIF (Controlled Numerical Diffusion with Internal Feedback) scheme of Runchal (1987). This is essentially a stabilized CD scheme, some false diffusion being introduced through a free parameter (**R**). The larger **R**, the less the false diffusion. For **R** = 1 CONDIF reduces to first order upwinding and with **R** >20, for most cells, essentially the CD scheme is used. Since CONDIF is less diffusive than HYBRID it requires significantly more convergence iterations (very roughly around a factor of 2 per time-step).

Prevalent oscillation removal techniques (*limiters*) are based on enforcing the *Total Variation Diminishing* (TVD) or *Local Extremum Diminishing* (LED) concepts, or *monotonicity conditions* of Van Leer. The Van Leer condition ensures that interpolated control volume face values are not greater or less than adjacent grid point values. The LED approach structures equations to ensure that local maxima cannot increase and local minima decrease (this is when considering the process described in *Figure 3-3.2* for an initially specified $\phi$ distribution). TVD's basis can be appreciated by considering Equation (3-4.1). For a correct solution, being integrated to a steady state, through time, over a domain of length $L$, a variable called the *total variation*

$$TV = \int_0^L \left| \frac{\partial \phi}{\partial x} \right| dx \tag{3-3.2}$$

should *not increase* with time (the scheme can also be formulated on an iterative basis). Based on this correct physical behaviour, TVD schemes are constructed so that

$$\int_0^L \left| \frac{\partial \phi^{new}}{\partial x} \right| dx \leq \int_0^L \left| \frac{\partial \phi^{old}}{\partial x} \right| dx \tag{3-3.3}$$

Or, if the above differentials are expressed as central differences, the following can be written

$$\sum_{x=0,L} \left| \phi_E^{new} - \phi_W^{new} \right| = \sum_{x=0,L} \left| \phi_E^{old} - \phi_W^{old} \right| \tag{3-3.4}$$

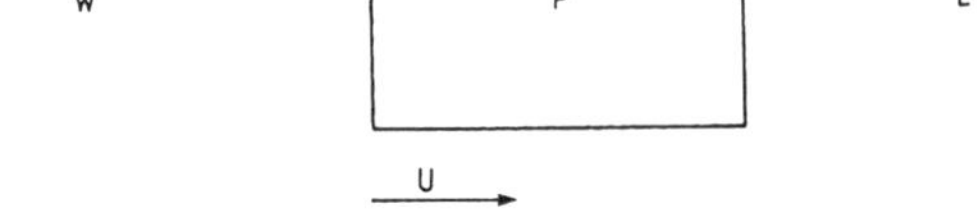


*Figure 3-3.3* Distributions of $\phi$ at $t$ and $t + \Delta t$ for a wave being convected at $u$

Enforcing the TVD constraint removes erroneous solution oscillations. Numerous studies on the performances of relatively simple convective schemes have been made by workers such as Huang, Launder and Leschziner (1985), Patel and Markatos (1986), Vanka (1987), Smith and Hutton (1982), Tamamidis and Assanis (1993), and Bo *et al.* (1995). It is not clear from these investigations which method is best. This is especially the case since, with the exception of Tamamidis and Assanis, they do not really address unsteady flows.

However, because for most industrial flows, cell Reynolds or Peclet numbers will be greater than two at virtually all grid points, clearly the use of the HYBRID is undesirable (despite this, it is still widely used in many industrial computations). Glass and Rodi (1982) illustrate the poor performance of HYBRID for unsteady scalar convection. Exceptions to the rule are simple parabolic boundary layer type flows where HYBRID can give reasonable results. Tamamidis and Assanis study unsteady scalar convection in specified velocity fields. The performances of first, second and third order upwinding, Van Leer's (1977) scheme and also QUICK are compared. Also, bounded (based on the Van Leer approach) versions the second and third order upwind, and QUICK schemes are assessed. As would be expected, results for the first order upwind scheme are poor. Overall, for cases without discontinuous gradients the unbounded schemes showed greatest accuracy. Hayase (1999) consider CD and QUICK scheme performances when using a second order accurate, three level, implicit time scheme to predict turbulent duct flows with no turbulence model. The focus of the work was grid convergence monotonicity (whether the convergence error always decreases with grid refinement), which is desirable for reliable estimation of grid convergence errors. For the flow considered, unlike QUICK, the CD convergence is found to be non-monotonic (one of its few defects). For Large Eddy Simulation work (see for example Yang and Ferziger (1993)) it is generally accepted that centrally based differences (such as CD) are superior to higher order offset schemes like QUICK.

*Unsteady Convective Schemes*

Temporal behaviour can display sensitivity to spatial discretization errors. Also, in transient flows, particularly high gradients of variables can arise. These factors (see Glass and Rodi (1982), Przulj and Younis (1993) and Tucker (2000)) place added demands on convective schemes. Therefore, for unsteady flows, for maximum accuracy, some of the above approaches can be modified. A CD scheme modification due to Leith (1965) is now described. With this $\phi_w$ is redefined as $\overline{\phi}_w$, where the overbar is used to represent a time averaged control volume face value for a time period $\Delta t$. The evaluation of the time average is best explained with reference to *Figure 3-3.3*. This shows $\phi$ distributions at $t$ and $t + \Delta t$ for a wave being convected at $u$. With reference to this figure, assuming the profile does not change, the following time averaged '$w$' control volume face expression can be written

$$\bar{\phi}_w = \frac{1}{2}\left[\left(\phi_W^{old} + \phi_P^{old}\right) - C_w\left(\phi_P^{old} - \phi_W^{old}\right)\right] \tag{3-3.5}$$

where $C_w$ is a Courant number. Equation (3-3.5) corresponds to

$$W_{fW} = \frac{1}{2}\left[1 + C_w\right]\phi_W^{old} \tag{3-3.6}$$

and

$$W_{fP} = \frac{1}{2}\left[1 - C_w\right]\phi_P^{old} \tag{3-3.7}$$

For other control volume faces, similar expressions can be written and this is the basis of Leith's method.

The accuracy of QUICK with first order forward Euler time differencing is essentially no better than when first order spatial upwinding is used (see Nasser and Leschziner (1985)). Therefore, it is best adapted in a similar fashion to Leith's CD scheme. This is the essence of QUICKEST (QUICK with Estimated Streaming Terms (EST)). With QUICKEST (Leonard (1979)) the control volume face time averaging is carried out in a more complex, non-linear fashion, consistent with the quadratic nature of QUICK. With Leith's method, since the CD interpolation is linear, this is not necessary. From the preceeding discussion it can be seen that the Leith and QUICKEST schemes, in a local sense, have similarities to the coordinate system adaptation procedure described in Section 3-2. However, Davies and Moore (1982) present a derivation of QUICKEST, starting with a 3rd order Taylor series expansion of the *time derivative* and finally arriving at Leonard's original equations. This enforces the point that QUICKEST is third order in both space and time (despite being formulated by Leonard around the 1st order forward Euler scheme) and that (since, $dx = udt$, space and time derivatives can be exchanged) schemes can be viewed in different ways. Davies and Moore also present a multi-dimensional version of QUICKEST, the original derivation being based on one-dimensional arguments. With relatively modest grids, this is successfully applied to the study of vortex shedding from rectangular cylinders.

Glass and Rodi (1982), Roache (1992), and Wallis and Manson (1997) present schemes which to varying degrees (the former loosely) are related to the Estimated Streaming Approach and are potentially (Glass and Rodi did not take advantage of this aspect) accurate for Courant numbers greater than unity. Consequently, in the latter two schemes, the convected wave shown in *Figure 3-3.3* can pass more than one control volume face per time-step. Since larger Courant numbers and hence time-steps can be used, a point in time can be reached using less time-steps and hence control volume face related interpolations. This gives the above three schemes the unusual property that their accuracies can increase with Courant number (they are also exact for integer values). In the two-dimensional, scalar transport work of Glass and Rodi, a cubic Hermitian interpolation (involving gradients of dependent variables, novely gained by the solution of transport equations) is used. The schemes of Roache, and Wallis and Manson use accuracy reducing flux limiters and are only applied to one-dimensional flow, with more work required for their extension to more dimensions.

The derivations of Glass and Rodi and Roache, and also to an extent that of Wallis and Manson are based around the *Method of Characteristics* (MOC). All the above Estimated Streaming Approach related methods are also members of a family called *semi-Lagrangian* methods. This name arises because they can be partly derived by considering the Lagrangian trajectories of scalars (these exactly correspond to pathlines and also characteristic lines) on an Eulerian grid. Staniforth and Cote (1991) give a thorough review of *semi-Lagrangian* approaches, noting that, in one-dimension, the Eulerian Lax-Wendroff scheme can be considered as a special case of the semi-Lagrangian schemes. They also show a highly challenging semi-Lagrangian test case involving the advection of a rotating slotted cylindrical scalar distribution. The schemes performance is impressive with little deterioration of the slotted cylinder form through the time integration (indicated by Glass and Rodi the standard CD scheme will not perform well for such cases).

An unusual, semi-Lagrangian related approach, called ICED-ALE, is presented by Hirt, Amsden and Cook (1974) (a similar technique called LINC is given by Butler (1970)). It is an extension to an earlier ALE (Arbitrary-Lagrangian-Eulerian) method devised by Hirt, in which grid nodes can be held fixed (Eulerian), move with the fluid (Lagrangian) or in any prescribed manner. Clearly, moving the grid with the flow, reduces the order of the problematic convective terms. However, severe grid distortion can occur. The ICED prefix (Implicit-Continuous-Fluid Eulerian (see Section 4-4)) signifies that, unlike the original ALE technique, the method is implicit. It seems appropriate to mention here the fully Lagrangian vortex method (see Khatir (2000)) where vortex blobs or elements are convected with the flow in a Lagrangian fashion. Such approaches remove the convective term modelling problem but accounting for diffusion then becomes problematic. Hence, the modelling difficulties are shifted from the convection to diffusion terms. An important class of high accuracy difference schemes (which can also be applied to temporal derivatives) that have small stencils are the *compact* schemes (see Lele (1992) and Hirsch and Ransbeeck (1999, p. 69)).

Worthy of note is the two-stage SHASTA (Sharp And Smooth Transport Algorithm) of Boris and Book (1973) extended by Book, Boris and Hain (1975). The first stage involves integrating the governing equations using a strongly diffusive method (incidentally, for this stage, Boris and Book's scheme has a semi-Lagrangian element) which gives rise to a monotone solution. The second involves correcting this solution (by correcting fluxes), reducing the diffusive error by introducing an anti-diffusive element. Ideally, the anti-diffusive component is made smaller than the initial diffusive component, leaving, if possible, a residual diffusive component large enough in principal to cancel any dispersive error. The method easily extends to three-dimensions. However, it is more intended for strong shock problems. Unlike SHASTA, to damp shock-induced oscillations, traditional methods can introduce levels of artificial viscosity, which based on diffusive stability limits, produce severe time-step restrictions. SHASTA has the key disadvantage that the initial diffusive stage can damp out maximas that will not be recovered in the anti-diffusion stage, thus giving rise to erroneous solution clipping. Motivated by this, Chapman (1981) presents the explicit FRAM (Filtering Remedy and Methodology) monotone solution approach. This has three stages. In the first, a higher order scheme is used to give a provisional time advanced solution. During the second stage, based on a local Lagrangian solution, bounds are calculated for the provisional solution. Where the initial solution is outside these bounds sufficient local dissipation is introduced to yield a monotone solution. For the first stage, several predictions presented by Chapman use the second order scheme of Crowley (1967) outlined further later.

If divergence is to be avoided, for flows without significant dissipation but requiring long time period integrations, special energy conserving convective term treatments should be used. Examples of these are given by Arakawa (1966), and Piacsek and Williams (1970). The divergence is related to *aliasing,* which is described in the next section. According to Staniforth and Cote semi-Lagrangian schemes do not generally formally conserve energy but are acceptably accurate with regard to this aspect.

*Current Convective Schemes*

For predictions presented here, either the CONDIF, CD or HYBRID schemes are used. Use of HYBRID is generally restricted to comparisons between different temporal schemes or concepts (where the spatial discretization errors will cancel). Alternatively, its use is restricted to parabolic boundary layer flows. With the increased diagonal dominance arising from unsteady flow computations (see Section 7-2) the CD scheme is mostly found to be quite stable. The derivation of the CONDIF scheme starts with an interesting discretization of the CD scheme. This gives unconditionally positive coefficients. For this reason and also because CONDIF is used for many cases presented here, a CONDIF derivation is included in Appendix B. The preference for CONDIF/CD schemes, for the unsteady flows modelled here is because they have accuracies consistent with the widely industrially used Crank-Nicolson scheme and relatively low *dispersion* errors (see later). Although QUICK is generally more popular, it has greater dispersion and an accuracy and form that is inconsistent with the Crank-Nicolson scheme. The form of the discretized HYBRID scheme coefficients is given in Appendix A. The nature of possible errors arising from the interaction between the chosen temporal and spatial differencing schemes will now be discussed more formally.

## 3-4 Dissipation, Dispersion and Aliasing

*The Modified Equation*

Normally a discretized equation is considered as modelling the original differential equation but having a certain truncation error. An alternative viewpoint is to say it is an 'exact' solution, but to a different equation. The form of the different equation (the *modified equation*) can be observed by replacing terms in the discretized equation with their Taylor series expansion and rearranging. For a simple illustrative example of this technique, the following one-dimensional equation is used

$$\frac{\partial \phi}{\partial t} + u\frac{\partial \phi}{\partial x} = 0 \tag{3-4.1}$$

Applying, to the above, a first order backwards difference in space and initially ignoring the time dependent term gives the following discretized equation

$$\frac{u(\phi_P - \phi_W)}{\Delta x_P} = 0 \tag{3-4.2}$$

Writing a Taylor series expansion for $\phi_W$ yields

$$\phi_W = \phi_P - \left(\frac{\partial \phi}{\partial x}\right)_P \Delta x_P + \left(\frac{\partial^2 \phi}{\partial x^2}\right)_P \frac{(\Delta x_P)^2}{2!} - \left(\frac{\partial^3 \phi}{\partial x^3}\right)_P \frac{(\Delta x_P)^3}{3!} + HOT \tag{3-4.3}$$

where *HOT* means Higher Order Terms. Replacing $\phi_W$ in Equation (3-4.2) with its Taylor expansion gives the following (dropping the *P* subscript because all terms are now for this point)

$$u\frac{\partial \phi}{\partial x} = C_0\frac{\partial^2\phi}{\partial x^2} + C_1\frac{\partial^3\phi}{\partial x^3} + HOT \tag{3-4.4}$$

where

$$C_0 = \frac{u\Delta x}{2} \tag{3-4.5}$$

and

$$C_1 = \frac{u\Delta x^2}{6} \tag{3-4.6}$$

Carrying out a similar procedure but also including the $\partial\phi/\partial t$ term, discretized using the explicit Euler scheme, yields,

$$C_0 = \frac{u\Delta x(1 - u\,\Delta t/\Delta x)}{2} \tag{3-4.7}$$

and

$$C_1 = \frac{u(\Delta x)^2(3u\,\Delta t/\Delta x - 2(u\,\Delta t/\Delta x)^2 - 1)}{6} \tag{3-4.8}$$

For $\Delta t = 0$, $C_0$ and $C_1$ are the same as for the steady analysis. For the backward Euler (see Gresho *et al.* (1984)) the unsteady discretization contribution to the false diffusion term changes sign hence giving greater stability.

*Dissipation and Dispersion*

From the above procedure it is clear that the equation actually being solving has additional terms (The exception to this is for a Courant number, *C*, (Equation (2-9.1)) of unity ($C = u\,\Delta t/\Delta x = 1$) where $C_0 = C_1 = 0$). The

$C_0\,\partial^2\phi/\partial x^2$ component has the same form as the dissipative term on the right hand side of the Navier-Stokes equations (i.e. $\mu\,\partial^2 u/\partial x^2$ ). Hence the equation actually being solved involves a dissipative *false* (or *numerical*) *diffusion* term. Physical intuition tells us that increasing a fluid's viscosity will damp flow unsteadiness and this (assuming $C_0$ is positive) is just the erroneous effect the $C_0\,\partial^2\phi/\partial x^2$ term has. The $C_1\,\partial^3\phi/\partial x^2$ term produces what is called a *dispersion* error. This error causes wiggles to appear in solutions and phase errors (waves will propagate at the wrong rate). Higher order terms (*HOT* in Equation (3-4.4)) also appear in the modified equation. Those with even order derivatives will cause dissipative behaviour (i.e. will damp flow unsteadiness). Odd ordered derivatives will cause dispersive behaviour, giving wiggles and phase errors. The omitted leading term, arising from the discretization, generally suggests the most striking behaviour of the scheme. *Table 3-3.1* gives the omitted leading terms for some common spatial discretization schemes.

Equation (3-4.7) suggests why the first order explicit forward Euler scheme is more unstable than the first order Implicit (backward Euler) scheme. In the explicit scheme, the stabilising spatial false diffusion term is reduced by $u\,\Delta t/\Delta x$ (the sign change for the backwards Euler has the reverse effect). Motivated by the above observation, Gresho *et al.* presented the Balancing Tensor Diffusivity (BTD) scheme (previously described by Dukowicz and Ramshaw (1979) and Crowley (1967)). This involves adding a contribution to the physical diffusion term that cancels the negative false contribution arising from the forward Euler scheme. In the present context, if Equation (3-4.1) is modified to include a left hand side diffusion term as below

$$\frac{\partial\phi}{\partial t}+u\frac{\partial\phi}{\partial x}=\Gamma\frac{\partial^2\phi}{\partial x^2} \tag{3-4.9}$$

the BTD scheme would correspond to discretizing Equation (3-4.9) with the forward Euler scheme and

$$\Gamma=\Gamma+u^2\frac{\Delta t}{2} \tag{3-4.10}$$

For two-dimensional flows, to apply BTD, cross-stream diffusion terms are necessary and so a 9 point difference stencil is required. In three-dimensions, 19 points are needed. Gresho *et al.* apply the BTD scheme to the convection of a rotating cone shaped passive scalar on a uniform grid (see Tamamidis and Assanis (1993)). BTD showed a similar performance to Arakawa's scheme. If $\Gamma = 0$, BTD becomes the Lax-Wendroff method described in Chapter 2. It is interesting to note that in the previous section the point is made that Lax-Wendroff can also be considered a relation to the semi-Lagrangian methods.

Approximate forms for the manifestation of dissipation and dispersion errors when a wave pattern propagates with a translational velocity $\tilde{U}$ are shown in *Figure 3-4.1*. As can be seen, dissipation results in wave form spreading with $A_2 < A_1$ and dispersion gives spurious oscillations and a phase error (i.e. $\Delta x < \tilde{U}\,\Delta t$ ). The extent of these errors is a function of the chosen temporal and spatial discretization schemes and the way they interact. Clearly, dissipative errors will have a profound effect on temporal behaviour. If the dissipative error is too high, steady flows will erroneously be predicted. Generally, high order convective schemes minimize dissipative errors and for accurate unsteady flow predictions are vital. The ideal minimum accuracy is second oˇrder.

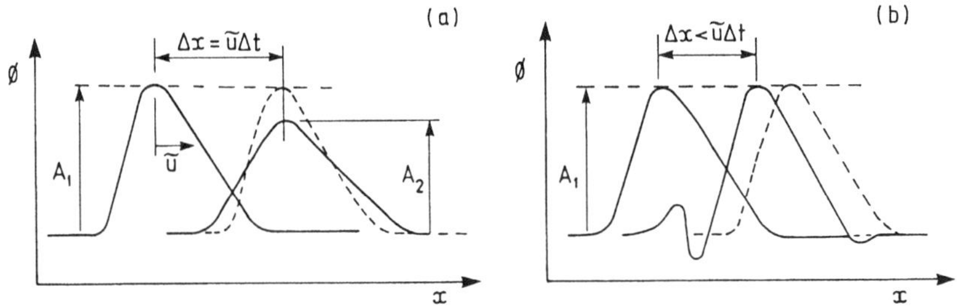


*Figure 3-4.1* Typical form of dissipation and dispersion errors:
(a) Dissipation error and (b) Dispersion error

*Grid Related Effects*

Non-uniform grids are valuable for resolving steep flow field gradients. However, for finite element, difference and volume methods non-smooth and discontinuous changes in adjacent grid spacings can produce dispersion and dissipation errors. Clearly this will affect temporal solutions and so smooth grid changes are recommended. For simple geometries, higher order spatial schemes can readily be implemented. Also, generally grids of acceptable quality are straightforward to generate. Hence good temporal accuracy is also expected. For complex geometry techniques, such as unstructured grids, more caution is required (see, for example, Yao *et al.* (2000)). Higher order schemes are less easy to implement. Also, for complex geometries, during mesh generation, poor cell types can be produced. These could affect temporal accuracy. It should be ensured that the order of all boundary conditions is consistent with the interior domain spatial discretization. With unstructured grids this can be less straightforward. For especially high spatial and hence temporal accuracy, *spectral* methods (generally involving a Fourier series based discretization, the order of which increases with the number of grid points) are recommended with high order temporal schemes. However, for complex geometries, these methods are of limited use. To extent these geometrical restrictions are overcome using *spectral element* techniques (see Patera (1984)) or compact difference schemes (see Lele. (1992)) that have spectral like behaviour.

*Stability Analysis*

It is not unusual to find, as a numerical solution progresses, rapidly growing oscillations and hence results that are totally unlike the expected physical behaviour. This is characteristic of the numerical scheme being *unstable.* The stability of linear, constant coefficient discretized equations with simple boundary conditions can be assessed using the Fourier series based method of von Neumann. However, since most practical problems do not have such a simple nature the use of this approach is limited to suggesting the likely behaviour.

To assist with overcoming the above difficulty Hirt (1968) presents a heuristic stability theory. This is in essence based about the modified equation approach presented earlier, where again the terms in the discretized equations are replaced with Taylor series expansions. The basic idea of the method can be appreciated by considering Equation (3-4.7). In this, when $u\Delta t/\Delta x > 1$ it can be seen that effectively any physical diffusion, which is well known to have a stabilizing effect on numerical schemes, is reduced. The condition $u\Delta t/\Delta x > 1$ corresponds to the well known CFL stability criterion. Hirt's approach involves making stability inferences from the modified equation form.

A relatively simple alternative for studying the dissipative and dispersive behaviour of numerical schemes is through a simplification of the von Neumann stability analysis (see Glass and Rodi, and Gresho *et al.* (1984)). This involves modelling Equation (3-4.1) for a constant $u$ with $\phi$ defined as single Fourier series components. Analytical solutions for these components can readily be derived. These can be used as a basis to assess the accuracies of predictions. For example, phase and amplitude errors can be plotted as a function of wavelength over grid spacing for different Courant numbers. Preferably for high Courant numbers, errors should decrease as the wavelength over grid spacing ratio increases. The rate of decrease depends on the scheme's quality. For an unstable scheme, errors will grow rapidly with time.

*Aliasing*

The shortest wavelength that can be resolved by a grid is $2\Delta\eta$, where for a Cartesian system $\eta = x, y$ or $z$ and for a cylindrical $\eta = r, r\theta$ or $z$. From Fourier analysis, flows can be seen to consist of a range of wave components. Because of the non-linear convective terms, without significant dissipation this range, with time, will grow. Energy from wavelengths that cannot be resolved by the grid (sub-grid scale wavelengths) can erroneously become combined with longer wavelengths. This is called *aliasing*. With most engineering flows, there is sufficient dissipation to damp out the shortest wavelengths (this is a similar concept to the turbulence energy cascade where the energy from smaller scales is absorbed by viscous dissipation). Also, generally, engineering flows do not involve long time period integrations. These longer integrations would give aliasing errors a chance to grow, and eventually result in computational instability and hence solution divergence.

Aliasing problems can be avoided by using special energy conserving convective term treatments. As noted earlier, examples of these are given by Arakawa (1966), and Piacsek and Williams (1970). According to Arakawa, aliasing divergence is characterized by a smooth fluid flow structure degenerating into eddies. Evidently, once formed, the eddies intensify without limit. As a result, there is an explosive growth of kinetic energy. This results in the solution diverging. The special energy conserving schemes noted above have mostly been applied to geophysical flows (see for example James, Jones and Farnell (1981) and Hignett *et al.* (1985)). For these, time integrations are relatively long and wave-like packets are convected with little dissipation. The schemes have also found use in large eddy simulation (see Mason and Cullen (1986)). For these, time integrations must be carried out for several eddy turnover periods, with the flow dynamics uncontaminated by inappropriate numerics. In Chapter 7 performances of the CD, CONDIF and HYBRID convective term treatments are explored for a geophysical related flow with, relatively little dissipation, long time period integrations and convected wave patterns.

## 3-5 Moving Boundaries

With moving boundary problems, the domain movement can either by specified or solved as part of the numerical solution. The latter approach can be far more numerically demanding. Examples of problems involving boundary movement include:

Orbiting bearings;
Piston engine and compressor cylinders;
Sloshing fluids in tanks;
Flow around ships and submerged objects;
Ruptured membranes or dams;
Crystal growth;
Electroplating;
Metal and glass forming;
Extrusion of polymers from nozzles and
Coating flows where fluid is deposited into solid surfaces.

With engines, as well as transients from the piston movement, there will also be rapid transients due to combustion. For journal and other bearings, dynamic instabilities can give rise to shaft orbiting and hence boundary movement. The later examples in the list above relate to free surface flows. Many types of interactions with regions having free surfaces (gas/fluid interfaces) will result in movement of the surface. Mechanisms for inducing this, include movement of boundaries (this could be as a result of either rupture or system oscillations) or objects through a fluid, fluid injection/pouring, external wind forcing and gravitational effects. Also, surface tension forces, themselves can drive flows as for example in Marangoni convection.

With a reciprocating compressor, it can be adequate to prescribe the piston movement. But, generally, with orbiting journal bearing and aero-elasticity problems the equations governing the dynamics of the fluid must be fully coupled with those of the solid. Similarly, with free surfaces the fluid dynamic equations should be coupled with equations for surface momentum conservation. The latter equations can include the fluid surface tension force, which itself can be a function of temperature and fluid contaminants such as chemicals or even dirt.

*General Modelling Techniques*

Ferziger and Peric (1997) present a moving boundary finite volume technique applicable to unstructured grids and introduce much background information on modelling moving boundary and free surface flows. Yang *et al.* (1997) give a Cartesian, trimmed cell method for moving body problems. However, the Chimera (a Chimera is a mythological creature with a goats body, snakes tail and lions head) grid method, described by Chesshire and Henshaw (1990), has seen more popularity for modelling such problems. With this, the flow domain is often mapped by a Cartesian grid and the body modelled by a moving curvilinear 'O' grid. The necessary interpolations between the significantly different grid systems makes conservation difficult to enforce. With other moving boundary schemes, conservation errors can also occur. However, these can be minimized by the use of small time-steps or the enforcement of what is called the *space conservation law* (see Demirdzic and Peric (1988)). Vortex methods, that generally do not require the use of grids, (see Khatir (2000)) naturally lend themselves to moving boundary problems. Martins and Ghoniem (1991), and Gharakhani and Ghoniem (1999) show their application to piston engine flows.

*Moving Boundaries with Body Fitted Grids*

Here a moving boundary technique, applicable to body fitted grids is briefly described. It is assumed that the boundary movement is prescribed. The governing equations in an $(x, y, z, t)$ coordinate system can be transformed into a body fitted $(\xi, \eta, \varsigma, \tau)$ using the chain rule of differential calculus. The transformation from steady cylindrical coordinates is given by Rayner (1993). Here it simply involves making the following substitutions in the governing $(x, y, z, t)$ equations, where $\xi = \xi\ (x, y, z, t)$, $\eta = \eta\ (x, y, z, t)$, $\varsigma = \varsigma\ (x, y, z, t)$ and $\tau = \tau(t)$

$$\frac{\partial}{\partial x} = \left(\frac{\partial \xi}{\partial x}\right)\frac{\partial}{\partial \xi} + \left(\frac{\partial \eta}{\partial x}\right)\frac{\partial}{\partial \eta} + \left(\frac{\partial \varsigma}{\partial x}\right)\frac{\partial}{\partial \varsigma} + \left(\frac{\partial \tau}{\partial x}\right)\frac{\partial}{\partial \tau} \tag{3-5.1}$$

$$\frac{\partial}{\partial y} = \left(\frac{\partial \xi}{\partial y}\right)\frac{\partial}{\partial \xi} + \left(\frac{\partial \eta}{\partial y}\right)\frac{\partial}{\partial \eta} + \left(\frac{\partial \varsigma}{\partial y}\right)\frac{\partial}{\partial \varsigma} + \left(\frac{\partial \tau}{\partial y}\right)\frac{\partial}{\partial \tau} \tag{3-5.2}$$

$$\frac{\partial}{\partial z} = \left(\frac{\partial \xi}{\partial z}\right)\frac{\partial}{\partial \xi} + \left(\frac{\partial \eta}{\partial z}\right)\frac{\partial}{\partial \eta} + \left(\frac{\partial \varsigma}{\partial z}\right)\frac{\partial}{\partial \varsigma} + \left(\frac{\partial \tau}{\partial z}\right)\frac{\partial}{\partial \tau} \tag{3-5.3}$$

and importantly in the present context

$$\frac{\partial}{\partial t} = \left(\frac{\partial \xi}{\partial t}\right)\frac{\partial}{\partial \xi} + \left(\frac{\partial \eta}{\partial t}\right)\frac{\partial}{\partial \eta} + \left(\frac{\partial \varsigma}{\partial t}\right)\frac{\partial}{\partial \varsigma} + \left(\frac{\partial \tau}{\partial t}\right)\frac{\partial}{\partial \tau} \tag{3-5.4}$$

The bracketed derivatives above are called *metrics*. Variables treated as fixed in the partial differentiations contained in Equations (3-5.1) to (3-5.4) are indicated using subscripts below

$$\left(\frac{\partial}{\partial x}\right)_{yzt} \left(\frac{\partial}{\partial y}\right)_{xzt} \left(\frac{\partial}{\partial z}\right)_{xyt} \left(\frac{\partial}{\partial t}\right)_{xyz} \left(\frac{\partial}{\partial \xi}\right)_{\eta\varsigma\tau} \left(\frac{\partial}{\partial \eta}\right)_{\xi\varsigma\tau} \left(\frac{\partial}{\partial \varsigma}\right)_{\xi\eta\tau} \left(\frac{\partial}{\partial \tau}\right)_{\xi\eta\varsigma}$$

Clearly $\partial\tau/\partial x = \partial\tau/\partial y = \partial\tau/\partial z = 0$ and so grid distortion from boundary movement does not affect transformations (3-5.1) to (3-5.3). Therefore, for simplicity a two-dimensional $(\xi, \eta, \tau)$ system is now considered. Generally $\tau = t$ and so Equation (3-5.4) can be re-expressed as

$$\frac{\partial}{\partial t} = \frac{\partial}{\partial t} + \left(\frac{\partial \xi}{\partial t}\right)_{xy} \frac{\partial}{\partial \xi} + \left(\frac{\partial \eta}{\partial t}\right)_{xy} \frac{\partial}{\partial \eta} \qquad (3\text{-}5.5)$$

When modelling moving boundary/grid problems the transformed grid stays fixed. It would therefore appear that the metrics $\partial\xi/\partial t = \partial\eta/\partial t = 0$ in Equation (3-5.5). But examination of their subscripts shows they are for a fixed $x$, $y$ location. Therefore, since the $x$, $y$ plane grid is moving these metrics are in fact non-zero. This can readily be demonstrated by revisiting the simple geometry shown in *Figure 3-2.1*. For this, the metrics can be easily specified analytically based on the following transformations

$$\xi = x\,, \quad \eta = y/h(x,t) \qquad (3\text{-}5.6)$$

The variable $h$ is the local instantaneous height between the upper and lower channel walls and so changes with $x$ and $t$. Clearly, since $\xi = x$, $\partial\xi/\partial t = 0$. From physical intuition this would be expected – movement of the top surface does not affect lines of constant $x$. However,

$$\frac{\partial \eta}{\partial t} = \frac{\partial\ (y/h)}{\partial t} = -\frac{y}{h^2}\frac{\partial h}{\partial t} = -\frac{\eta}{h}\frac{\partial h}{\partial t} \qquad (3\text{-}5.7)$$

Clearly, this metric is non-zero giving rise to an additional term in the transformed governing equations. Examples, showing the application of moving curvilinear grids to flows in piston/cylinder assemblies are given by Gosman and Watkins (1977), and Gosman and Johns (1978). In both these studies, the grid movement, as in the simple example presented above, is restricted to one direction.

*Space Conservation Laws*

For moving grid problems, in essence (either explicitly or implicitly), grid velocities are required in the discretized equations. If these velocities are not carefully specified, the grid movement can result in mass conservation errors. To avoid these, the grid velocities must be such that the *Space Conservation Law* (SCL) is obeyed. For an arbitrary shaped domain, containing a constant density fluid, the SCL can be expressed (see Demirdzic and Peric (1988)) in the following integral form

$$\frac{V^{new} - V^{old}}{\Delta t} = \oint \boldsymbol{v}_g \, . \boldsymbol{ds}^{new} \tag{3-5.8}$$

where $\boldsymbol{v}_g$ are grid velocities, $V$ cell volumes and $\boldsymbol{ds}$ surface areas. For a region containing rectangular, geographically notated, elements the following form of Equation (3-5.8) is most convenient

$$\frac{V^{new} - V^{old}}{\Delta t} = \sum v_{gNB} \cdot s_{NB}^{new} \tag{3-5.9}$$

To enforce the SCL, grid velocities must be chosen so that equations (3-5.8) or (3-5.9) are automatically satisfied. This implies

$$\left(\frac{\Delta V}{\Delta t}\right)_{SCL} = \left(\frac{\Delta V}{\Delta t}\right)_{A} \tag{3-5.10}$$

the subscript $A$ referring to an actual value. First it is instructive to consider the case where the temporal grid movement is in just one direction. This is shown in *Figure 3-5.1a*, where the grid is stretched vertically at a velocity $v_{gn}$ such that

$$v_{gn} = \frac{\delta y_n}{\Delta t} \tag{3-5.11}$$

Equation (3-5.9) then gives (since $u_{ge} = u_{gw} = v_{gs} = 0$)

$$\left(\frac{\Delta V}{\Delta t}\right)_{SCL} = v_{gn}\Delta x \tag{3-5.12}$$

Also, it can be readily seen that for this simple case

$$\left(\frac{\Delta V}{\Delta t}\right)_{A} = \frac{\delta y_n \Delta x}{\Delta t} \tag{3-5.13}$$

Using (3-5.11), Equation (3-5.13) can be re-expressed as

$$\left(\frac{\Delta V}{\Delta t}\right)_{A} = v_{gn}\Delta x \tag{3-5.14}$$

Comparison of equations (3-5.12) and (3-5.14) shows the SCL is automatically satisfied. It will now be considered for the simple unidirectional coordinate transformation embodied in Equations (3-5.6). The moving control volumes relating to this are shown in *Figure 3-5.1b*, $h^{old}$ and $h^{new}$ corresponding to the temporal height change of the cell. The SCL (Equation (3-5.9)) is again as given by Equation (3-5.12), where (assuming $h$ is centrally located within the cell, which is not always the case)

$$v_{gn} = \frac{\delta h}{\Delta t} \tag{3-5.15}$$

Also, it can be seen, since the volumes of the left and right hatched areas in the $h^{new}$ frame will cancel out (this is again assuming $h$ is centrally located in the cell)

$$\left(\frac{\Delta V}{\Delta t}\right)_{A} = \frac{\delta h \Delta x}{\Delta t} \tag{3-5.16}$$

Substituting Equation (3-5.15) into (3-5.16) gives Equation (3-5.12), showing again that the SCL is satisfied.

Finally, the more general case will be considered where two perpendicular surfaces move ( the *n* and *e* control volume faces) as shown in *Figure 3-5.1c*. For this case we have the additional velocity equation

$$u_{ge} = \frac{\delta x_e}{\Delta t} \tag{3-5.17}$$

The space conservation law (Equation (3-5.9)) then becomes

$$\left(\frac{\Delta V}{\Delta t}\right)_{SCL} = v_{gn}\Delta x^{new} + u_{ge}\Delta y^{new} \tag{3-5.18}$$

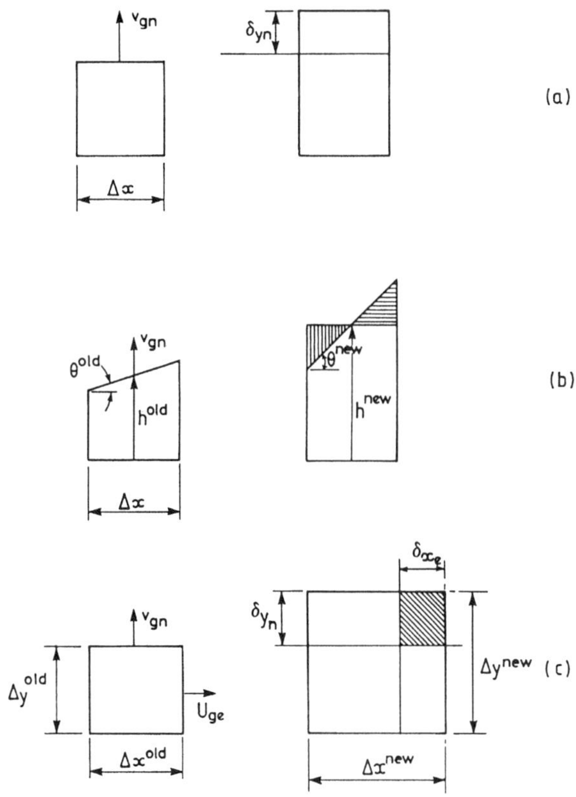


*Figure 3-5.1* Some basic temporal control volume deformations: (a) deformation just in one direction; (b) deformation for unidirectional coordinate transformation and (c) deformation in two directions.

Using Equations (3-5.11) and (3-5.17) (the obvious grid velocity expressions) gives the following artificial continuity equation mass source error expression

$$\left(\frac{\Delta V}{\Delta t}\right)_{ERROR} = u_{ge}\left(-v_{gs}\right)\Delta t \tag{3-5.19}$$

If equations (3-5.11) and (3-5.17) are to be used in computations, it must be ensured that the error term $\varepsilon = (\Delta V)_{ERROR}/(\Delta V)_{SCL}$ (this assumes $(\Delta V)_A \cong (\Delta V)_{SCL}$) is small, i.e.

$$\varepsilon \approx \frac{u_{ge}\left(-v_{gs}\right)\Delta t}{u_{ge}\Delta y^{new} + v_{gn}\Delta x^{new}} \tag{3-5.20}$$

Equation (3-5.20) implies that to make the error negligible $\Delta t$ must be small or the grid spacing large! On the other hand, (3-5.20) shows that keeping one set of grid lines fixed ensures $\varepsilon = 0$. The SCL time-step constraint (i.e. Equation (3-5.20)) can be more severe than the $\Delta t$ constraint for temporal accuracy, hence the importance of its consideration. A possible grid velocity specification that automatically satisfies the SCL, in the example above, is as follows

$$u_{ge} = \frac{\Delta y^{old} + \Delta y^{new}}{2\Delta y^{new}} \frac{\delta x_e}{\Delta t} \tag{3-5.21}$$

$$v_{gn} = \frac{\Delta x^{old} + \Delta x^{new}}{2\Delta x^{new}} \frac{\delta y_n}{\Delta t} \tag{3-5.22}$$

Further expressions, for arbitrary moving grids, are presented by Demirdzic and Peric (1988).

*Inertial Coordinate Systems with Moving Boundaries*

For certain geometries, moving coordinate systems can be chosen to minimize physical plane grid movement thus minimizing the temporal metrics in Equation (3-5.5) (an example suggesting this is noted in Section 3-5.2). This gives the potential for greater computational *accuracy* and economy. The *Figure 3-2.1* geometry can be used to illustrate this point. If the top wall has purely transitional motion $\tilde{U}$ in the $x$ direction with $\tilde{V} = \tilde{W} = 0$, the minimum in $h$ (the local vertical height) will propagate downstream at a velocity $\tilde{U}$ giving an expensive unsteady flow problem. However, if the solution is made in a coordinate system moving at $\tilde{U}$ there is no grid distortion. The problem effectively becomes steady. Even if say $\tilde{U} \neq 0$ and $\tilde{V} = A \sin kt$ (where $A$ and $k$ are amplitude parameters), for small $A$, solution in a coordinate system moving at $\tilde{U}$, although not steady should yield computational benefits. An example showing the use of a moving coordinate system to minimize temporal grid distortion for an unsteady hydrodynamic journal bearing problem is shown in Chapter 9.

*Free Surface Modelling*

Essentially there are the following three key free surface flow modelling techniques:

(A) Use a grid that extends beyond the free surface. The fluid boundary, within the extended domain, is then tracked. At cells cut by the free surface, modified flow governing equations are used. This is the basis of the relatively well-known marker and cell (MAC) technique of Harlow and Welch (1965). There are many variants to the basic original explicit MAC scheme. For example, to overcome the viscous stability limit of the original explicit scheme, Pracht (1970) presents MACRL, an implicit version.

(B) Have a deforming mesh that conforms to the free surface region and moves with it. Initially, the free surface shape is prescribed. Neglecting a boundary condition, the governing equations are solved based on the prescribed shape. The neglected boundary condition is then used to update the free surface shape (see Tanner, Nickell and Bilger (1975)). This iterative procedure is repeated until convergence. Variables on the old mesh must be repeatedly interpolated onto the new mesh.

(C) Extend (B) above by solving for the free surface position simultaneously with free surface equations (see Engelman and Sani (1986)). Interpolation between different meshes is avoided. This technique is used in the FIDAP commerical CFD program.

With Approach (C), care is required when considering the Eulerian frame time derivative in the governing equations. The grid is moving, but not with the flow, in a Lagrangian frame. Instead, the movement is in a mixed frame. Hence, the transport equation time derivatives must be transformed using the following substantial derivative type expression

$$\frac{\partial \phi}{\partial t} = \frac{\partial \phi}{\partial t} + \frac{\partial \mathbf{x}}{\partial t} \cdot \nabla \phi \qquad (3\text{-}5.23)$$

where **x** are coordinates of moving grid points.

*Free Surface Tracking Techniques*

There are several free surface-tracking techniques. The simplest is to define a free surface height, $h(x,t)$, relative to a datum and solve for this variable (see *Figure 3-5.2a*). For a two-dimensional $x$-$y$ plane system the following equation is sufficient

$$\frac{\partial h}{\partial t} + u\frac{\partial h}{\partial x} = v \qquad (3\text{-}5.24)$$

Miyata and Nishimura (1985) use the above technique for modelling surface waves arising from the passage of ships. Since relatively little storage and computational effort is required, use of Equation (3-5.24) is attractive. However, for the same values of $x$ and $t$, $h(x,t)$ can become multivalued and then the method fails. In certain situations, the coordinate system in which $h$ is defined can be altered to overcome this problem. *Figure 3-5.2b* shows a technique resorted to by Sato and Scriven (1981). In this, a polar coordinate description of $h$ is used.

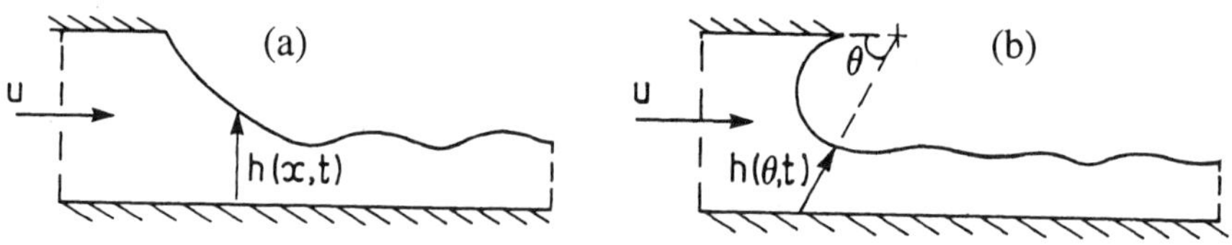


*Figure 3-5.2* Description of free surface profile using a height variable: (a) profile where $h(x,t)$ is not multivalued and (b) profile where $h$ can be multivalued.

*Figure 3-5.3* shows more flexible variant, used by Engelman and Sani (1986) and related to the height function method above. With this, a more complex fixed reference frame, indicated by the heavy line, is used. Essentially, a local 'height' variable is implemented. Through parametric equations this is subsequently related to $x$ and $y$ coordinates.

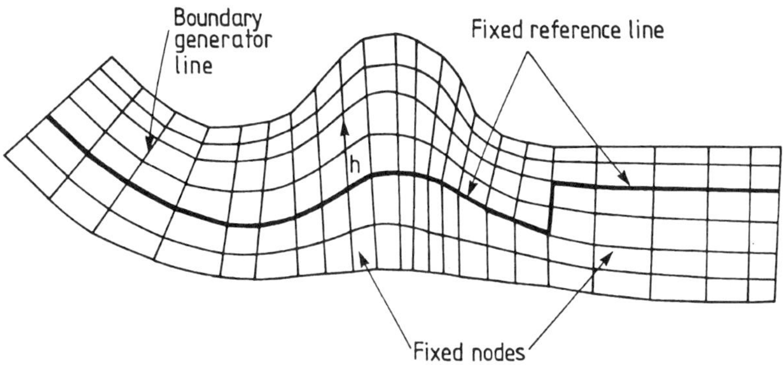


*Figure 3-5.3* More flexible 'height' function variant (from Engelman and Sani (1986))

An alternative to the use of Equation (3-5.24) is to seed the surface with weightless particles and solve the following Lagrangian equations for their advection (see Castrejon and Spalding (1988))

$$\frac{\partial x_P}{\partial t} = u_P$$

$$\frac{\partial y_P}{\partial t} = v_P \tag{3-5.25}$$

Using an explicit solution procedure the following particle location equations can be written

$$x_P^{new} = x_P^{old} + u_P \Delta t$$

$$y_P^{new} = y_P^{old} + v_P \Delta t \tag{3-5.26}$$

Alternatively, not just the surface, but all the fluid region can be seeded with particles and Equations (3-5.25) applied, as demonstrated, for a cylindrical geometry, in Section 2-11 (see *Figure 2-12.2*). The seeding of the entire fluid region with weightless particles is used in the popular MAC approach. Relative to Equation (3-5.24), the use of Equation (3-5.25) requires more storage and complex logic to recover the free surface unambiguously from particle positions.

A further alternative is to solve a transport equation for, $F$, the fractional volume of cell occupied by the fluid (see Liu Jun (1986))

$$\frac{\partial F}{\partial t} + u\frac{\partial F}{\partial x} + v\frac{\partial F}{\partial y} = 0 \qquad (3\text{-}5.27)$$

$F = 1$ corresponds to a cell completely full of fluid and $F = 0$ the reverse.

The free surface form can be recovered from the nodal point distribution of $F$. Equation (3-5.27) is simple to implement, no complex logic is required and the storage need is relatively low. However, to avoid smearing of the free surface, low false diffusion convective term discretizations must be implemented in the solution procedure. A key disadvantage is that Equation (3-5.27) cannot resolve sub-grid scale free surface features. *Figure 3-5.4* shows a solution of the $F$ transport equation cast in polar coordinates for the *Figure 2-12.1* geometry. The boundary condition $F = 1$ is enforced for the cylinder. A 65 x 120 $(r,\theta)$ grid with CONDIF is used. This scheme and grid is ample to get velocity profile grid independence. However, the significant smearing of the scalar interface is clear (cf. *Figure 2-12.2* using the MAC type interfacing method).

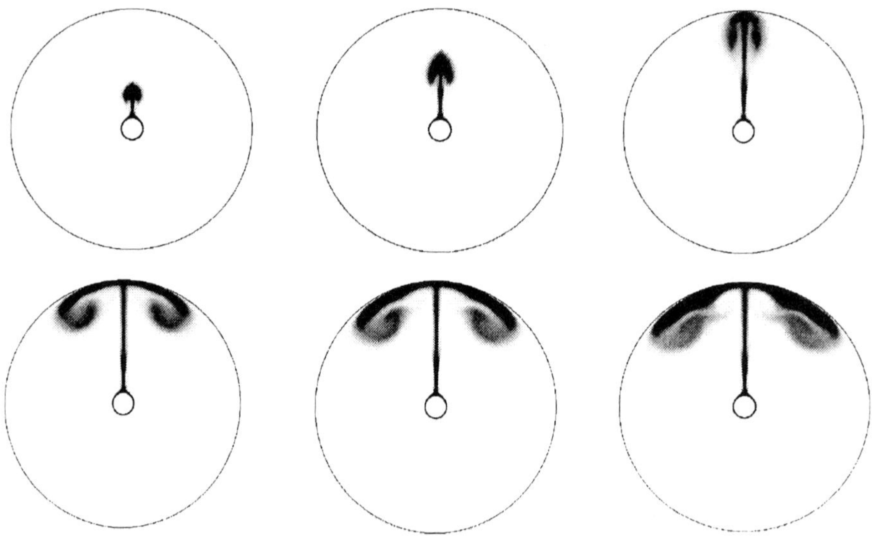

*Figure 3-5.4* Solution of $F$ transport equation in polar coordinates for *Figure 2-11.1* geometry at: (a) $t \approx 9.5$ s; (b) $t \approx 13.5$ s; (c) $t \approx 19.5$ s; (d) $t \approx 31.0$ s; (e) $t \approx 37$ s and (f) $t \approx 55$ s.

## 3-6 Time Adaptive Solutions

Adaptive solutions, where space or time-steps or even the order of numerical algorithms are automatically adjusted during computations to minimize errors are attractive. Schonauer, Glotz and Raith (1980) perform boundary layer flow calculations where, based on error estimates, space and time-steps, and also algorithm orders are all automatically adapted. Wolfsberg and Freyberg (1994) locally adapt both space and time-steps for species transport problems. Generally, with adaptive space step methods, either the number of grid points can be fixed but simply redistributed or alternatively new grid points can be introduced. Forms of the latter (involving what is called adaptive grid embedding) are given by Wolfsberg and Freyberg, Davis and Dannenhoffer (1994) and Coelho *et al.* (1991). The application of this grid point insertion technique to time dependent shock hydrodynamics is shown by Berger and Colella (1989). Importantly, multilevel (or multigrid) methods (a powerful convergence acceleration technique in itself, involving overlaying grids of different densities) discussed in Section 4-2, lend themselves well to grid adaptation (see Brandt (1980)) and embedding (see Thompson and Ferziger (1989)). A grid embedding example is given in Section 5-3. Spatial grid refinement can either be carried out after various numbers of steady state solution iterations or alternatively as a solution temporally evolves. A simple version of the latter is now briefly outlined. Adaptive refinement of time-steps is discussed in Chapter 5.

*Spatial Grid Adaptation with Time*

The time moving grid transforms discussed in the previous section can be used as part of a solution procedure to cluster grid points in areas of high gradients (see Anderson (1995)). A simple adaptive grid strategy for a fixed number of grid points is to define relationships such that

$$\Delta x = C_x \Delta\xi\, f\left(\frac{1}{\left|\partial\phi/\partial x\right|}\right)$$

$$\Delta y = C_y \Delta\eta\, f\left(\frac{1}{\left|\partial\phi/\partial y\right|}\right) \qquad (3\text{-}6.1)$$

where $C_x$ and $C_y$ are arbitrary scale factors. The transformed grid and hence $\Delta\xi$ and $\Delta\eta$ is fixed. Grid refinement could be based on for example, $\phi = p, T$ or $\rho$. In regions of high $\phi$ gradient, as can be seen from Equations

(3-6.1), the grid spacing in the physical plane will be made closer. Once $\Delta x$ and $\Delta y$ have been established the temporal metrics $\partial\xi/\partial t$ and $\partial\eta/\partial t$ are evaluated, giving contributions to the governing equations. The adaptive process is carried out continuously through time. The final form of the adaptive grid can in itself be helpful in interpreting the physical nature of the predicted flow. Steger and Kutler (1977) successfully use spatial transformed grid adaptation with time to capture aircraft vortex wakes. The adaptation is based on changes in vorticity.

## 3-7 Conclusions

Generally, for flows classified as unsteady it is meant that variables undergo temporal changes with respect to the grid. For certain systems (classified as exhibiting unsteady flow), the fluid structure may stay exactly or much the same, but simply translate or rotate in space, relative to the coordinate system/grid. In the interests of computational accuracy and speed, minimizing temporal variations with respect to the grid is valuable. Also, selecting the coordinate system velocity to minimize temporal grid distortions is worthwhile.

As part of the convective term discretization, dependent variables are interpolated from grid points to control volume faces. For example, the third order QUICK scheme involves a quadratic fit to three grid point values. When used with ADI the involvement of three grid points detracts from its efficiency. Compact spline techniques offer the potential of high accuracy without involving extra grid points. This is advantageous for ADI type methods. Compact difference schemes can also be used to give high accuracy with small differencing stencils.

Oscillations arising from the use of higher order schemes are associated with *dispersion* errors, which cause phase errors. Generally introducing numerical or false diffusion damps oscillations. A more superior oscillation removal technique is the Total Variation Diminishing (TVD) approach. The accuracy of QUICK with first order forward Euler time differencing is poor. Therefore, it is best modified using the Estimated Streaming Approach to give QUICKEST. Schemes related to this approach can be derived that are accurate for Courant numbers greater than unity. But these have only applied to one-dimensional flow with more work required for their extension to two and three-dimensions. For systems without significant dissipation but requiring long time period integrations, energy from wavelengths that cannot be resolved by the grid can erroneously become combined with longer wavelengths. This is called *aliasing*. Aliasing problems can be avoided by using special energy conserving convective term treatments.

For moving boundary problems, the domain movement can either by specified or solved as part of the numerical solution. The latter approach can be far more numerically demanding. Cartesian, trimmed cell methods can be used for problems involving bodies moving through fluid domains. However, the Chimera grid method is more popular, despite interpolations between the significantly different grid systems making conservation difficult to enforce. With other moving boundary schemes, involving cell distortion, conservation errors can also occur. However, these can minimized by the use of small time-steps or the enforcement of the space conservation law.

A possibility when modelling free surfaces is to have the grid conform to the fluid region and move with it. Another strategy is to use a grid that extends beyond the free surface. The fluid boundary, within the extended domain, is then tracked. This is the basis of the relatively well-known MAC technique. There are several free surface-tracking techniques. The simplest is to define a free surface height relative to a datum and solve for this variable. Since relatively little storage and computational effort is required the method is attractive. However, for complex free surface configurations the height variable can become multivalued and then the method fails. An alternative is to seed the surface with weightless particles and solve Lagrangian equations for their advection. This approach requires more storage and complex logic. A further alternative is to solve a transport equation for the fractional volume of cell occupied by the fluid. The free surface form can be recovered from the nodal point distribution of the fractional volume variable. This technique is simple to implement and the storage needed relatively low. However, to avoid smearing of the free surface, low false diffusion convective term discretizations must be implemented. Adaptive solutions, where space steps or even the order of numerical algorithms used to resolve these are automatically adjusted during computations to minimize errors, are attractive. Time moving grid transforms can be used to cluster grid points in areas of high gradients.

CHAPTER

# 4

# SOLVER ALGORITHMS AND GENERAL SOLUTION PROCEDURES

## 4-1 Introduction

Once transport equation temporal derivatives have been discretized as in Chapter 2 and the spatial terms as outlined in Chapter 3 the following stages still remain:

(I) The discretized equations must be arranged in a form that will be both *stable* in the chosen solution framework and not be subject to error build up mechanisms.

(II) The discretized equation sets must be solved in an efficient manner that has compatibility with the nature of the fluid flow and the architecture of the computer being used. Also, the accuracy to which the equations are solved must be sufficient and consistent with other solution errors.

(III) A pressure field (pressure is an unknown in the momentum transport equations) that is consistent with a mass conserving solution must be produced. Alternatively, pressure can, in some way, be removed from the governing equations.

The compatibility of the solution method with the flow and also computer architecture (Point (II)) generally involves some compromise. Computers can be broadly classified into *serial*, *vector* or *parallel* machines. Parallel machines, with a smaller number of powerful processors, can also utilize an element of vector processing. Emerson (1996) and Lacor (1996) describe different computer architectures and compatible solution algorithms. For parallel and vector machines data dependencies must be removed. However, the removal of these can result in significant algorithm modifications or the choice of completely different algorithms.

In this chapter, common solvers and solution procedures are outlined along with their relationship to particular computer architectures. Multilevel (or multigrid) convergence is presently, arguably, the most efficient simultaneous equation solver technique and at the cutting edge of CFD technology. Therefore, some attention will be paid to this and its potential benefits. These include: circumvention of the Courant-Friedrichs-Lewy (CFL) condition (when time marching to steady states); computational savings for time evolving flows; savings with adapted time-step solution procedures (see Chapter 5) and, when parallel processing, increased coupling of spatially decomposed domains. Some attention is also paid to Newton methods, which for unsteady flows, can be viable. Also, parallel processing in the time domain is discussed.

The following pressure elimination or prediction techniques are outlined in relation to unsteady flows: streamfunction vorticity; SIMPLE including AVPI and CTS modifications; fractional step; artificial compressibility; penalty function and ICE. Also, techniques for avoiding oscillatory pressure fields and increasing numerical precision of the pressure field are noted. The chapter finishes with a review of algorithms used in some relatively well known commercial CFD codes. Importantly, the contents of this chapter focus around commercially used algorithms.

## 4-2 Simultaneous Equation Solvers

Explicit temporal schemes do not require a simultaneous equation solver and can be easily parallelized. However, they have more severe time-step stability restrictions. Consequently they are only suitable when high temporal resolution is required. Implicit techniques require a simultaneous equation solver. This is generally a major solution overhead. Approximate factorization or splitting methods, of which ADI is a family member, successfully combine explicit and implicit techniques only requiring a one-dimensional solver. With these, when time marching to a steady state, best efficiency is gained by using a series of increasing $\Delta t$ 's in a cyclic fashion (see Ferziger and Peric (1977, p.122)).

A simple technique such as Gaussian elimination is an effective solver in one-dimension. However, in two or three-dimensions, for sparse non-linear matrix systems it is inefficient. Since solvers are a major solution overhead, there has been much work aimed towards establishing efficient methods and so a wide range are available. Due to the non-linear nature of

CFD problems, solution procedures that are essentially *iterative* in nature are preferred. *Jacobi* and *Gauss-Siedel* are two of the simplest solvers but are relatively inefficient and seldom used on their own. A popular CFD solver (see Patankar (1980)) is the *Tri-diagonal Matrix Algorithm* (TDMA). With this, Gaussian-elimination is used in one coordinate direction. The other directions are treated using Gauss-Siedel. The direction that Gaussian-elimination is applied is generally varied through the solution. For vector and parallel processing, to avoid data dependencies the *red-black* procedure is popular. This can be considered as a Jacobi/Gauss-Siedel hybrid. If a structured grid is viewed as a chessboard, with red and black squares, the technique involves applying the Jacobi method to all the red and then finally all the black squares. If a sufficiently good initial guess is available *Newton-Raphson* (Newton) methods have quadratic convergence (i.e. the convergence error is proportional to the inverse square of the number of iterations). This property is very compatible with unsteady flows, where, if time-steps are sufficiently small, the solution from the previous time-step is a good approximation to the solution for the next. Because of this unsteady flow compatibility, the Newton method is considered further later.

More sophisticated and popular solvers include Stone's (Stone (1968)) method (also called the *Strongly Implicit Procedure* – SIP), *biconjugate* solvers and variants, and *multilevel* (*multigrid*) methods. Unlike the simpler solvers, the SIP and biconjugate are less sensitive to violation of the *Scarborough* criterion (Equations (4-2.5) and (4-2.6)) relating to *diagonal dominance*. For most solvers, the number of iterations to convergence increases with approximately the square of the number of grid points. However, with multilevel methods, in theory, the number of iterations stays constant. Consequently it is perhaps the most impressive of solution techniques and so some attention will be paid to it here. It should be noted that, in practice, multilevel methods are best viewed as a convergence acceleration technique, needing a base solver such as Gauss-Siedel or TDMA. For satisfactory convergence, many solvers require diagonal dominance of the governing equations. Therefore, in this section, ways in which this can be achieved or increased are discussed first. For all Case Studies to be presented either TDMA or multilevel convergence with TDMA is used. Therefore, next details of the present TDMA implementation are given along with, most importantly, convergence criteria. Multilevel convergence will then be discussed followed by the Newton method.

*Diagonal Dominance Issues*

For predictions with no source terms, at the steady state, the diagonal coefficient $A_{P,\phi}$ (see Equation (2-3.1)) should equal $\sum A_{NB,\phi}$. If this were not the case, for a uniform property domain with a constant boundary value, say $\phi = 0$, and no source terms the wrong $(\phi \neq 0)$ steady state solution would be returned. To make $A_{P\phi} = \sum A_{NB\phi}$, following Patankar (1980), in discrete form

$$\phi_P^{new}\left[\frac{\partial \rho}{\partial t} + \nabla \cdot (\rho \mathbf{u})\right] \tag{4-2.1}$$

can be subtracted from the governing convection-diffusion transport equations. The terms in the square brackets constitute the continuity equation. Correctly, if at each time-step continuity is satisfied, the original modelled equation will be unaffected by the subtraction. Clearly it is important to ensure this is the case. Therefore, with SIMPLE (see Section 4-4) related iterative pressure correction methods, to avoid errors growing with time, an adequate number of continuity enforcing pressure correction sweeps is desirable.

Subtracting the discrete forms of the time derivatives in the transport

$$\frac{\partial(\rho\phi)}{\partial t} \cong \frac{\rho_P{}^{new}\phi_P{}^{new} - \rho_P{}^{old}\phi_P{}^{old}}{\Delta t} \tag{4-2.2}$$

and continuity multiplied by $\phi_P{}^{new}$

$$\phi_P^{new}\frac{\partial \rho}{\partial t} \cong \phi_P^{new}\frac{\left(\rho_P{}^{new} - \rho_p{}^{old}\right)}{\Delta t} \tag{4-2.3}$$

equations, gives the following two-level scheme discretized equation time derivative

$$\frac{\rho_P{}^{old}\left(\phi_P{}^{new} - \phi_P{}^{old}\right)}{\Delta t} \tag{4-2.4}$$

As can be seen, $\rho_P^{new}$ no longer features in the temporal derivative. Following on from the $A_{P\phi} = \sum A_{NB\phi}$ steady state, no source physical constraint, a criterion for convergence of Gauss-Siedel related iterative solvers is that of Scarborough. This states that for at least one equation

$$\frac{\sum |A_{NB\phi}|}{|A_{P\phi}|} < 1 \tag{4-2.5}$$

with, for all others,

$$\frac{\sum |A_{NB\phi}|}{|A_{P\phi}|} \leq 1 \tag{4-2.6}$$

The criterion echoes a relatively well-known fact that it is desirable, where possible, to make $|A_{P\phi}|$ large. This is called *diagonal dominance* and will give solution stability by lessening changes in $\phi$ between iterations. For steady flows, diagonal dominance is mostly improved by directly incorporating under-relaxation in the discretized equations using the expression below

$$\phi_P = \alpha_\phi \phi_P^{new} + (1 - \alpha_\phi)\phi_P^{old} \tag{4-2.7}$$

The 'new' and 'old' subscripts are used here in an iterative sense and $\alpha_\phi$ is an under-relaxation parameter. Equation (4-2.7) is amalgamated with the coefficients of Equation (2-3.1). The amalgamation results in contributions to both $A_{P\phi}$ and the source. For small $\alpha_\phi$ values, under-relaxation can cause excessively slow convergence. In this situation, when convergence is based on changes in variables, convergence can be erroneously assumed. Clearly, for unsteady flows to increase diagonal dominance the $\rho_P^{old}/\Delta t$ component multiplying $\phi_P^{new}$ can be combined with $A_{P\phi}$ (the $\rho_P^{old}\,\phi_P^{old}/\Delta t$ term is committed to the source term). Hence as $\Delta t$ becomes smaller, $A_{P\phi}$ will increase in size and even extremely unstable equation systems will become stable (for example, the CD scheme, which sometimes produces negative coefficients and hence of loss of diagonal dominance can be used (see Section 3-3)). *False time-stepping* is a popular approach where the superscripts 'old' and 'new' in the above are used in an iterative sense and $\Delta t$ becomes a false time scale (see Section 2-8 for more details). In fact, there are strong similarities between the equation structures for the 1st order implicit Euler temporal scheme and also those where under-relaxation is applied to the steady equations. Both strategies produce contributions to $A_{P\phi}$ and $S_{1\phi}$. A key difference is that a constant time-step for each cell essentially corresponds to a spatially varying under-relaxation factor. On the other hand using a constant $\alpha_\phi$ corresponds to a spatially varying $\Delta t$.

With unsteady flows, it is desirable to avoid under-relaxation as much as possible. Instead, time-steps should be reduced, giving the bonus of increased temporal accuracy. For unsteady solutions, when convergence judgement parameters based on changes in variables between iterations are used, caution is required. This is especially the case when significant additional forms of under-relaxation are implemented. For small time-steps variable changes will already be small and this effect, combined with strong under-relaxation, can be misleading.

For certain flows, rotating at $\Omega$, an important relaxation/diagonal dominance enhancement is that of Gosman *et al.* (1976). The enhancement's existence arises because for rotating systems, a dominant radial momentum equation term is the tangential velocity. Momentum equation coupling is improved by addition of the following to the radial, *u,* momentum equation:

$$\frac{\alpha_0 \rho_P \left|v+\alpha_1 \Omega r\right| \left(u_P{}^{old} - u_P{}^{new}\right)}{r} \tag{4-2.8}$$

The parameters $\alpha_0$ and $\alpha_1$ are relaxation factors and in the original formulation $\alpha_1 = 1$. Here, the superscripts '*old*' and '*new*' are in an iterative sense and $u_P{}^{new}$ can be combined with $A_{Pu}$, thus increasing diagonal dominance. The term of Gosman *et al.* is used in some rotating flow predictions presented later and as shown by Vaughan (1989) is highly beneficial.

For SIMPLE type methods (Section 4-4), the coefficients of the pressure correction equation are an inverse function of the momentum equation diagonal coefficients. When the time contribution to the latter is expressed as $\rho_P^{old}/\Delta t$, for small $\Delta t$ pressure correction equation coefficients will become small. Relative to the transport Equation (2-3.1) the convergence rate for pressure correction equation can then be disappointing. However, the dominance of the $\rho_P^{old}/\Delta t$ term, in the momentum equation's diagonal elements gives the potential for a storage saving simplification to the SIMPLE method, discussed later.

*TDMA Solver*

As noted above, the multigrid method requires a simultaneous equation solver. This must produce smooth error distributions (i.e. TDMA acts partly as a smoother). In all predictions presented here (which involve structured grids) TDMA is used. Therefore, the form of this algorithm is briefly described.

TDMA is applied with the discretized equations arranged in the following tri-diagonal form

$$A_{P\phi}\,\phi_P = A_{Q1\phi}\,\phi_{Q1} + A_{Q2\phi}\,\phi_{Q2} + S \qquad (4\text{-}2.9)$$

where

$$S = A_{Q3\phi}\,\phi_{Q3} + A_{Q4\phi}\,\phi_{Q4} + A_{Q5\phi}\,\phi_{Q5} + A_{Q6\phi}\,\phi_{Q6} + S_{\phi}. \qquad (4\text{-}2.10)$$

Equation (4-2.9) is then solved with the subscripts, *Q1, Q2 ..., Q6* arranged in the following three ways:

$$\begin{aligned} &Q1 = E,\ Q2 = W,\ Q3 = S,\ Q4 = N,\ Q5 = F,\ Q6 = B;\\ &Q1 = N,\ Q2 = S,\ Q3 = E,\ Q4 = W,\ Q5 = F,\ Q6 = B;\\ &Q1 = F,\ Q2 = B,\ Q3 = S,\ Q4 = N,\ Q5 = E,\ Q6 = W. \end{aligned} \qquad (4\text{-}2.11)$$

Effectively, the terms in the source are solved using Gauss-Siedel and the others (the three diagonals in Equation (4-2.9)) with Gaussian elimination. It is worth noting that the red-black is an acceptable multilevel convergence solver/smoother. This enables multilevel convergence to be used on vector and parallel computers. Such an approach is tried with one of the programs included in this text on a 16 Processor SGI Origin machine. Good scaling is found.

*Convergence Criteria*

For predictions presented here, the convergence is based on either normalized residual levels or monitoring root mean square (*rms*) changes of the solution variables, where

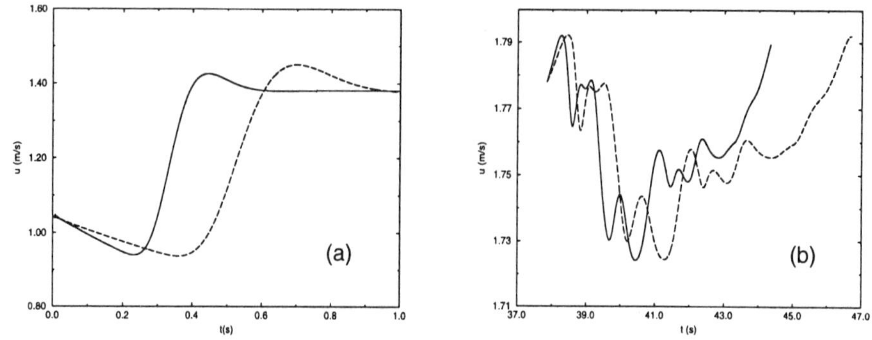


*Figure 4-2.1* Turbulent flow velocity time traces illustrating the effects of poor convergence: (a) Transient flow predictions with convergence based on *rms* and (b) Cyclic flow predictions with convergence based on normalised residuals ( — Converged results, ---- Poorly converged results)

$$rms_{\phi} = \sqrt{\frac{\sum\left(\phi^{new} - \phi^{old}\right)^2}{\sum\left(\phi^{new}\right)^2}} \tag{4-2.12}$$

and the subscripts '*old*' and '*new*' are considered in an iterative sense. As noted previously, for unsteady flows, with strong under-relaxation the judgement of convergence using *rms* changes can be dangerous. The effects of poor convergence are illustrated in the *Figure 4-2.1* velocity time traces. Frame (a) shows results for a transient turbulent two-dimensional flow prediction. Convergence is based on *rms* changes. The full line is the prediction assuming convergence for about $rms_{u,v} < 1 \times 10^{-5}$ and the dashed $1 \times 10^{-4}$. As can be seen from the velocity agreement at $t = 1$ s, the latter looser condition is certainly accurate enough for the steady state solution. However, when trying to model the time history the looser criterion gives clear solution errors. The error is mostly in phase. Frame (b) shows results for a cyclic turbulent three-dimensional flow prediction. Convergence is based on normalised residuals. The full line is for a 0.5 % residual error and the dashed a 2% error. If first order diffusive upwinding is used to force a steady flow solution it can be readily seen (results are not presented here) the 2 % criterion is sufficient. However, again, for the unsteady flow results a tighter convergence criterion is required. As with the transient case, the poor convergence mostly results in a *phase error*.

### *The Multilevel Algorithm*

Standard iterative methods effectively smooth high frequency errors with a wavelength proportional to the grid spacing, but are inefficient at reducing low frequency errors. Consequently computing times vary approximately as the square of the number of grid points. By utilizing coarser grid structures, the multilevel method (see Brandt (1977)) enhances the reduction of low frequency errors and enables computing times to vary approximately linearly with the number of grid points. In a SIMPLE (see Section 4-4) framework (to be used in later Case Studies) there are many examples of its application. For example, Lonsdale (1988) and Vaughan *et al.* (1989) apply the technique to laminar and turbulent rotating cavity flows, respectively. Also, Rayner (1991, 1990) and Shy *et al.* (1992, 1993) use multilevel convergence for solutions involving body fitted coordinate systems. An *algebraic* method, suitable for unstructured grids, is presented by Ruge and Stuben (1986) and Raw (1995). The multilevel convergence technique is now outlined.

The semi-discretized equation sets to be solved can be expressed symbolically as:

$$\left[\dot{\phi}_m\right]+\left[\tilde{A}_m\right]\left[\phi_m\right]=\left[\tilde{S}_m\right] \tag{4-2.13}$$

where $\phi_m$ corresponds to the specific variable being solved for (eg, pressure, a velocity component or temperature), $\dot{\phi}_m$ its temporal gradient, $\left[\tilde{A}_m\right]$ the semi-discretized equation coefficients and $\left[\tilde{S}_m\right]$ their source terms. For the completion of the discretization of Equation (4-2.13) and solution, an assumption must be made as to how $\phi_m$ changes with time. It is then convenient here to re-express it in the following form

$$\left[A_m\right]\left[\phi_m\right]=\left[S_m\right] \tag{4-2.14}$$

where both $\left[A_m\right]$ and $\left[S_m\right]$ contain contributions from the discretized time derivative. For a fully explicit forward Euler temporal discretization $\left[A_m\right]$ will have just diagonal elements. Despite this, as will be seen later, multilevel convergence still has its uses. The subscript $m$ indicates the current grid level and $m$ = *1* corresponds to the finest. It is assumed the exact solution to Equation (4-2.14), $[\phi_m]$, can be expressed as the sum of approximate values $[\phi_m^a]$ and corrections $[\phi_m^c]$ to the approximate values. So we can re-state Equation (4-2.14) as:

$$[A_m][\phi_m^a+\phi_m^c]=[S_m] \tag{4-2.15}$$

and write the following residual equation

$$[R_m]=[S_m]-[A_m][\phi_m^a] \tag{4-2.16}$$

Equations (4-2.15) and (4-2.16) can be combined to eliminate $[S_m]$ giving

$$[A_m][\phi_m^a+\phi_m^c]=[R_m]+[A_m][\phi_m^a] \tag{4-2.17}$$

For a coarser mesh denoted by $m+1$ Equation (4-2.17) can be stated as

$$[A_{m+1}][\phi_{m+1}^{a,f}+\phi_{m+1}^c]=[R_m^f]+[A_{m+1}][\phi_{m+1}^{a,f}] \tag{4-2.18}$$

where the superscript $f$ indicates the value originates from a finer mesh. As can be seen, Equation (4-2.18) solves the *Full Approximation* $\left(\phi_{m+1}^{a,f}+\phi_{m+1}^c\right)$. If this and the right hand side of the equation are expressed as

$$\begin{aligned}&[\phi'_{m+1}]=[\phi_m^{a,f}+\phi_{m+1}^c]\\&[S'_{m+1}]=[R_m^f]-[A_{m+1}][\phi_{m+1}^{a,f}]\end{aligned} \tag{4-2.19}$$

(since $S'_{m+1}$ forces the coarse grid solution to have the accuracy of the fine grid solution it is generally called the *forcing function*) Equation (4-2.18) can be restated as:

$$[A_{m+1}][\phi'_{m+1}]=[S'_{m+1}] \tag{4-2.20}$$

giving, for $m > 1$, an equation similar in form to Equation (4-2.14). Corrections for lower grid levels can then be obtained by subtracting $[\phi_m^{a,f}]$ from $[\phi'_{m+1}]$. For the method to work, values originating from level $m$ to level $m+1$ must be sufficiently smooth. For implicit temporal methods, this can be achieved using the SIMPLE type methods to be described later in conjunction with a reasonable simultaneous equation solver such as TDMA. For explicit methods the residual smoothing technique (see Blazek *et al.* (1991)) outlined in Section 2-8 can be used. However, generally, for highly stretched grids (which can be most economical when resolving relatively high-speed flow in boundary layers), implicit methods make better smoothers than their explicit counterparts.

Importantly, if the equation to be solved is linear $[A_{m+1}][\phi^{a,f}_{m+1} \ldots]$ cancels either side of Equation (4-2.18). Therefore, Equation (4-2.20) simplifies to the following *Correction Storage* or *Linear* multilevel equation:

$$[A_{m+1}][\phi^{c}_{m+1}]=[R^{f}_{m+1}] \tag{4-2.21}$$

For a two grid method $\phi^{c}_{m+1}$ is interpolated to grid level $m$ where

$$[\phi_{m}]=[\phi^{a,f}_{m}]+\alpha[\phi^{c}_{m+1}] \tag{4-2.22}$$

and $\alpha$ is an under or over-relaxation parameter. However, there is no reason why a correction equation on grid level $m+2$ should not be written for Equations (4-2.20) and (4-2.21) themselves (which have the same form as Equation (4-2.14)). Importantly, this procedure could be repeated on higher grid levels, until no further coarsening can take place. This is the basis of the multilevel method. For computations here, as recommended by Brandt (1977), grid coarsening is achieved by missing out every other grid line in the different co-ordinate directions. There are several possible paths that can be used between the different grid levels. Here a basic V cycle, illustrated in *Figure 4-2.2,* is used. As can be seen for this, Equations (4-2.20) and (4-2.21) are represented on successively coarse grids to make corrections to solutions on the next lower grid. Generally, for predictions here, as indicated in *Figure* 4-2.2, two solution sweeps are used on all levels except the coarsest, where three are used.

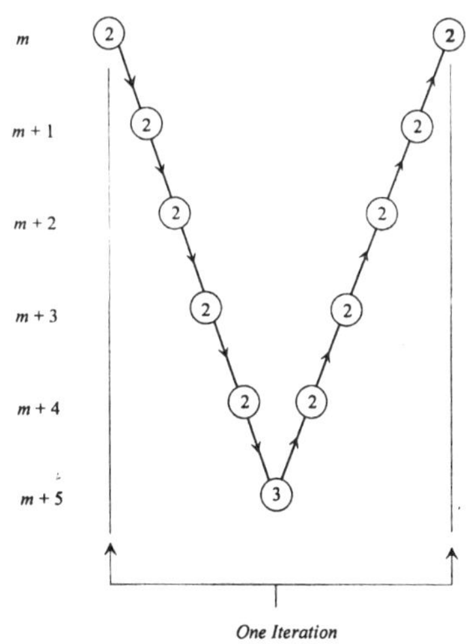


*Figure 4-2.2* Multilevel V cycle procedure (Number of sweeps at each level are represented by the circled numbers)

To represent variables on say level $m+1$ from level $m$ (restriction), a nine point distance weighted averaging is used for main grid point variables. For staggered variables, six points are used. Values of $\rho^{new}$ and $\mu$ are not restricted but calculated at each level. However, $\mu_t$ is restricted from the finest grid, since as found by Vaughan *et al.* (1989), evaluation at each level produces numerical difficulties. For unsteady flows, since $[S_m]$ contains a contribution from the time derivative, $\phi^{old}$ values must also be restricted onto the coarser grids. For variable density flows $\rho^{old}$ values must also be restricted, these giving a contribution to $[A_m]$ and $[S_m]$. Once corrections ($\phi^{c}_{m+1}$) have been calculated, they are bilinearly interpolated onto the next finer grid level. This procedure is called *prolongation.* For later Case Studies, simultaneous equation sets are solved using a tri-diagonal matrix algorithm. Under-relaxation is carried out implicitly through the coefficients of the discretized equations. Full prolongation and restriction expressions are given in Appendix A.

*Specific Multilevel Convergence Temporal Applications*
Brandt (1980) describes a multilevel based strategy for modelling flows evolving with time. This has the potential to yield significant computing time savings. The basis of the method is Brandt's assumption that for evolution processes the high and low frequency (in a spatial sense) components can be very different. Evidently, during early time-steps, the high frequency components change rapidly but then settle down. Then, the lower frequency components only need be resolved. The evolution of these can be modelled using coarser grids, the higher frequency component being accounted for by contributions to the $S'_{m+1}$ terms and $\phi_m^{a,f}$. Therefore, after an initial period, the solution can mostly be integrated through time on a coarser grid. The fine grid is periodically revisited to update the relationship between the newer low frequency and older high frequency components. No examples applying this technique to fluid flow problems are presented by Brandt. However, some concepts embodied in the approach are exploited for solution adaptive time-stepping in Chapter 5.

Explicit solutions, being time marched to a steady state, can greatly benefit from multilevel convergence acceleration (see Dimitriadis and Leschziner (1989)). This is because, the method circumvents the CFL condition. To utilize multilevel algorithms, time-steps are considered equivalent to iterative relaxation steps and coarsened with mesh. Hujeirat and Rannacher (1998) apply what they refer to as a multilevel time-stepping technique to evolve both time dependent and steady flow problems related to astrophysics. The combination of the technique with adaptive time-stepping yields dramatic savings in computing time. However, it does not, in the conventional sense, appear to be a true multilevel convergence method.

The impressive efficiency and potential of the multilevel method is illustrated for some wall and buoyancy driven, and also channel type flows in Appendix C. Also, more details of the method can be gained.

*The Newton Method*

As commonly known, the method solves algebraic equations of the form $A(\phi)=0$ (or in the more traditional nomenclature $f(\phi)=0$). The function can be expressed as the following Taylor series

$$A(\phi)=A(\phi^{old})+A'(\phi^{old})(\phi-\phi^{old}) \tag{4-2.23}$$

where the superscript 'old' is used in an iterative sense. Setting $A(\phi)=0$ in Equation (4-2.23) yields the well known Newton solution equation, found in many elementary texts

$$\phi=\phi^{old}-\frac{A(\phi^{old})}{A'(\phi^{old})} \tag{4-2.24}$$

The form of Equation (4-2.24), when solving a set of simultaneous matrix CFD equations, is given by Engelman and Sani (1986). One important change is that $A(\phi)$ is replaced by $A(\phi)-S$ where $S$ represents a source term. A point to note is that the procedure does not dispense with the need for, and thus additional expense of, a simultaneous equation solver. A further and key snag with the method, implied in (4-2.23) and (4-2.24), is that ideally $\partial A(\phi^{old})/\partial\phi$ (a *Jacobian*) must be repeatedly re-evaluated. The function $A(\phi)$ corresponds to the discretized equations. Generally, these are of significant size and complexity. Therefore, if the Jacobian differentiation can be evaluated at all, it is generally expensive. Orkwis and Vanden (1994) compare the performances of numerically and analytically evaluated Jacobians. The analytical Jacobian expressions are evaluated using a symbolic mathematics software package. Better convergence is found for the numerically evaluated Jacobians. It is worth noting here the well known *Secant* method. With this, to avoid the need for analytical differentiation, the Jacobian can be approximated as a finite difference. Therefore,

$$\frac{\partial A(\phi^{old})}{\partial \phi} \approx \frac{A(\phi^{old}) - A(\phi^{older})}{\phi - \phi^{older}} \tag{4-2.25}$$

giving the requirement for extra storage of 'older' values. Evidently, this approach has yet to find use in CFD (see Ferziger and Peric (1997, p. 114)).

For unsteady flows, the Newton method is used in the well-known, commercial, FIDAP finite-element CFD program. To avoid the Jacobian re-evaluation expense, FIDAP has two approaches (see Gresho and Lee (1980) and Engelman and Sani (1986)). These are as follows:

a) Use adaptive time-stepping, with a small user defined time truncation error. This results in the need for only one Newton iteration and hence Jacobian up-date per time-step.

b) Use a *quasi-Newton* approach, where the Jacobian is only re-evaluated every two or three time-steps.

## 4-3 Parallel Processing

*Parallel Processing and Multilevel Convergence*

As noted earlier, for parallel and vector processors multilevel convergence can, when combined with the red-black method, give an efficient solver. With parallel processing, the computational domain may be decomposed spatially and also more interestingly in time. Decoupling, caused by spatial domain decompositions, can make convergence deteriorate. However, coarser levels, from the multilevel method can improve this coupling (see Jenssen and Weinerfelt (1995)). A multilevel method that involves parallel processing in both the space and time domains is given by Burmeister and Horton (1991). The time domain decomposition aspect will now be outlined.

*Domain Decomposition in Time*

The method is explained by Seidl, Peric and Schmidt (1995), and Burmeister and Horton (1991). The latter combine the method with multilevel convergence acceleration. Ideally the decomposition technique should be used to enhance the performance of spatially decomposed parallel solutions. It should be implemented after the parallel performance significantly drops off due to use of a large number of spatial domains. To illustrate the idea of the method, Equation (2-3.1) can be rewritten for a constant density flow as

$$A_{P\phi}^{n}\ \phi_{P}^{n}+\frac{(-1-\eta)\rho_{p}\phi_{p}^{n-1}}{\Delta t}+\frac{\eta\rho_{p}\phi_{p}^{n-2}}{\Delta t}=\sum A_{NB}^{n}\ \phi_{NB}^{n}+S_{\phi}^{n} \tag{4-3.1}$$

A notable modification in the above is that the 'new', 'old' and 'older' time levels are denoted as $n$, $n$-$1$ and $n$-$2$, respectively. Also, depending on its sign, $S_{2\phi}$ is merged with $A_{p\phi}$ or $S_{\phi}$.

If $m$ successive time-steps are to be solved concurrently on different processors, from Equation (4-3.1) we have the following

$$A_{\rho\phi}^{n}\ \phi_{\rho}^{n}+\frac{(-1-\eta)\rho_{p}\phi_{p}^{n-1}}{\Delta t}+\frac{\eta\rho_{p}\phi_{p}^{n-2}}{\Delta t}=\sum A_{NB}^{n}\ \phi_{NB}^{n}+S_{\phi}^{n}$$

$$A_{\rho\phi}^{n-1}\ \phi_{\rho}^{n-1}+\frac{(-1-\eta)\rho_{p}\phi_{p}^{n-2}}{\Delta t}+\frac{\eta\rho_{p}\phi_{p}^{n-3}}{\Delta t}=\sum A_{NB}^{n-1}\ \phi_{NB}^{n-1}+S_{\phi}^{n-1}$$

$$\cdot \tag{4-3.2}$$

$$A_{\rho\phi}^{n-m}\ \phi_{\rho}^{n-m}+\frac{(-1-\eta)\rho_{p}\phi_{p}^{n-m-1}}{\Delta t}=\sum A_{NB}^{n-m}\ \phi_{NB}^{n-m}+S_{\phi}^{n-m}-\frac{\eta\rho_{p}\phi^{n-m-2}}{\Delta t}$$

$$A_{\rho\phi}^{n-m-1}\ \phi_{\rho}^{n-m-1}=\sum A_{NB}^{n-m}-1\,\phi_{NB}^{n-m-1}+S_{\phi}^{n-m-1}-\frac{(-1-\eta)\rho_{p}\phi_{p}^{n-m-2}}{\Delta t}-\frac{\eta\rho_{p}\phi^{n-m-3}}{\Delta t}$$

Terms such as $\phi_{p}^{n-1}$ and $\phi_{p}^{n-2}$ in Equation (4-3.1) can be combined with $S_{\phi}^{n}$. Also, depending on the form of the discretization (whether linearization is used or not) $\sum A_{NB}$ may be a function of just $\phi^{n}$. These terms will be periodically updated after new data becomes available.

Unlike with spatial decompositions, load balancing with time parallelization is automatic. However, communication requirements are much higher. With spatial domain decompositions only boundary data needs to be exchanged between processors. However, for time parallelization, values at all grid points must be transferred. Therefore, for the method to be efficient, this information transfer must be optimized as much as possible. Burmeister and Horton (1991), and Seidl *et al.* (1995) apply the time parallelization technique to two and three-dimensional oscillating wall driven cavity flows, respectively. For these cases, the method is shown to be efficient. However, its success relies on the work at higher time levels (which is based only approximate prior time level information) being of sufficient benefit. With moving wall flows, the higher time level solutions are likely to be of some benefit. This is because, if nothing else, they will sense the walls new higher or lower velocity and start computing the consequences of this boundary condition change. In the author's opinion, it is likely that the success of time parallelization is, amongst other factors, problem dependent. This is because, the notion of concurrently solving elements in a parabolic time domain seems to an extent at odds with physical reality.

### 4-4 Evaluation of the Pressure Field

In this section, evaluation of the pressure field ($p$) with the emphasis on *incompressible* flows is addressed. For strongly *compressible* flows, the continuity equation (Equation (1-2.4)) yields the fluid density and then $p$ can be recovered from an equation of state. For constant density flows the approach is less clear. Pressure does not appear as the dependent variable in any of the flow governing equations. Perhaps the most obvious method is to substitute the momentum equations for $u$, $v$ and $w$ (Equations (1-2.1)) into the continuity equation and then manipulate the result to give a pressure equation. However, the arising Poisson equation (that can either be discretized and solved in an explicit or implicit time framework) is rather large and unwieldy and is seldom implemented. To overcome this difficulty, simplifications can be made and there are many Poisson equation based solution techniques for pressure. Some of these are detailed later. In the early days of CFD the streamfunction vorticity approach, that eliminated pressure from the governing equations, was popular. To start with, this method is briefly outlined below.

*Streamfunction Vorticity Method*

For two-dimensional, incompressible, iso-viscous flows the need to evaluate pressure can be avoided. For a Cartesian $x$-$y$ system this can be achieved by differentiating the $u$ momentum equation with respect to $y$ and the $v$ with respect to $x$. Subtraction of the two resulting equations gives the following single vorticity equation

$$\rho\frac{\partial \omega}{\partial t}+\rho u\frac{\partial \omega}{\partial x}+\rho v\frac{\partial \omega}{\partial y}=\mu\left(\frac{\partial^2 \omega}{\partial x^2}+\frac{\partial^2 \omega}{\partial y^2}\right) \tag{4-4.1}$$

The stream function is defined using the following : $\partial\psi/\partial x=-v$ and $\partial\psi/\partial y=u$. Similarly vorticity can be expressed as : $\omega=\partial v/\partial x-\partial u/\partial y$. Combination of the stream function and vorticity definitions gives

$$\frac{\partial^2 \psi}{\partial x^2} + \frac{\partial^2 \psi}{\partial y^2} = -\omega \tag{4-4.2}$$

The solution procedure involves specifying initial velocities and converting these into a vorticity field. Equation (4-4.1) is then advanced in time to get a new vorticity field. This is then used as a source term in Equation (4-4.2) to calculate $\psi$. The stream function values are then differentiated to give velocities. The time integration procedure involving Equation (4-4.1) is then repeated. The method in two-dimensions (see Fromm (1970)) has the attraction that it involves solving for just two variables instead of the usual three (*u, v* and *p*). However, since for complex geometries the boundary condition specification is awkward, the method is no longer popular. Furthermore, in three-dimensions there are other complications and drawbacks.

*Basic SIMPLE Type Methods - Including PISO*

SIMPLE (Patankar and Spalding (1972)) type methods have been popular for around three decades. With these, the general approach is to establish a pressure field that is compatible with a mass conserving solution. This method, along with the SIMPLEC (Van Doormaal and Raithby (1984)) and AVPI (Jones and Marquis (1985)) variants will be explained below.

The discretized momentum equations can be written in the form of Equation (4-4.3), where the pressure gradient term has been separated from the primary source term.

$$A_{P,\phi}\phi_P = \sum A_{NB,\phi}\, \phi_{NB} + \frac{\Delta p}{\Delta n_\phi} + S_{1,\phi} \tag{4-4.3}$$

The term $\Delta p/\Delta n_\phi$ is the driving pressure gradient for the variable $\phi$, taken in discrete form across the control volume faces. Therefore, for example, in a Cartesian system when $\phi = u$, $\Delta p/\Delta n_u = \Delta p/\Delta x$, and similarly $\Delta n_v = \Delta y$ and $\Delta n_w = \Delta z$. If $p^*$ is an approximate initial pressure field and $u^*$, $v^*$ and $w^*$ approximate three-dimensional velocity components, Equation (4-4.3) can be written as Equation (4-4.4) for *u, v* and *w*:

$$A_{P,\phi}\phi^*_P = \sum A_{NB,\phi}\, \phi^*_{NB} + \frac{\Delta p^*}{\Delta n_\phi} + S_{1,\phi} \tag{4-4.4}$$

It is assumed that correct pressures and velocities will be obtained if exact pressure and velocity corrections are added to the guessed velocity and pressure fields. Correction equations for the velocities and pressure can then be written using Equation (4-4.5), where $\phi'$ is the correction, that when added to the approximate value $\phi^*$ will give the correct solution, $\phi$.

$$\phi'_P = \phi_P - \phi_P^* \tag{4-4.5}$$

Subtracting Equation (4-4.4) from Equation (4-4.3) gives a correction Equation (4-4.6) for $u', v'$ and $w'$, (the source terms are assumed to cancel).

$$A_{P,\phi}\,\phi'_P = \sum A_{NB,\phi}\,\phi'_{NB} + \frac{\Delta p'}{\Delta n_\phi} \tag{4-4.6}$$

The next key stage involves simplifying the correction Equation (4-4.6). For the SIMPLE method, drastically $\sum A_{NB,\phi}\,\phi'_{NB}$ is neglected. For SIMPLEC a more refined approach is taken by assuming $\sum A_{NB,\phi}\,\phi'_{NB} \cong \sum A_{NB,\phi}\,\phi'_P$. Equation (4-4.6) for $\phi = u, v$ and $w$ can then be expressed as

$$A_{P,\phi}\,\phi'_P = \sum A_{NB,\phi}\,\phi'_P + \frac{\Delta p'}{\Delta n_\phi} \tag{4-4.7}$$

giving equations of the form of Equation (4-4.8),

$$\phi'_P = d_{P,\phi}\,\frac{\Delta p'}{\Delta n_\phi} \tag{4-4.8}$$

where $\phi = u, v$ or $w$ and

$$d_{P,\phi} = \frac{1}{\left(A_{P,\phi} - \alpha_\phi \sum A_{NB,\phi}\right)} \tag{4-4.9}$$

The term $\alpha_\phi$ has been introduced. This is an under-relaxation parameter $\left(0 < \alpha_\phi \le 1\right)$ which improves the stability of the iterative solution. If $\sum A_{NB\phi}$ is ignored in Equation (4-4.9) the SIMPLE algorithm is gained.

Using (4-4.5) and (4-4.8) enables the following velocity equations to be established

$$\phi_P = \phi^*_P + d_{P,\phi} \frac{\Delta p'}{\Delta n_\phi} \tag{4-4.10}$$

Equations (4-4.10) for *u, v* and *w* are now substituted into the discretized continuity equation $-\sum_{\phi=u,v} \Delta\phi_P^{new} / \Delta n_\phi = 0$ (Equation (1-3.4) is the non-discretized continuity equation). This gives the *pressure correction*, Equation (4-4.11).

$$A_{P,p}\, p'_P = \sum A_{NB,p}\, p'_{NB} + S_{1,p'} \tag{4-4.11}$$

Equation (4-4.11) coefficients, for a structured staggered grid finite volume/difference solver are given in Appendix A. The pressure at the grid point $P$, $p_P$, is evaluated as

$$p_P = p^*_P + \alpha_P\, p'_{P,} \tag{4-4.12}$$

where $\alpha_P$ is an under-relaxation parameter $(0 < \alpha_P \leq 1)$. For SIMPLE and unsteady flows, the optimal $\alpha_P$ value is a strong function of time-step size. For steady flows, the best $\alpha_P$ is highly dependent on momentum equation under-relaxation parameters. The velocity components are updated using (4-4.10).

Generally, two types of boundary condition are required for the pressure correction equation. When the velocity is specified, the condition is $\Delta p'/\Delta n_\phi = 0$. This boundary condition is implemented by setting the appropriate coefficient in the discretized pressure correction equation to zero. When *p* is given, no pressure correction is required, and so $p' = 0$ at the boundary giving a Dirichlet condition. For flows where unsteady vortex motions cross outlet boundaries the latter condition can be inappropriate. This is because it tends to act like a pseudo-physical barrier from which waves are reflected. Under such circumstances *non-reflecting* outlet boundary conditions are preferable (see for example Amano (1984), and Breuer and Rodi (1994)). Reflection at inlet Dirichlet boundaries can also occur (see Barton (1998b)). This can be remedied by inlet repositioning.

One relaxation sweep using the SIMPLE, SIMPLEC and AVPI (see later) algorithms takes the following form: (i) Guess the pressure field $p^*$; (ii) Calculate the coefficients for the momentum equations and solve for the velocities $u^*$, $v^*$ and $w^*$; (iii) Calculate the coefficients in the pressure correction equation and solve for $p'$; (iv) Using Equation (4-4.12), calculate $p$ from $p'$ and $p^*$; (v) Correct velocity components using Equation (4-4.10); (vi) If required, calculate the coefficients for the energy equation and solve, and (vii) Unless solution converged return to (ii).

Further, closely related variants of SIMPLE are the SIMPLER and PISO schemes. SIMPLER is similar to SIMPLE but also solves a Poisson equation for pressure rather than just $p'$ i.e. both $p$ and $p'$ equations must be solved for each iteration. PISO (see Issa (1986)), which as originally proposed involves the Crank-Nicolson scheme, extends SIMPLE with an additional velocity correction step and Poisson based pressure correction equation. For steady state flows, there have been many comparisons of these SIMPLE type methods. Generally these have been inconclusive regarding which scheme is best. More recent, PISO/SIMPLE unsteady flow performance comparisons are given by Barton (1998a,b), a key finding being that PISO is stable for larger time-steps.

*SIMPLE2 and SIMPLE**

Barton (1998a) implies that as originally proposed the SIMPLE formulation does not involve the updating of, for example, $d_{P,\phi}$ from the 'old' time level. Barton also gives a slightly different argument for neglecting the $\sum A_{NB\phi}$ terms in Equation (4-4.6). To introduce the former noted coefficient update Barton proposes SIMPLE2. This involves the application of SIMPLE twice in a time-step thus giving a coefficient update. Barton (1998b) introduces SIMPLE*. With this, SIMPLE is repeatedly applied over $N$ relaxation sweeps. For the momentum equations, Equation (4-2.7) is indirectly employed i.e. it is used as a separate auxiliary equation to be applied after the solution of each momentum equation. Also, $\alpha_\phi = \min[I/(N-1),1]$ where $\phi = u, v$ or $w$ and $I$ is the relaxation cycle number. Equation (4-4.12) is gain applied but with $\alpha_P = \min[(I/(N-1))^2,1]$. No under-relaxation is used on the final two cycles. It should be noted that the quadratic term in $\alpha_P$ ($\alpha_{u,v,w}$ uses a linear function) gives strong initial stage under-relaxation. This is because, as suggested by Barton, it is generally the pressure equation that is the most troublesome. In the present author's opinion, the original derivation of SIMPLE does not preclude the unsteady flow update of coefficients. The computations presented in the current text span many years and all involve repeated updates of the SIMPLE coefficients, this process being continued until convergence. The key difference between SIMPLE* and the application of SIMPLE used here is the manner of under-relaxation (as noted earlier the preference, if possible, here is to avoid under-relaxation). As noted by Barton, the easier to implement SIMPLE approach can with careful application give similar computational performance to PISO and does not require the storage of velocity correction terms.

*AVPI Scheme*

A simplification to Equations (4-4.9), for time dependent flows, is described by Jones and Marquis (1985) and Henkes (1990). This modification, called, by the former authors, the AVPI scheme, reduces core memory requirements by avoiding the need to store $d_{pu}$, $d_{pv}$ and $d_{pw}$. From Appendix A it can be seen that generally

$$A_{P,\phi} = \sum A_{NB,\phi} + \frac{\Delta t}{\rho_P^{\,old}} + S_{2,\phi} \tag{4-4.13}$$

where $S_{2,\phi}$ is a contribution to $A_{P,\phi}$ to improve diagonal dominance. Examining Equations (4-4.9) and (4-4.13) shows that, when $\alpha_\phi = 1$ and $\rho_P^{\,old}/\Delta t >> S_{2\phi}$ , Equation (4-4.9) can be approximated as (4-4.14).

$$d_{P,\phi} \cong \frac{\Delta t}{\rho_P^{\,old}} \tag{4-4.14}$$

When the pressure correction scheme is derived using Equation (4-4.14), the coefficients of the discretized momentum equations need not be stored (they can quickly be evaluated when required – *'on the fly'*) as they must with Equations (4-4.9). When solving three-dimensional flows, this storage saving can be valuable. However, with rotating flows, for the tangential and radial momentum equations (containing Coriolis and Centripetal acceleration terms), $S_{2,\phi}$ can be significant. This gives the restriction that $\Delta t << \rho^{old}/(S_{2,\phi})$, if Equation (4-4.14) is to be a reasonable approximation to (4-4.9). However, hybrid pressure correction schemes are possible, where for example $d_{P,w}$ (the coefficient associated with axial momentum, for which generally in rotating flows $S_{2,\phi} = 0$) is evaluated using the AVPI simplification (where $\Delta t$ can be relatively large) and $d_{P,u}$ and $d_{P,v}$ (radial and tangential momentum related coefficients) using SIMPLEC. This idea is tried in Chapter 8. For the 1[st] order backwards Euler scheme and a constant density two-dimensional system, AVPI gives the following relatively simple Poisson equation for $p'$

$$\frac{\Delta^2 p'^{new}}{\Delta x^2}+\frac{\Delta^2 p'^{new}}{\Delta y^2}=\frac{\rho}{\Delta t}\left(\frac{\Delta u^*}{\Delta x}+\frac{\Delta v^*}{\Delta y}\right)$$

The above can be verified by applying the AVPI assumptions to discretized terms in Appendix A.

*Consistent 'Time-Step' variant of SIMPLE*
Using the AVPI assumption Equation (4-4.10) becomes

$$\phi_P=\phi^*{}_P+\frac{\Delta t}{\rho_P{}^{old}}\frac{\Delta p'}{\Delta n_\phi} \tag{4-4.15}$$

which for $\phi = u$ is a finite difference analog to

$$\rho\frac{\partial u}{\partial t}=-\frac{\partial p'}{\partial x} \tag{4-4.16}$$

Equation (4-4.16) gives (ignoring viscous forces) the change in $u$ over a time period $\Delta t$ due to $\partial p'/\partial x$ alone. It implies a time scale is embodied in Equation (4-4.15) and its SIMPLE equivalent. Raithby and Schneider (1979,1980) argue that, for SIMPLE, inconsistencies between time scales in Equation (4-4.15) and the discretized momentum equations can give rise to the general need for small values of $\alpha_P$ in Equation (4-4.12) and hence slow convergence. As a remedy, they propose the Consistent 'Time-Step' variant of SIMPLE. This is also derived by Ferziger and Peric (1997, p. 169) using different arguments and nomenclature. Using this nomenclature, it just involves defining the relaxation parameter for Equation (4-4.12) as

$$\alpha_P=1-\frac{\sum A_{NB,\phi}}{A_{P,\phi}} \tag{4-4.17}$$

The coefficients in the above originate from the momentum equations.

Interestingly, as can be seen from Equation (4-4.13), as $\Delta t \to 0$, diagonal dominance of the momentum equations is improved. However, for the pressure correction equation, the coefficients tend to zero as $\Delta t \to 0$, and since $A_{P,p'} = f\left(\sum d_{NB}\right)$, $A_{P,p'}$ tends to (without action) zero. This point was mentioned in Section (4.2). Slightly more recent than the above SIMPLE related approaches are the *fractional step* methods. A second order version is outlined below. Fractional step methods appear to be especially popular for Large Eddy Simulations (see Chapter 6) but unlike SIMPLE related approaches have found relatively little use in commercial CFD programs.

*Fractional Step Methods*

As the name fractional step suggests, the solution procedure is carried out as a series of steps in predictor corrector like manner. The Crank-Nicolson based variant of Choi and Moin (1994) will now be described (a hybrid Adams-Bashforth/Crank-Nicolson variant is presented by Kim and Moin (1985)). For simplicity, this is done for a constant density fluid, two-dimensional Cartesian system. First, Equation (2-3.1) is divided by $\rho$ with $\eta = 0$ giving a two-level scheme. The pressure gradient $\left(\partial p/\partial n_\phi\right)$ is removed from $S_{1\phi}$ and expressed discretely as $\Delta p/\Delta n_\phi$ where $\Delta n_u = \Delta x$ and $\Delta n_v = \Delta y$. The terms $A_{P\phi}$, $A_{NB\phi}$, $S_{1\phi}$ and $S_{2\phi}$ are grouped into a single variable $A(\phi)$. To give a Crank-Nicolson form, the discretization of the function $A$ is expressed as a half and half weighted average of '*' and 'old' time levels. The discretization for the momentum equations is then

$$\frac{\phi_P^* - \phi_P^{old}}{\Delta t} = \frac{1}{2}\left(A\left(\phi^*\right) + A\left(\phi^{old}\right)\right) - \frac{1}{\rho}\frac{\Delta p^{old}}{\Delta n_\phi} \tag{4-4.18}$$

Equations (4-4.18) are solved as usual with a choice of any simultaneous equation solver. Since the pressure gradients are based on known 'old' values, this should be relatively fast. Next, half the old time level pressure gradient $\left(-\Delta p^{old}/2\Delta n_\phi\right)$ is subtracted from the $\phi_P^*$ estimates to give $\phi^{**}$

$$\frac{\phi_P^{**} - \phi_P^*}{\Delta t} = \frac{1}{2\rho}\frac{\Delta p^{old}}{\Delta n_\phi} \tag{4-4.19}$$

It is then desired to add half the new time level pressure gradients $\left(-\Delta p^{new}/2\Delta n_\phi\right)$ to gain expressions for $\phi_P^{new}$. The form of these is shown below.

$$\frac{\phi_P^{new}-\phi_P^{**}}{\Delta t}=-\frac{1}{2\rho}\frac{\Delta p^{new}}{\Delta n_\phi} \tag{4-4.20}$$

Since $p^{new}$ is not known, Equation (4-4.20) cannot be directly evaluated. Instead, the expressions for $\phi = u$ and $v$ given by (4-4.20) are substituted into the discretized continuity equation. The following Poisson equation for pressure is gained

$$\frac{\Delta^2 p^{new}}{\Delta x^2}+\frac{\Delta^2 p^{new}}{\Delta y^2}=\frac{2\rho}{\Delta t}\left(\frac{\Delta u^{**}}{\Delta x}+\frac{\Delta v^{**}}{\Delta y}\right) \tag{4-4.21}$$

Equation (4-4.21) is solved for $p^{new}$ enabling $\phi^{new}$ to be evaluated from Equation (4-4.20). The equation computed from this procedure is not Equation (2-3.1) divided by $\rho$ with $\eta = 0$ but in fact

$$\frac{\phi_P^{new}-\phi_P^{old}}{\Delta t}=\frac{1}{2}\left(A(\phi^{*})+A(\phi^{old})\right)-\frac{1}{2\rho}\left(\frac{\Delta p^{new}}{\Delta n_\phi}+\frac{\Delta p^{old}}{\Delta n\phi}\right) \tag{4-4.22}$$

As can be seen, this is not entirely the correct equation for the Crank-Nicolson scheme. For the correct equation $\phi^{*}$ should be replaced by $\phi^{new}$. However, this error is second order and so acceptable. Adaptation of a SIMPLE type code to use the fractional step method appears relatively straight forward. A Crank-Nicolson solution of Equation (2-3.1) is made using $p^{old}$ values. The resulting $\phi^{new}$ values equal $\phi^{*}$ in Equation (4-4.18). Equation (4-4.19) is then used to evaluate $\phi^{**}$ (this equation could be viewed as a velocity correction equation where the correction $\phi' = (\Delta t/2\rho)(\Delta p^{old}/\Delta n_\phi)$). This is followed by solution of the Poisson pressure Equation (4-4.21) (if desired, by subtraction of Equations (4-4.18) and (4-4.22), this Equation could also be expressed in a pressure correction form (see Ferziger and Peric (1997 p.172)). Finally, Equation (4-4.20) is applied. This could be viewed as a second velocity correction ($\phi'$) stage where $\phi' = -(\Delta t/2\rho)(\Delta p^{new}/\Delta n_{\phi})$.

*Artificial Compressibility Methods*

A distinctive feature of the compressible flow continuity Equation (1-3.4) is the presence of the density time derivative, which provides a direct link to pressure. For *incompressible* flow, the subject of this text, this derivative is zero. With artificial compressibility methods the link is restored through the approximation that $\rho = C\ p$ where $C$ is a constant. The smaller $C$, the more 'incompressible' the fluid. Using the approximate 'equation of state,' Equation (1-3.4) can be re-expressed as

$$\frac{\partial p}{\partial \tau} = -\frac{1}{C}\left(\frac{\partial \rho u}{\partial x} + \frac{\partial \rho v}{\partial y} + \frac{\partial \rho w}{\partial z}\right) \tag{4-4.23}$$

where $\tau$ is a pseudo time scale. The above can be directly solved explicitly but, for stability, small time-steps are required. Importantly, *the artificial nature (and hence the use of* $\tau$ rather than $t$) *of Equation (4-4.23) makes its use for time accurate solutions impractical* (see Kim and Moin (1985)). If an attempt is made to minimize temporal errors by minimizing the artificial compressibility effect a highly stiff set of equations results. Therefore, without modification, its only real application is for steady state solutions.

A *dual-time method* that overcomes the aforementioned temporal accuracy problems is proposed by Gaitonde (1998). With this, the transport equations for $\phi$, written below in a simplified one-dimensional Euler form, have, like Equation (4-4.23), a pseudo time derivative appended to them giving

$$\frac{\partial \phi}{\partial \tau} + \frac{\partial \phi}{\partial t} + \frac{\partial u\phi}{\partial x} = -\frac{\partial p}{\partial x} \tag{4.4.24}$$

The equation (4-4.23) and (4-4.24) system are integrated to a steady state in pseudo time (where $\partial\phi/\partial t, \partial p/\partial t \to 0$) giving a time accurate solution in real time.

*Penalty Function Method*

The penalty function approach is popular in Finite Element programs. It has similarities to the artificial compressibility method. The continuity equation is modified along similar lines to give the following pressure relationship

$$p = -\frac{1}{C}\left(\frac{\partial u}{\partial x} + \frac{\partial v}{\partial y} + \frac{\partial w}{\partial z}\right) \tag{4-4.25}$$

where $C$ is a small number, typically $10^{-9} \le C \le 10^{-4}$. In a sense, the $C$ parameter corresponds to a weak compressibility of the fluid. The penalty function method involves some error, but provided $C$ is small, this will be insignificant. However, the requirement for small values of $C$ means that, for stability, explicit methods need prohibitively small time-steps. Equation (4-4.25) can be substituted directly into the momentum equations and then the equations discretized. Alternatively, a discrete form of (4-4.25) can be used in the discretized equations. Using the latter approach (see Engelman and Sani (1986)), the penalty function method is adopted in the successful commercial FIDAP CFD code. When starting solutions, the method can give rise to an initial non-physical transient component (Engelman and Sani (1986)). This problem can be dealt with by using a few (3 or 4) initial dissipative first order backward Euler steps. A higher order scheme can then be switched to.

*Implicit Continuous-fluid Eulerian (ICE) Scheme*

The ICE scheme of Harlow and Amsden (1971) (also see Harlow, Amsden and Hirt (1970)), like SIMPLE, yields a Poisson equation for pressure requiring an efficient simultaneous equation solver. It is suitable for flows at all speeds (when used for supersonic flows SIMPLE, ideally, requires modification). An example of a single flow problem exhibiting all flow speeds is an explosion where supersonic flow will be followed by subsonic flow (to avoid poor convergence, most schemes, depending on the direction, would require modification around subsonic/supersonic velocity transitions). ICE, relative to SIMPLE, has the following distinctive features:

- Its derivation is based around the continuity Equation (1-2.4) expressed in the following time centered form

$$\frac{\partial \rho}{\partial t} + \lambda\left(\frac{\partial}{\partial x}(\rho u) + \frac{\partial}{\partial y}(\rho v) + \frac{\partial}{\partial z}(\rho w)\right)^{new} + (1-\lambda)\left(\frac{\partial}{\partial x}(\rho u) + \frac{\partial}{\partial y}(\rho v) + \frac{\partial}{\partial z}(\rho w)\right)^{old} = \tau\left(\frac{\partial^2 \rho}{\partial x^2} + \frac{\partial^2 \rho}{\partial y^2} + \frac{\partial^2 \rho}{\partial z^2}\right) \tag{4-4.26}$$

where $0 \leq \lambda \leq 1$. In some circumstances, the right hand side mass diffusion term can enhance stability ( $\tau > 0$ ) or by cancellation of discretization errors numerical accuracy ( $\tau < 0$ ), but generally $\tau = 0$.

- Equation (4-4.3) (the discretized momentum equations) is solved in the following explicit form, with the pressure gradient expressed in a similar time centered form to that used for the continuity equation.

$$A_{P,\phi}\phi_P^{new} = \sum A_{NB,\phi}\,\phi_{NB}^{old} + \lambda\frac{\Delta p^{new}}{\Delta n_\phi} + (1-\lambda)\frac{\Delta p^{old}}{\Delta n_\phi} + S_{1,\phi} \tag{4-4.27}$$

- The pressure at the 'new' time level is evaluated using the following equation

$$p_P^{new} = p_P^{old} + \left(\frac{\partial p}{\partial \rho}\right)_P^{old}\left(\rho_P^{new} - \rho_P^{old}\right) \tag{4-4.28}$$

where $\partial p/\partial \rho$ is related to the speed of sound. ICE, like the fractional step methods, appears to have found little application in the most widely available commercial CFD programs.

*Fully Coupled Solvers*

The latest commercial CFX program uses a fully coupled solver with multilevel convergence acceleration. A description of this approach is given by Raw (1996). All the pressure evaluation methods described above relate to what are called *sequential* solvers. With these, the momentum equations are solved for the complete domain one at a time. Then perhaps the pressure evaluation equation is solved for the complete domain and finally say an energy equation (again for the complete domain). Coupled solvers basically take the reverse approach; cells being visited one at a time with all the above equations solved concurrently. This dispenses with the need for algorithms like SIMPLE and also evidently the grid staggering approach (or other related techniques) to be described later. Hence coupled solvers are highly attractive. When time marching to steady states, as with multilevel convergence, higher CFL numbers can be used and hence savings relative to sequential solvers made. However, solvers that can deal with *block matrices* are required, i.e. the array elements we are normally used to considering are arrays in themselves, the size of which depends on the number of differential equations required to model the problem. This means that array elements for all the flow governing equations must be stored at the same time. Hence coupled solvers need more storage than their sequential counterparts.

The basic nature of coupled solvers is illustrated for the model equation below cast in a form compatible with a 1$^{st}$ order backwards Euler temporal approach.

$$\frac{\bar{\phi}_P^{new} - \bar{\phi}_P^{old}}{\Delta t} + \left(\frac{\partial \bar{J}}{\partial x}\right)^{new} = 0 \tag{4-4.29}$$

The above can be considered as a one-dimensional Euler type equation system. The overbars are used to indicate the terms in the above are arrays. The elements of these could have the following form

$$\bar{\phi} = \begin{bmatrix} \rho \\ \rho u \\ \rho T \end{bmatrix}, \quad \bar{J} = \begin{bmatrix} \rho u \\ \rho uu + p \\ \rho uT \end{bmatrix} \tag{4-4.30}$$

where $T$ is temperature. The first row terms give the continuity equation, the second momentum and third an energy equation. Employing a similar approach to that used in several other places in this text (see for example Section (2-4)) the following can be written

$$\bar{J}^{new} = \bar{J}^{old} + \frac{\partial \bar{J}}{\partial \bar{\phi}} \Delta \bar{\phi} + HOT \tag{4-4.31}$$

and hence

$$\left[\frac{\partial \bar{J}}{\partial x}\right]^{new} = \delta_x \bar{J}^{old} + \delta_x \frac{\partial \bar{J}}{\partial \bar{\phi}} \Delta \bar{\phi} + HOT \tag{4-4.32}$$

where $\delta_x$ is some form of say finite difference operator, $\partial \bar{J} / \partial \bar{\phi}$ a Jacobian matrix and $\Delta \bar{\phi} = \bar{\phi}^{new} - \bar{\phi}^{old}$. Substituting equations (4-4.31) and (4-4.32) into (4-4.29) yields the following 3 x 3 'submatrix' Δ form system

$$\left[\frac{I}{\Delta t}+\delta_x\left(\frac{\partial \bar{J}}{\partial \bar{\phi}}\right)\right]\left[\Delta\bar{\phi}\right]=-\delta_x \bar{J}^{old} \tag{4-4.33}$$

where $I$ is a unit matrix. Equation (4-4.33) can then be assembled into a global solution matrix, each element of which is a 3 x 3 submatrix. When using coupled solvers, relative to a sequential, the number of iterations to convergence should decrease. However, this does not necessarily mean that the actual computing time will be lower. This is a significant function of the equation set being solved. For example, with turbulent flows, the momentum and turbulence model equations are generally only weakly linked through diffusion terms. Then it can be better (see Wilcox (1998)) to keep the momentum and turbulence model equation link sequential (the momentum equations being still solved in a coupled fashion).

*Reduced Pressures*

Pressure, $p$, can be solved for as the hydrostatically or centrifugally reduced value $\tilde{p}$ below

$$\tilde{p}=p-\rho_o gz-\frac{1}{2}\rho\Omega_g{}^2 r^2 \tag{4-4.34}$$

where $\rho_o$ is a reference density, $g$ the earth's acceleration due to gravity, $z$ depth, $r$ radial coordinate and $\Omega_g$ the angular velocity of the coordinate system. This will increase numerical precision, which can be important for certain flows. A classic example is systems of significant depth involving relatively high-density fluids such as water. For these the hydrostatic pressure can dominate and swamp out smaller flow driving pressure differences. Examples, involving reduced pressure are used are given in Chapters 7 and 8. For the case in Chapter 7 solutions without the reduced pressure will erroneously be axisymmetric and steady.

*Pressure/Velocity Locations*

For many years, a popular technique devised by Harlow and Welch (1968) has been to arrange the grids such that the relevant pressure nodes are coincident with the control volume faces of the momentum variable being solved for. To achieve this staggered multiple grids or control volumes must be used. *Figure 4-4.1* shows a typical staggered arrangement that is used for cylindrical polar coordinate predictions presented here. Arranging the flow driving pressure difference to be coincident with the momentum equation control volume faces avoids the prediction of nonphysical oscillatory

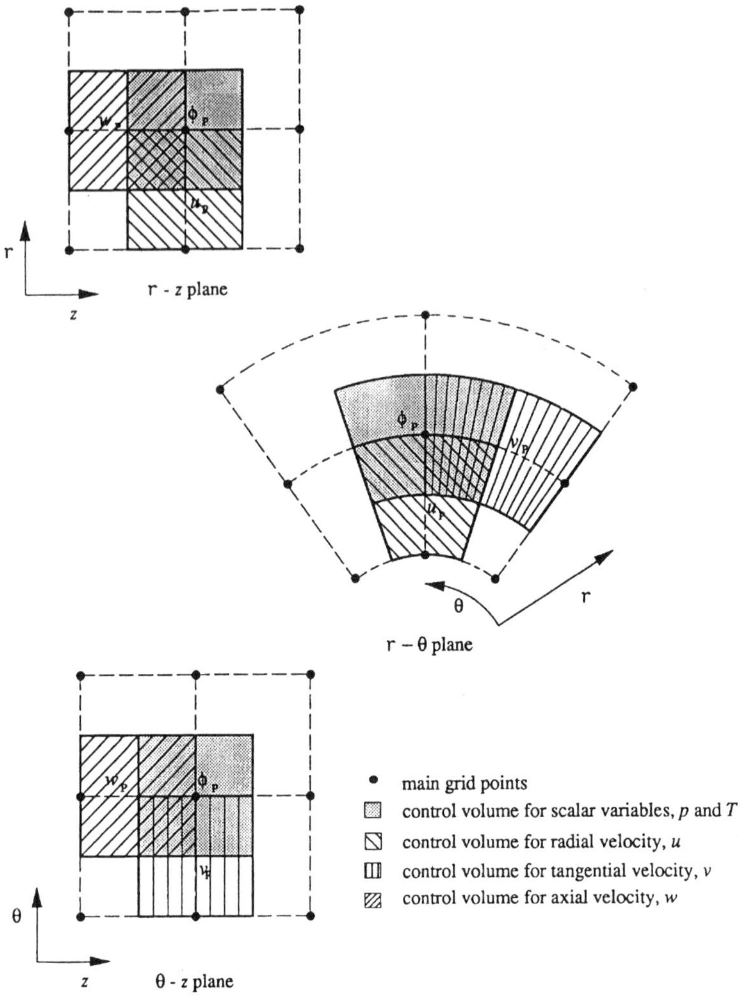


*Figure 4-4.1* Staggered grid arrangement for a cylindrical polar coordinate system

pressure fields. It also has the following advantages: more compact discretization expressions; avoidance of the need to use artificial pressure boundary conditions; compatibility with porous media modelling and (in the absence of time differencing errors and viscosity) global conservation of kinetic energy and momentum (see Kim and Moin (1985)).

More recently staggered grids, which are inconvenient to implement in general geometry codes have become less popular with collocated variable arrangements being preferred. The most well known collocated variable technique to avoid unrealistic pressure field oscillations is due to

Rhie and Chow (1983).The smoothing properties of this approach, which applies dissipation to the pressure field, depend on time-step and under-relaxation values (see Barton and Kirby (2000)). Other more modern approaches are discussed by Roychowdhury (2000) who advocates the *Pressure Weighted Interpolation Method – Corrected* (PWIMC) of Kobayashi and Pereira (1991). Evidently, PWIMC gives convergence rates that are not so strongly dependent on under-relaxation parameter values. Barton and Kirby also present a method where performance is de-sensitized to time-step and under-relaxation values.

*Pressure Subcycling*

As previously noted, sometimes, explicit scheme stability requirements dictate that much smaller time-steps must be used than needed to resolve actual physical processes. In this situation, to lessen computational expense a procedure called subcycling, described by Gresho *et al.* (1984) can be used. The premise for the procedure is that, for the momentum equations (and other transport equations), explicit scheme stability is not linked to pressure and continuity constraints. The subcycling procedure, illustrated in *Figure 4-4.2*, is as follows:

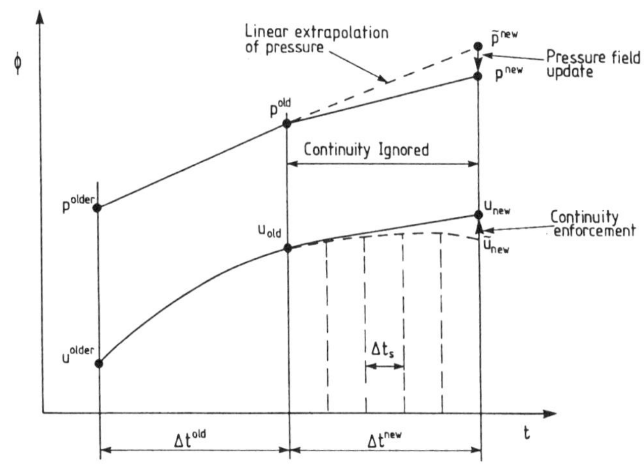


*Figure 4-4.2* Illustration of the subcycling procedure

(I) Over a time period $\Delta t^{new}$, solve the momentum equations explicitly using minor steps, $\Delta t_s$, governed by stability limits. During this period use extrapolated pressure values.

(II) At the end of the period $\Delta t^{new}$, by correcting the velocity field, re-enforce continuity.

(III) Finally, use the continuity satisfying velocity field to update pressures.

By not requiring a pressure update for every minor time-step the above procedure saves computational effort. Subcycling, is offered in the commercial FIDAP finite element program discussed more below.

## 4-5 Review of Commercial Programs and their Unsteady Flow Capabilities

*Codes Considered*

In this section the capabilities of some commercial programs are briefly reviewed. Due to the rapid advances in CFD technology such information can become quickly dated. However, it is hoped the information here will give a flavour of the presently available commercial CFD technology. Also, comparisons of the features offered by the different codes may be indicative of future trends. The following codes are considered: FLOTHERM (Flomerics Plc); FLUENT 4 and 5 (Fluent Inc.); FLOTRAN (ANSYS Inc.); FIDAP (Fluent Inc.); CFX 4, 5 and TASCflow (AEA Technology – United Kingdom Energy Authority); PHOENICS/HOTBOX (Concentration, Heat and Momentum Ltd) and STAR-CD (Computational Dynamics Ltd). HOTBOX is a customized version of PHOENICS for electronics cooling. Fluent Inc. also offer a customized electronics-cooling program called ICEPAK. The FLOTHERM program is designed just for electronics cooling. Flomerics Plc, also market a program called FLOVENT, intended specifically for room ventilation modelling. The development of customized codes appears to be an increasing trend. Until recently the FLUENT and AEA Technology codes used structured grids. However, now unstructured grid versions of these programs are available. This perhaps indicates an increasing market for unstructured solvers. In the past, such solvers were in a minority.

The range of codes discussed here is far from exhaustive and many (in the author's opinion) impressive codes are omitted. Most commercial CFD codes are tremendous pieces of human endeavour, which people have dedicated much of their lives to. However, the complex nature of the physical processes which it is attempted to model and other constraints mean that in use they can at times be frustrating and the results, and time scales in which they are achieved sometimes disappointing.

*General Code Features*

*Table 4-5.1* summarizes some general features for the codes noted above. Reference to Chapter 6 may be required for details of some of the listed turbulence models.

| Code name | Discreti-zation | Nature of grids | Variable positions | Convective term treatments | Pressure treatment | Turbulence models | Equation solvers |
|---|---|---|---|---|---|---|---|
| STAR-CD | Finite volume | Unstructured | Collocated | Extensive range | SIMPLE, PISO | $k$-$\varepsilon$ variants, RNG and multi-layer $k$-$\varepsilon$ based model | Conjugate gradient |
| CFX4 | Finite volume | Curvilinear with multiblocking, blocking off and sliding mesh capability | Collocated | Extensive range including QUICK, HYBRID, Upwind (first order and above) CONDIF, TVD/MUSCL | SIMPLE, PISO | $k$-$\varepsilon$, RNG, low Reynolds number $k$-$\varepsilon$, $k$-$\omega$, Algebraic stress, Differential Reynolds stress | Range including algebraic multilevel convergence |
| CFX-TASCflow | Finite volume | Structured, non-orthogonal curvilinear multiblock with multi-frame of reference capability[7] and grid embedding | Collocated | 1st order upwind, mass weighted skew upwinding, linear profile skew upwinding | Coupled solver | Low and high Reynolds number $k$-$\varepsilon$, Kato and Launder modification and RNG, zonal $k$-$\varepsilon$, /$k$-$l$ model | Algebraic multilevel convergence coupled solver |
| CFX-5 | Finite volume | Unstructured with solution adaptive meshing capability | Collocated | Second order NAC (Numerical Advection Correction) scheme – a free parameter is used to increase/decrease false diffusion | Coupled solver | Zero equation, $k$-$\varepsilon$, RNG $k$-$\varepsilon$, Reynolds stress model | Algebraic multilevel convergence coupled solver |

*Table 4-5.1* Summary of general code features

*Continued on next page...*

*Continued...*

| Code name | Discreti-zation | Nature of grids | Variable positions | Convective term treatments | Pressure treatment | Turbulence models | Equation solvers |
|---|---|---|---|---|---|---|---|
| FIDAP | Finite element | Unstructured | Velocities reside at node points and pressure over entire element | Streamline upwind | Penalty function method | Mixing length, standard $k$-$\varepsilon$, RNG, $k$-$\omega$, and non-linear $k$-$\varepsilon$ models | Range of coupled and segregated solvers including Newton based methods[2] |
| FLOTRAN | Finite element | Unstructured | Collocated | Streamline upwind | SIMPLER | Standard $k$-$\varepsilon$ | Segregated solver with TDMA, conjugate gradient and Newton methods |
| FLUENT 4.5 | Finite volume | Curvilinear with blocking off and multiple reference frame and sliding grids | Collocated | 1st order upwind, power law, second order upwind, QUICK | SIMPLE, SIMPLEC PISO | Extensive range | TDMA with multilevel convergence |
| FLUENT5 | Finite volume | Unstructured with multiple reference frame and sliding meshes | Collocated with solution adaptive meshing including hanging nodes. Can do adaptation on $y^+$ values | As above | As above | Extensive range including Spalart-Allmaras model and Smagorinsky LES model and zonal models | Multilevel convergence |

*Table 4-5.1* Summary of general code features

*To minimize cross-stream spatial diffusion, hexagonal cells can be subdivided by connecting diagonally opposite corners. The original cell appears as if an X has been drawn on it.

[7]CFX-TASCflow is intended for turbomachinery design. The multi-frame of reference option is to allow grids to rotate with different angular velocities. This is important in, for example, regions where a set of stationary blades are upstream of a set of rotating blades.

[2]The Newton based methods, which have small convergence radii, are efficient for unsteady flows. The solution from the previous step is a good starting approximation for the next time-step.

*Capabilities Specific to Unsteady Flows*
*Table 4-5.2* Summarizes flow modelling capabilities specific to unsteady flows for the codes above.

**4-6 Features of Codes used for Case Studies**
For the later Case Studies presented three programs are used. These are given here the following names:

*(I)* *FLO_CAVITY;*
*(II)* *FLO_BEARING* and
*(III)* *FLO_ELEC.*

These programs have an easy to understand structure and are of a training nature. Hence, they can be readily further developed making useful algorithm development platforms. All of the above codes use the finite-difference based finite volume approach of Patankar (1980) with structured staggered grids and TDMA or Gauss-Siedel solvers. Except for *FLO_BEARING,* convergence can be accelerated using non-linear multilevel convergence. The *FLO_CAVITY* program uses a cylindrical polar coordinate system. *FLO_BEARING* is a bearing flow program utilizing a customized coordinate transformation. For conjugate heat transfer to define the fluid, solid shaft and bearing regions separate coupled meshes are used. *FLO_ELEC* uses a Cartesian coordinate system. Its structure is such that electronic and room geometry configurations can be readily meshed.

In all of the above structured mono-block programs, rather than use high viscosities to define solid regions, a Boolean, marker array $[n]$ is used. At fluid, solid and differential boundary condition regions individual entries of $n$ take the following, respective, values 1, 0, -1. This marker array technique requires extra storage. However, for non-conjugate problems, it prevents wasted computations on redundant cells. Also, importantly, it easily allows for second order wall diffusion term treatments, which are consistent with the discretization order inside the domain. At solid walls, for velocities, the no-slip and impermeability conditions are applied. Also, when solving for scalar transport normal scalar velocity gradients are set to zero. At open boundaries, $p$ is fixed and generally the gradients of other variables are set to zero in a second order fashion. Generally, at inlets, Dirichlet velocity and scalar boundary conditions are used. More specific algorithm options for the above codes are identified in Chapters 3 and the present.

| Code name | Time differencing | Adaptive time-stepping ? | Time moving grid capability ? | Free surface modelling capabilities |
|---|---|---|---|---|
| STAR-CD | First order fully implicit, Crank-Nicolson | No | Yes, with space conservation law and cell deletion | MAC Method |
| CFX4 | First order fully implicit backward difference, Second order backward difference, Crank-Nicolson | Yes[++] | Yes, no cell insertion or deletion | Volume of fluid (VOF) method (see Liu Jun (1986)) |
| CFX-TASCflow | First order fully implicit backward difference, Second order backward difference. | No | No | No |
| CFX5 | First order fully implicit backward difference. | No | No | No |
| FIDAP | First order fully explicit/implcit[1], 2$^{nd}$ order trapezoidal | Yes | Yes, with space conservation law | Deforming or fixed mesh with VOF, Surface tension can be included. 2-D, axisymmetric and 3-D free surfaces can be modelled |
| FLOTRAN | Crank-Nicolson | Unknown | Unknown | Unknown |
| FLUENT4.5 | First order fully implicit | No | Yes, but has no cell insertion or deletion | VOF |
| FLUENT5 | First and second order implicit, explicit Runge-Kutta | No | As above | VOF |
| PHOENICS/ HOTBOX | First order implicit and X-cell in time** | Yes, Just for combustion modelling | Yes*** | VOF |
| FLOTHERM | None | None | None | None |

*Table 4-5.2* Summary of flow modelling capabilities specific to unsteady flows

**The spatial X-cell strategy can be extended to the temporal domain.

[++]The time-step adaptation is not based on solution error and so the method is relatively crude.

***For solid bodies moving through the fluid domain a Chimera grid like technique is used. The technique employs a rectangular cage around the periphery of the solid moving object's mesh (this could be a valve moving into the cylinder of a piston engine). This lessens the amount of interpolation required relative to the Chimera method. Also, for moving domain problems grid transformations can be used along with a cell deletion strategy (this could be used to model piston movement in an engine).

[1]The penalty function approach suffers problems with the explicit scheme. Therefore, pressure is advanced implicitly through time and velocities, temperature etc, explicitly.

## 4-7 Conclusions

Explicit schemes do not require a simultaneous equation solver and can be easily parallelized. However, they have severe time-step restrictions. Consequently they are only suitable if high temporal resolution is required. Implicit techniques require a simultaneous equation solver. This is generally a major solution overhead. Approximate factorization or splitting methods of which ADI is a family member successfully combine explicit and implicit techniques and only require one-dimensional simultaneous equation solvers. For vector and parallel processing, to avoid data dependencies the red-black procedure is popular. For most solvers, the number of iterations to convergence increases with about the square of the number of grid points. However, with multilevel methods, (that also work well with red-black) in theory, the number of iterations stays constant. Explicit Euler solvers, when used to time march to steady states, can greatly benefit from multilevel convergence acceleration. This is because the Courant-Friedrichs-Lewy condition is circumvented. Multilevel algorithms also have the potential to yield computational enhancements for evolution problems and also solution adaptive time-step approaches. For unsteady flows, Newton methods can be viable. Also, when parallel processing, temporal domain decomposition techniques can be used. However, the success of this appears, amongst other factors, problem dependent.

For satisfactory convergence, many solvers require the diagonal elements of the discretized equations to be of sufficient size. To help with this, the continuity equation, in discrete form can be subtracted from the governing convection-diffusion transport equations. If at each time-step, continuity is satisfied, the original modelled equation will be unaffected by the subtraction. It is important to ensure this is the case. Therefore, with iterative pressure correction methods, to avoid temporal error growth, an adequate number of continuity enforcing pressure sweeps is desirable. For unsteady simulations, as the time-step becomes smaller, transport equation diagonal dominance increases and even extremely unstable equation systems can become stable. There are strong similarities between the equation structures for the implicit Euler scheme and those when under-relaxation is applied. Both strategies produce contributions to the diagonal coefficient and source term. A key difference is that a constant time-step, for each cell, essentially corresponds to a spatially varying under-relaxation factor. On the other hand, using a constant under-relaxation parameter corresponds to a spatially varying time-step. For SIMPLE type methods, the coefficients of the pressure correction equation are an inverse function of the momentum equation diagonal coefficients. Relative to the momentum equations, the convergence rate for pressure correction equations can then be disappointing. The dominance of the time derivative term in the momentum equation diagonal elements allows use of the storage saving AVPI scheme.

With unsteady flows, it is desirable to avoid under-relaxation. Instead, generally, to gain stability, time-steps should be reduced giving the bonus of increased temporal accuracy. For unsteady solutions, convergence level care is required. For each time-step, errors will grow. Hence, over a long integration period, relatively small local convergence errors can become highly significant.

Usefully, the streamfunction vorticity approach eliminates pressure from the governing equations but does not naturally extend to flows in three-dimensional complex geometries. Therefore, for industrial CFD SIMPLE related methods are more popular approaches. The AVPI scheme is essentially an unsteady flow storage saving simplification to SIMPLE. AVPI can be combined with other pressure correction schemes to form hybrid solution approaches. Fractional step methods are an alternative, unsteady flow, pressure field prediction approach. They have found little use in industrial CFD. The use of artificial compressibility methods to gain pressures for time accurate solutions is dubious. The penalty function method has similarities to the artificial compressibility method. It is mostly used in Finite Element programs, has a sound theoretical basis and so can be used for time accurate solutions. Fully coupled solvers naturally overcome pressure prediction problems.

CHAPTER

# 5

# SOLUTION ADAPTED TIME-STEPS

## 5-1 Introduction

For many years, when solving general Ordinary Differential Equations (ODEs), temporal type solution adaptive approaches have been used in many scientific fields. Also, when solving the fluid flow equations spatial adaptations have seen significant use (such methods are outlined in Section 3-6). Both approaches can involve the variation of step lengths, discretization orders or a combination of the two. Where there are relatively high dependent variable gradients, either step lengths must be reduced or discretization orders increased. For low solution gradients the opposite action must be taken. In this way, the best computational efficiency can be achieved.

Generally, whatever numerical resolution approach is used, some form of truncation error estimate is required. The sophistication of this can vary significantly. For example in the spatial adaptation work reported by Anderson (1995) (see Chapter 3), grid refinement is crudely expressed (via simple scaling factors) as a function of local variable gradients. Thus, the error estimate is quite indirect. More direct approaches compare solutions for example on two grids (i.e. solutions for different step sizes) or schemes of different orders. Multilevel convergence acceleration (see Section 4-2), which uses a hierarchy of grids, with generally a one to two-step coarsening ratio, naturally lends itself to spatial step adaptation (see Brandt (1980)). It also has the desirable properties that it can be used to circumvent temporal stability restrictions (see Dimitriadis and Leschziner (1989)) and coarser levels may more economically model evolving flows – the finer levels being periodically revisited (see Brandt). Therefore, temporal adaptive approaches compatible with multilevel convergence are attractive.

Schonauer *et al.* (1980) describe a variable order spatial adaptation for high-speed boundary layer flows. Ascher and Petzold (1998) give the temporal equivalent to this. In practical cases, due to higher order scheme dispersion, this latter approach is likely to also involve some time-step ($\Delta t$) adaptation. For example, when instability is detected $\Delta t$ can be reduced.

With the exception of the FIDAP finite-element program (as can be seen from the Section 4-5 review), in most major commercial codes, little use has been made of adaptive time-stepping. Predictor-corrector methods naturally lend themselves to time adaptation (see Ascher and Petzold (1998)). Gresho and Lee (1980) successfully apply this FIDAP approach, with its second order explicit Adams-Bashforth predictor step, to the modelling of von Karman vortex streets. The method is also used by Tu *et al.* (1992) when studying pulsatile arterial flows, Reindl *et al.* (1991) for transient buoyancy drive channel flows, and Hujeirat and Rannacher (1998) for astrophysics related flows. For the latter work, when combined with a crude grid overlaying through time procedure, significant computing time savings are found. As with the indirect spatial approach reported by Anderson, indirect temporal approaches can be used. For example, the commerical CFX4 (Glanfield (2000)) program adapts time-steps with respect to convergence rates. Here the assumption is made that if a time-step is requiring a relatively large number convergence iterations, significant temporal variable changes are occurring. For combustion problems, PHOENICS also has an *ad hoc* indirect adaptation approach.

When modelling dissolving gas bubbles, Gupta and Kumar (1983), adapt $\Delta t$ so that the bubble interface is always coincident with grid points. This has some similarities with the Case Study presented in Chapter 9, where $\Delta t$ is chosen so that sliding spatial grid points are always exactly coincident. Muramatsu and Ninokata (1992) apply a novel fuzzy logic based $\Delta t$ adaptation procedure with an algebraic stress turbulence model. Bell and Surana (1994) use special finite-elements that map both the time and space domains. Based on integral error estimates, element sizes in the time domain are adapted (this is only done after spatial accuracy refinements fail to reduce errors below a specified level). Im (2000) successfully applies an adaptive time integration based around a variable-order, variable-step backwards differentiation formula to the modelling of a transient flame flow.

Through the method of lines, the discretized flow governing transport equations can be cast as temporal ODEs. Then, in principle, any standard time adaptive ODE approach (see Press *et al.* (1988), Ascher and Petzold (1998), and Brenan *et al.* (1996)) can be applied to the solution of the unsteady flow governing equations. Possible standard approaches have many similarities to spatial adaptations already more widely used in CFD. Temporal truncation error estimates are here conveniently grouped into the following four classes:

(I) Gradient Based Method (*GBM*) - this is the temporal equivalent of an indirect error estimate spatial adaptation approach presented by Anderson;

(II) Altered Scheme (*ASc*) - solutions for schemes of the same order but different natures are made. By analysis of truncated Taylor series terms solution errors can be inferred;

(III) Altered Step ($A\Delta t$) - solutions with one long and two short time-steps are compared and

(IV) Altered Order (*An*) - solutions of different orders are compared.

Versions of these four approaches are discussed in more detail later. As will be seen from the above discussion, especially in a commercial solver environment, with a view to industrial applications, more consideration could be paid to solution adaptive time-steps. This is considered in the present text and later Case Study chapters. In an industrial setting, as part of the design optimisation process, many simulations with minor parameter changes can be carried out (such repetitive simulations may also take place in more academic research). It is then worth investing a small amount of time selecting and then refining parameters in an optimal solution approach. As will be seen later, although the introduction of automatic adaptive time-stepping removes the need to directly select $\Delta t$, other parameters, on which the solutions success is generally more weakly dependent must be set. Therefore, the efficiency of the adaptive time-stepping approach, as with many other CFD algorithms, should partly be viewed in the light that a series of quite similar computations will often be carried out. This warrants the investment of some parameter tuning time.

The scope here is restricted to adaptive schemes that can most readily be incorporated into commercial codes with minimal changes. Three-dimensional, unsteady, turbulent, non-isothermal industrial flows require significant storage. Since most adaptive time scheme approaches require comparison of two solutions, various storage saving measures are considered. More mathematical details of adaptation approaches (I) to (IV) are given in Section 5-2. Solution adaptive time-step strategies are tested for the following cases:

(A) Transient wall driven flow in a square box;
(B) Transient buoyancy driven flow between cylinders (introduced in Section 2-12);
(C) Transient wall driven free disc flow;
(D) Cyclic buoyant flow in a rotating cavity (see Chapter 8);
(E) Transient flow in a complex system (see Chapter 11) and
(F) Cyclic flow in a complex system (see Chapter 10).

(a)

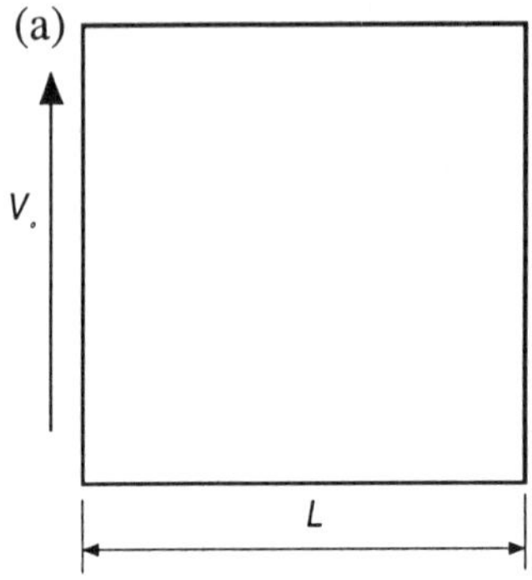


(b)

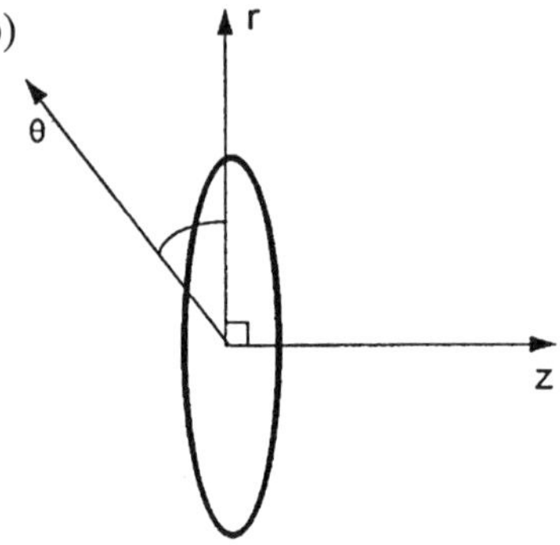


*Figure 5-1.1* Case (A) and (C) Schematics: (a) Cavity with suddenly moving wall and (b) Suddenly rotating disc

Schematics for Cases (A) and (C) are given in *Figure 5-1.1*. *Figure 2-12.1* is the Case (B) schematic. For Cases (D)-(F) adaptive results and schematics are given in Chapters 8, 11 and 10, respectively, where other temporal aspects are also addressed.

## 5-2 Numerical Methods

*(I) Crude Time-step Adaptation with no Direct Error Estimate*

A simplistic adaptive time-step approach can be gained by mimicking the spatial Equations (3-6.1) of Anderson, and expressing the time-step as

$$\Delta t^{new} = C_t f\left( \frac{1}{\frac{1}{N_\phi}\sum_{i=1}^{N_\phi}\left|\partial\phi/\partial t\right|_{ave}^{old}} \right) \tag{5-2.1}$$

where $C_t$ is a scale factor and $N_\phi$ is the total number of variables solved for (eg. $u$, $v$, $w$, $p$, $T$, $k$, $\varepsilon$ .....). Since, for time accurate solutions, the $\Delta t$ must be spatially constant, for all variables, $|\partial\phi/\partial t|$ should be expressed as some characteristic spatially averaged value, hence the subscript '*ave*' is used. As noted earlier, for the above scheme, $\Delta t$ is not directly based on the actual solution error. Also, the method would appear only really suited for flows where $|\partial\phi/\partial t|_{ave}^{old}$ is monotonic and decreasing (Case (A) is such an example). Then, the $\Delta t$ evaluation, based on the temporal gradient for the previous step will be reasonable – as the gradient decreases a time-step increase will be suggested. If $|\partial\phi/\partial t|$ increases with $t$, the $\Delta t$ estimate, which is based on the gradient for the previous step will be erroneous – an increase in time-step being suggested where a reduction is required. More refined and sensible approaches are now outlined.

*(II) Time-step Adaptation with Altered Scheme (ASc) Error Estimate*

The following 1st Order Forward/Backward Euler procedure is used by Gresho *et al.* (1984). From Taylor's series, for a scheme of order $n - 1$, the equation below can be written

$$\phi = \phi_1 + \frac{(\Delta t)^n}{n!}\frac{\partial^n \phi}{\partial t^n} + HOT \tag{5-2.2}$$

where $\phi_1$ is the solution for the backwards Euler scheme, $\phi$ the exact answer and here $n = 2$. Similarly the following can be written

$$\phi = \phi_2 - \frac{(\Delta t)^n}{n!}\frac{\partial^n \phi}{\partial t^n} + HOT \tag{5-2.3}$$

where $\phi_2$ is the forward Euler scheme solution. Combining Equations (5-2.2) and (5-2.3) gives

$$E = \frac{\Delta t^n}{n!}\frac{\partial^n \phi}{\partial t^n} = \frac{1}{2}(\phi_1 - \phi_2) + HOT \tag{5-2.4}$$

In all of the above $HOT = O(\Delta t)^{n+1}$. From Equation (5-2.4) it can be stated that

$$\frac{E^{new}}{E^{old}} = \left(\frac{\Delta t^{new}}{\Delta t^{old}}\right)^n \frac{\partial^n \phi^{new} / \partial t^n}{\partial^n \phi^{old} / \partial t^n} \tag{5-2.5}$$

where the superscripts 'new' and 'old' refer to solutions for different step sizes, 'new' being the latest. Setting $E^{new} = \varepsilon$, a pre-set normalised error input value, enables the maximum temporal solution error to be specified ($\varepsilon = 0.001$ corresponds to a 0.1 % solution error).

Assuming $\partial^n \phi^{new} / \partial t^n \approx \partial^n \phi^{old} / \partial t^n$ Equation (5-2.5) can be expressed as

$$\Delta t^{new} = C_t \Delta t^{old} \left| \frac{\varepsilon}{E^{old}} \right|^{1/n} \tag{5-2.6}$$

where $C_t$, this time, is a safety factor set here between 0.45-0.9. Since $\Delta t$ must be spatially constant for all $N_\phi$, $E^{old}$ needs to be an averaged (or more conservatively a maximum) normalised (to be consistent with the $\varepsilon$ definition) value. Forms of averaging will be discussed later.

*(III) Time-step Adaptation with Altered Step (AΔt) Error Estimate*

An alternative approach, readily applicable to schemes of any order (see Press *et al.* (1988)) is described below. The following two Taylor series based (forward difference) expressions are used

$$\phi = \phi_1 + \frac{(2\Delta t)^n}{n!} \frac{\partial^n \phi}{\partial t^n} + HOT \tag{5-2.7}$$

$$\phi = \phi_2 + \frac{2(\Delta t)^n}{n!} \frac{\partial^n \phi}{\partial t^n} + HOT \tag{5-2.8}$$

The second, right hand side terms in equations (5-2.7) and (5-2.8) are leading term discretization truncation errors for a scheme of order *n-1*. Equation (5-2.7) corresponds to a solution $\phi_1$, from one time-step of length $2\Delta t$. Equation (5-2.8) corresponds to a solution $\phi_2$, from two time-steps of $\Delta t$. Combining Equations (5-2.7) and (5-2.8) we can write the error $E$ for the increment involving two steps as

$$E = \frac{2(\Delta t)^n}{n!}\frac{\partial^n \phi}{\partial t^n} = \frac{1}{2^{n-1}-1}(\phi_2 - \phi_1) + HOT \tag{5-2.9}$$

Following previous arguments, again from Equation (5-2.9)

$$\frac{E^{new}}{E^{old}} = \left(\frac{\Delta t^{new}}{\Delta t^{old}}\right)^n \frac{\partial^n \phi^{new} / \partial t^n}{\partial^n \phi^{old} / \partial t^n} \tag{5-2.10}$$

and so, as before

$$\Delta t^{new} = C_t \Delta t^{old} \left|\frac{\varepsilon}{E^{old}}\right|^{1/n} \tag{5-2.11}$$

For the above procedure, three solutions are required, two of these being used in the error estimate. Whether the benefits of being able to adapt time-steps outweigh the significant computational overhead is likely to be problem dependent. As noted by Ascher and Petzold (1998) step doubling approaches, although easy to implement, can be relatively expensive.

*(IV) Time-step Adaptation with Altered Order (An) Error Estimate*

The previous error estimates have used schemes of the same order. Skelboe (1977) use schemes with different orders. In this situation (see Schonauer *et al.*) the following can be written

$$\phi = \phi^n + \frac{(\Delta t)^n}{n!}\frac{\partial^n \phi}{\partial t^n} + HOT \tag{5-2.12}$$

$$\phi = \phi^{n+1} + \frac{(\Delta t)^{n+1}}{(n+1)!}\frac{\partial^{n+1} \phi}{\partial t^{n+1}} + HOT \tag{5-2.13}$$

If the truncation error for the higher order (forward difference) scheme is assumed negligible, the lower order scheme solution error can be expressed as

$$E = \frac{(\Delta t)^n}{n!}\frac{\partial^n \phi}{\partial t^n} = (\phi^{n+1} - \phi^n) + HOT \tag{5-2.14}$$

Equation (5-2.11) can again be used to modify the time-step.
Schonauer *et al.* plot error values for a function of the form $\phi = 1/t$ (in the original paper $t = x$) for different values of $n$ and $\Delta t$. The results of this are reproduced in *Figure 5-2.1*. As can be seen, for lower order schemes with small time-steps, the simple error estimate is reliable.

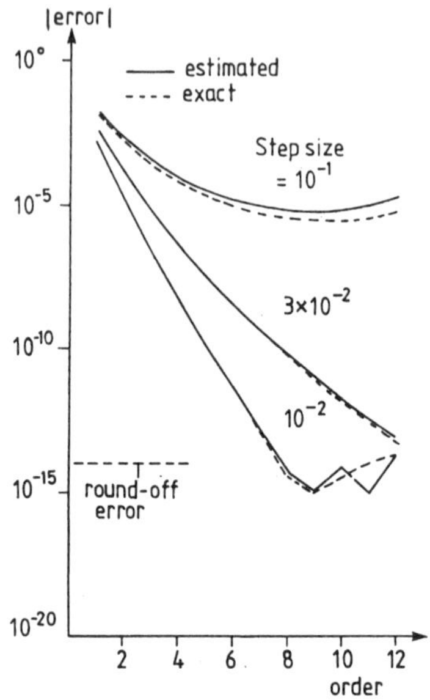


*Figure 5-2.1* Error values using (5-2.14) for different values of $n$ and $\Delta t$ (Schonauer *et al.*)

When using higher order schemes with adaptive step strategies, instabilities and hence oscillations can occur. To help with this, the order of schemes can be limited (see Brenan *et al.* (1996)). In the present work, as an alternative, the damping factor of Hujeirat and Rannacher (1998) is used. This introduces an additional tuning parameter but is required for just one case (see Chapter 10).

*Evaluating a Representative Solution Error*

In all of the above approaches some form of error estimate ($E^{old}$) is required. As will be seen through later Case Studies the sensible and economical evaluation of this is key to making adaptive methods both accurate and efficient. For time accurate solutions spatially constant time-steps are needed. This suggests the error estimate should be spatially averaged. Gresho *et al.* propose the following average

$$E^{old} = \left[ \frac{1}{N_\phi} \sum_{i=1}^{N_\phi} \left( \frac{1}{N_{P\phi}} \sum_{j=1}^{N_{P\phi}} \left( \frac{E_{P\phi}^{old}}{\left|\phi_P^{old}\right| + \phi_o} \right)^2 \right) \right]^{1/2} \quad (5\text{-}2.15)$$

where $N_{P\phi}$ is the number of nodal points for each variable (this is generally the same for all variables) and $\phi_o$ a reference scale. Evaluation of Equation (5-2.15) requires storage of two full solutions. Hence for a three-dimensional system it gives a significant storage burden. Also, it would seem a more refined volume weighting would be better. Therefore, in the present work the following volume weighted storage saving expression is tested

$$E^{old} = \frac{1}{N_{\phi m}} \sum_{j=1}^{N_{\phi m}} \left[ \frac{\int^{R'} \left|E_{P\phi}^{old}\right|_m dv}{\int^{R'} \left(\left|\phi_P^{old}\right| + \phi_o\right)_m dv} \right] \quad (5\text{-}2.16)$$

where $\phi$ is a, mostly reduced, selected range of variables and $R' \le R$ the full integration region. The $m$ subscripts are used to indicate that variables can be restricted in some way to a coarser grid $m+1$ (saving storage) and then Equation (5-2.16) evaluated.

Tests are made where Equation (5-2.16) is evaluated for just $N_{\phi m} = 1$. When this is done, it might, at first, seem best to automatically evaluate the above for the variable with the greatest average temporal gradient (this can be readily done with no extra storage). However, if the absolute value of this variable is large, significant temporal errors may arise in the parameter of true interest (in the context of the calculation). Therefore, here, when using (5-2.16) with $N_{\phi m} = 1$, the simple approach of basing the error estimate on the key variable of solution interest is tested.

Another factor is that for certain larger scale systems, flow unsteadiness can be local to the region $R'$. An example would be a large domain where the only unsteadiness arose due to vortex shedding from say an isolated cylinder. Then, use of equations (5-2.15)-(5-2.16) over $R$ would result in a diluted error estimate. Under these circumstances, it would be better to apply (5-2.15)-(5-2.16) over a reduced region $R'$. The extent of $R'$ can be ascertained through initial crude coarse space and time-step solutions or alternatively physical intuition (the correctness of which could be verified on completion of the prediction). As well as potentially improving accuracy, $R/R' > 1$ saves storage.

Another means of reducing storage is to restrict or inject values over the region $R$ or $R'$ onto a coarser grid $m + 1$. When using multilevel convergence this naturally occurs. When not, a virtual grid is used and the approach corresponds to data thinning. The impact on accuracy of using injection or restriction is explored later. The staggered variables used here (see Section 4-4) prevent direct velocity injection. For this reason here the term *psuedo-injection* instead of injection is used.
The parameter $\phi_o$ will have a significant effect on solution accuracy. This improves with decreasing $\phi_o$. A sensible $\phi_o$ value is not always easy to estimate. Therefore, in the present work generally the parameter is just used to avoid divisions by zero, i.e. $\phi_o \approx 1 \times 10^{-9}$.

*Relating Error Estimate to New Time-steps*
The error estimate is related to $\Delta t$ through Equation (5-2.6). Again, there are many ways in which (5-2.6) can be used. For stability reasons (see Skelboe (1977)), ideally, the maximum *swing* ($\left(\Delta t^{new}/\Delta t^{old}\right)_{\max}$) should be restricted. However, the minimum ($\Delta t$ decreases) need not be limited. According to Ascher and Petzold (1998), $\Delta t^{new}$ should also be limited to avoid round-off error. *Table 5-2.1* gives higher order implicit scheme stability results of Skelboe.

| Schemes Order (*n-1*) | 2 | 3 | 4 | 5 | 6 |
|---|---|---|---|---|---|
| Maximum swing $\left(\Delta t^{new}/\Delta t^{old}\right)_{\max}$ | 2.6 | 1.9 | 1.5 | 1.2 | 1.0 |

*Table 5-2.1* Maximum swing against schemes order

It shows that permitted swing is inversely proportional to $n$. Also, for a second order scheme, as used here, $\left(\Delta t^{new}/\Delta t^{old}\right)_{\max} \approx 2.5$. With Skelboe's predictions, provided the error estimate is satisfied, $\Delta t^{new}$ is fixed for $n$-$1$ (the scheme order) steps. Therefore, computations use a piecewise constant $\Delta t$ distribution. Gresho *et al.* limit $\Delta t$ changes by setting $\Delta t^{new} = \Delta t^{old}$ when $\Delta t^{new}$ does not change by a certain fraction of $\Delta t^{old}$. Also, for initial time-steps, as here, Gresho *et al.* use the first order backwards Euler fixed step method. This is because adaptive stepping cannot be applied on the first step (there is no error estimate). Also, initial non-physical penalty function solution oscillations are damped by the dissipative first order Euler scheme. Generally, for the present flow computations, the swing is found to be naturally well within the bounds suggested by Skelboe. Therefore, it is not explicitly limited.

An important consideration is sudden transients from quiescent conditions. At the start, Equation (5-2.6) can become highly dependent on $\phi_o$. If $\phi_o$ is negligible, excessively small time-steps may be produced. On the other hand, as the steady state is reached, although lower order stable schemes will give $\Delta t^{new} \to \infty$ monotonically, higher order schemes can give $\Delta t$ oscillations. For economy, under-relaxation factors should be as large as possible. However, in many practical situations as $\Delta t^{new} \to \infty$ an unusually stable solver would be required to perform well with little under-relaxation. Therefore, here, for stability and sensible time-steps, simple $\Delta t$ clipping is used. Equation (5-2.6) is constrained between specified limits so that $\Delta t_{\min} \le \Delta t^{old} \left|\varepsilon / E^{old}\right|^{1/n} \le \Delta t_{\max}$. Also, the use of temporally adapted under-relaxation is considered. This is discussed later.

With the *An* approach (IV), for larger steps the higher order scheme can introduce the dominant solution error. Therefore, a conservative step selection measure, tested in later Case Studies, for $\left|\varepsilon / E^{old}\right| \ge 1$, is to base $n$ in Equation (5-2.6) on the higher order scheme. The larger $n$ value produces smaller $\Delta t$ increases.

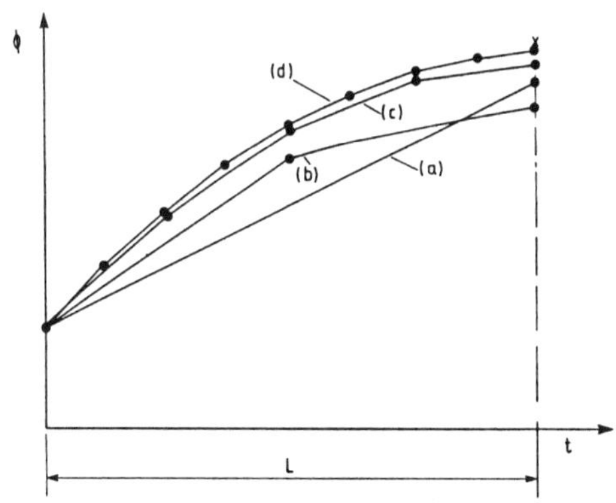


*Figure 5-2.2* Solutions for an integration over the time period *L* with four time-steps sizes

*Solution Correction*

Equation (5-2.9) can be used to effect a Richardson (see Richardson (1910)) type extrapolation. If the time-steps are sufficiently small, this should yield an improvement to the solution. However, the radius of convergence of the Taylor (power) series is small. Some points relating to the Taylor series convergence radius are illustrated in *Figure 5-2.2.* This shows possible solutions for a time integration over a period $L$, using four $\Delta t$ sizes. For profiles (a) to (d) $\Delta t$ progressively decreases. The sketched profiles show that if $\Delta t$ is not sufficiently small, convergence to the correct solution may not even be monotonic (cf. profiles (a) and (b)). It is possible (if time-steps are too large) that the comparison of two solutions falsely suggests $\Delta t$ independence. Therefore, ideally convergence should be based on solutions for three step sizes. When the step is small enough, two solutions can be used to extrapolate to the exact solution. This is indicated by the x symbol. Solutions from profiles (c) and (d) may be suitable for making this extrapolation. With adaptive time-stepping the error estimate could be used to correct solutions thus, in theory, enhancing accuracy. However, bearing in mind the above discussion this is perhaps not worthwhile (see also Press *et al.*).

*Specific Implementation Details for Schemes (I)-(IV)*
In the present work just implementations of (I)-(IV) that can be readily incorporated into available commercial programs are considered. Most commercial codes have the first order forward and backwards Euler schemes with Crank-Nicolson and perhaps ADI.

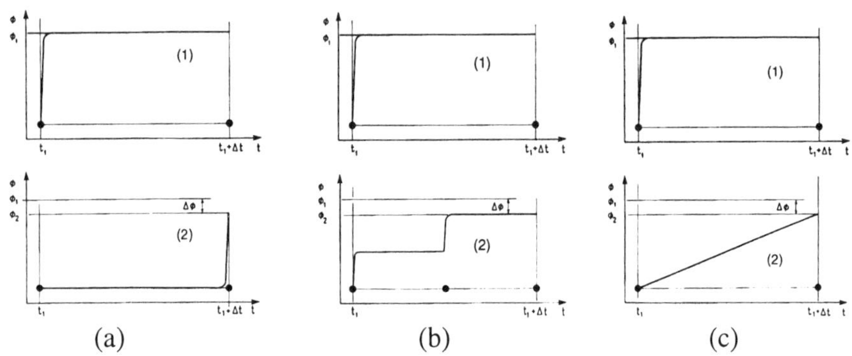


*Figure 5-2.3* Schematics of the temporal finite difference stencils and solution variable variations for approaches (II-IV): (a) Approach (II); (b) Approach (III) and (c) Approach (IV)

Essentially, these can all naturally be coded through the use of the simple substitution embodied in Equation (2-2.1) and variation of the parameter $\lambda$ in Equation (2-2.3). Schematics of the temporal finite difference stencils and solution parameter variations for approaches (II-IV) are given in *Figure 5-2.3.* In essence, the implementation of all the methods described here requires the addition of some simple intrinsic arguments to: appropriately vary $\lambda$ (thus obtaining two solutions for comparison); activate appropriate storage and Equation (5-2.15)-(5-2.16) evaluation subroutines and apply Equation (5-2.6). When using multilevel convergence acceleration additional calls for existing restriction subroutines must be made.

The crude gradient based method *(GBM)*, lacking a direct error estimate gives no additional storage burden. Its implementation just requires that an appropriate Equation (5-2.1) functional relationship is devised. Some conceptual drawbacks of *GBM* were discussed earlier. For the other approaches, to evaluate $E^{old}$ two solutions sets at the 'new' time level should be stored ready for later comparison. Even, for a basic three-dimensional, turbulent, non-isothermal flow prediction this introduces a significant extra storage burden. Since the present interest is industrial CFD, where storage is still an important consideration, this matter is further addressed below.

For the present Altered Scheme (*ASc*) approach, illustrated in *Figure 5-2.3a*, the forward Euler solution ($\lambda$ = *0*), which is subject to severe stability restrictions, is carried out first. This means the more stable and hence reliable backwards Euler solution ($\lambda$ = *1*) is fully stored ready as initial conditions for the next time-step. This should prevent temporal error growth. The error $E^{old}$ is evaluated using Equation (5-2.4). If $E^{old} < \varepsilon$, the solution is continued using the fully stored $\lambda$ = *1* solution. If $E^{old} \geq \varepsilon$, $\Delta t$ is re-evaluated and the solution recomputed.

For ease of implementation, with the Altered Order (*An*) method, mostly, the first order backwards Euler ($\lambda$ = *1*) and second order Crank-Nicolson ($\lambda$ = *0.5*) schemes are used. This approach is illustrated in *Figure 5-2.3b*. For the second order solution component, another relatively easy to implement choice is the ADI method of Peaceman and Rachford (1955). This *An(ADI)* approach is tested for one case. With these *An* methods, the lower order computation is generally made first and then selected solution error comparison variables stored. The second order computation is then made and finally the solution error $E^{old}$ evaluated using Equation (5-2.6). If $E^{old} < \varepsilon$, the solution is continued using the fully stored second order solution. If $E^{old} \geq \varepsilon$, $\Delta t$ is re-evaluated and the time-step recomputed. For complex flows, where the second order component is observed to be unstable it is just used, first, as part of the error estimate (i.e. not carried forward to the next step). Also, the stabilised form of $\lambda$ (Equation (2-4.30)), specified by Hujeirat and Rannacher (1998), is employed. Although the stabilisation parameter $\alpha$ embodied in Equation (2-4.30) still maintains second order accuracy, in practice, if $\alpha$ is too large $E^{old}$ will become artificially small. Therefore, caution is required.

The Altered Step ($A\Delta t$) approach (see *Figure* 5-2.3c) is based around two first order backward Euler solutions. Perhaps, for best convergence (if $\Delta t$ small enough), the ideal would be to compute the two half steps ($\Delta t$) first and then use the final solution to this as a guess to the $2\Delta t$ step (approach 1). However, this would require full data storage at the 'old' time levels for the full step and second half step (this would be on top of the need to store two final solutions for the error estimate). Here, to save storage, generally the long step ($2\Delta t$) is carried out first (at the end of this the selected error evaluation solution variable field is stored) and then the $\Delta t$ steps. At the end of the first short step the 'old' time level data for the long step is over written by the short step data. Multilevel restriction of the new short step 'old' time level variables is then performed.

To save computing time, as a starting guess for the first half time-step, half the values for the 'old' and 'new' time levels of $2\Delta t$ step is taken. Later tests for a box with a suddenly accelerated wall suggest this practice can give around 10 % computing time savings. The problem arises if the error criterion is violated. There is no long step, 'old' time level data available for a solution re-run. To overcome this, for cases with few (less than 10) time-step fails, the previous half step solution is assumed valid and the computation continued from this. The above approach seems reasonable - for most cases tested here there are few time-step failures. This is because generally the $\Delta t$ estimate from the previous step is quite conservative and holds well for the next. Obviously, if storage is not a concern, the above uncertainty is easily remedied by storing all the old time level data (then computation of the half steps followed by the full would probably be best). This latter approach is used for solutions with around more than 10 failed steps. To yield a further time savings, the velocity field stored from the one-long-step solution (for the later error evaluation) could be directly used to assist forming an initial guess to the final half step solution. However, partial velocity field, Case (A) tests indicated this not to be helpful.

*Novel Sub-cycling with Adaptive Time-stepping*

The adaptive $\Delta t$ approaches considered here are summarised in *Figure 5-2.4.* As can be seen, there are two solution elements ((1) and (2)) where $\Delta t \propto \phi_2 - \phi_1$. The subcycling approach, proposed by Gresho *et al.* (1984), applies an extrapolated pressure field to minor time-steps inside a larger interval (mass conservation is enforced at the larger interval end). The motivation for this is that the pressure field evaluation is a major solution expense. Also, the transport equation time-steps can be more dictated by stability than temporal resolution requirements. Stability, like the avoidance of non-physical oscillations, can dictate that time-steps are excessively small. Subcycling and the past practice of using coarser spatial step pressure fields suggests that perhaps $p$ for solution (1) (see *Figure 5-2.4*) could be interpolated to the comparative solution element (2) without resulting in significant errors.

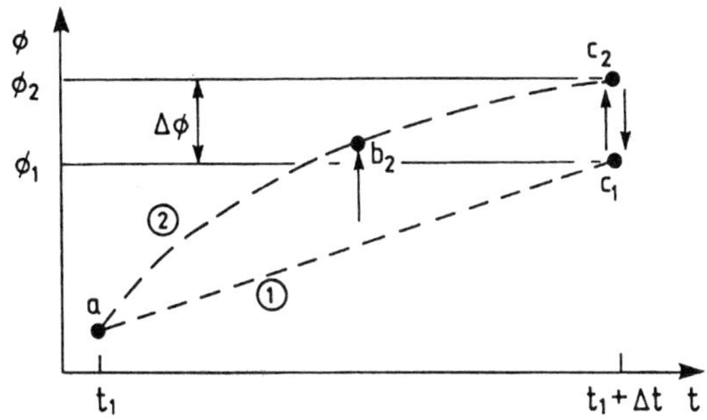


*Figure 5-2.4* Summary of adaptive approaches considered here

Generally, SIMPLE (see Section (4-4) and Patankar and Spalding (1972)) type pressure evaluation methods, give rise to many (typically 5) extra iterative solver calls. This relatively expensive solution element can then be followed by a pressure and velocity field correction stage. Omission of these in element (2) could yield over 50 % computational effort savings (in *just* element (2)). This makes the consideration of a 'subcycling' approach attractive. To perform the $p$ interpolation, additional pressure storage is required.

Several 'subcycling' possibilities exist. For example, there are the following:

($a_{12}$) Solution (2) using solution (1) pressures with mass conservation being enforced at $c_2$;

($b_{12}$) Approach ($a_{12}$) but mass conservation enforced through discarding solution (2);

($c_{12}$) Approach ($a_{12}$) with $\varepsilon$ small enough to give a negligible conservation error. Hence $\phi_I$ is used without mass conservation enforced.

Obviously, scheme (2) need not go before (1) and the reverse can occur. This gives schemes designated here as $a_{21}$, $b_{21}$ and $c_{21}$.

Approaches (a) are classic subcycling, but, in the present context, the number of inner interpolated pressure stages is too small. Unless the $A\Delta t$ scheme is used, the relatively expensive pressure and velocity correction stages are still a dominant feature. Therefore, Approaches (a) are not considered further.

Approaches (b) are conservative, but require storage of the full $\phi_l$ solution. The $b_{21}$ approaches, relative to the $b_{12}$ have the attraction that the higher order solutions are not discarded. However, with the $A\Delta t$ approach an extra pressure evaluation at $b_2$ (see *Figure 5-2.4*) is required. This is replaced in the $b_{12}$ method with an interpolation from points a and $c_1$.

Approaches (c) avoid the need to discard solutions and hence offer the possibility of storage savings and require minimal additional programming. However, if $\varepsilon$ is not sufficiently small there is the possibility of error growth. Assuming solution (2) is more accurate than (1) the $b_{12}$, unlike $b_{21}$, approaches perhaps undesirably discard the higher order solution. The reverse occurs for the (c) approaches. As can be seen, numerous model arrangements exist. To give an indication of potential 'subcycling' savings and the accuracy impact attention is focused here on the following approaches:

$An(p=b_{12})$, $An(p=c_{12})$, $A\Delta t(p=b_{12})$, $A\Delta t(p=b_{21})$, $A\Delta t(p=c_{12})$ and $A\Delta t(p=c_{21})$.

Although many engineering turbulence models greatly impact temporal accuracy, this is in an inconsistent fashion (see Tucker (2000), Franke and Rodi (1993), Johansson, Davidson and Olsson (1993), Chapter 10 and also Section 6-2). Therefore, in the context of real industrial computations, it might seem excessive to demand high turbulence variable temporal resolution (the temporal accuracy is swamped by the turbulence model trait – for example the $k$-$\varepsilon$ model is too dissipative). This suggests, like pressure, certain turbulence parameters could be subcycled. Since here, the $k$-$l$ model is used, this implies a 'subcycling' of $k$. Here the influence of $k$ 'subcycling' is explored in the

$An(p,k=c_{12})$, $A\Delta t(p,k=c_{12})$

frameworks, $p$ and $k$ being 'subcycled' together. For this work the transient complex system turbulent flow Case (E) presented in Chapter 11 is used.

*Novel Forcing Function Element Adaptation*

A typical multilevel convergence cycle (a V cycle) is shown in *Figure 4-2.2.* Brandt (1980) proposed that for evolving flows the finest grid level ($m$=1) can be periodically revisited, most time advancement effort being more economically focused on the coarser levels ($m$>1). The finer level information is sensed on coarse grids through the *forcing function* (see Equation (5-2.19)). This non-adaptive solution strategy suggests that the first adaptive solution element (1) could be used to generate a forcing function for element (2), the latter being advanced on computationally cheaper coarser grids ($m$>1). For element (2), the probably more dominant lower spatial frequency components will be supported on coarser grids and higher frequencies accounted for through the forcing function. The question of whether the accuracy of element (2) is sufficient to give reasonable time-steps is addressed by later computational tests. *Figure 5-2.5* is a schematic of the new novel forcing function element adaptation. The schematic is directly applicable to the $An$ ($An_m$) and $ASc$ ($ASc_m$) approaches. Since the element (1) solution will be close to that of (2), just one element (2) outer iteration is used. For the $A\Delta t$ approach ($A\Delta t_m$), here, both half steps are advanced at $m$ = 2, adding another single iteration stage with forcing function assembly to *Figure 5-2.5*. Other arrangements are possible. These are not explored here. Also, no $ASc_m$ implementation is made. This is because, it is considered that the explicit forwards Euler element may not always be stable enough to assemble a reliable forcing function. Also, generating of a forcing function from the stable backwards Euler element, for use in the forwards Euler would yield little computational benefit. The explicit forward Euler component is already extremely cheap.

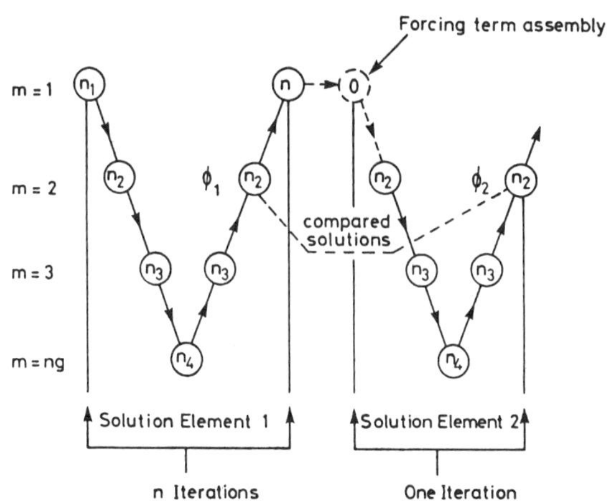


*Figure 5-2.5* Schematic of new novel forcing function element adaptation

| Case | A | B | C | D | E | | F |
|---|---|---|---|---|---|---|---|
| | | | | | 2D | 3D | |
| Grid type | Non-uniform $x$-$y$ | Radially refined $r$-$\theta$ | Axi-symmetric non-uniform $r$-$z$ | Radially and axially refined $r$-$\theta$-$z$ | Uniform $x$-$y$ | Non-uniform $x$-$y$-$z$ | Non-uniform $x$-$y$-$z$ |
| Density | 33x33-121x121 | 17x40-65x120 | 33x33-65x65 | 21x40x17 | 101x101 | 88x77x53 | 101x89x45 |
| Adaptation Variables | $u$ and $v$ and just $v$ | $v$ | $v$ | $u$, $v$, and $w$ | $u$ and $v$ and just $v$ | NA | $u$ |
| $\Delta t$ (s) | 0.001 | 0.0001 | 0.001 | 0.05 | 0.001 | 0.001 | 0.001 |
| $\Delta t_{min}$ (s) | 0 | 0.01 | 0.001 | 0.0001 | 0.0001 | NA | 0.001 |
| $\Delta t_{max}$ (s) | $\infty$ | 0.5-1 | 0.5 | 0.5 | 0.5 | NA | NA |
| $C_t$ | 0.9 | 0.9 | 0.45 | 0.45 | 0.45 | NA | 0.45 |
| $\alpha_{u,v,w}$ | 0.3 | 0.8 | 0.3 | 0.5 | 0.5 | 0.3 | 0.4 |
| $\alpha_{k,\varepsilon,\mu t}$ | NA | NA | NA | NA | 0.5 | 0.3 | 0.3 |
| $\alpha_{p'}$ | 0.7 | 1 | 0.7 | 1 | 1 | 0.5 | 0.7 |
| $n_g$ | 4 | 2 | 2 | 3 | 1 | 1 | 1 |
| $n_p$ | 6 | 8 | 8 | 6 | 16 | 6 | 6 |
| $y^+_{ave}$ | NA | NA | NA | NA | - | 100 | 15 |
| $\rho_o$ (kg/m$^3$) | 1.2 | 998.66 | 1.2 | 1.2 | 1.2 | 1.2 | 1.2 |
| $\mu_o$ (kg/ms) | $1.83\times10^{-5}$ | $1.15\times10^{-2}$ | $1.82\times10^{-5}$ | $1.82\times10^{-5}$ | $1.83\times10^{-5}$ | $1.83\times10^{-5}$ | $1.83\times10^{-5}$ |
| $V_o$ (m/s) or $\Omega$ (rad/s) | $1.52 \times 10^{-4}$ | - | 1.45 | 17.017 | 3.8 | 3.8 | 4.5 |
| $v_o$ (m/s) | $1\times10^{-10}$ | $1\times10^{-10}$ | $1\times10^{-10}$ | $1\times10^{-10}$ | 3.8 | NA | $1\times10^{-10}$ |
| $rms_{max}$ | $2.5\times10^{-5}$ | $2.5\times10^{-6}$ | $2.5\times10^{-6}$ | $2.5\times10^{-5}$ | $1\times10^{-4}$ | NA | NA |
| Residual Error (%) | NA | NA | NA | NA | NA | 2 | 0.5 |

*Table 5-3.1* Nominal parameter values for Cases (A)-(F)

*Diagonal Dominance*
When performing adaptive time-step solutions, the $\Delta t$ changes can cause large diagonal dominance (see Section 4-2) variations (this is especially true for Case (E)). This would suggest, that for best convergence, under-relaxation factors (see Section 4-2) should change through the solution, with high values where $\Delta t$ is small and low values elsewhere. The following *ad hoc* adaptive under-relaxation expression is tested

$$\alpha_{u,v,w} = \max\left[\min\left[\frac{S|A_{P\phi}|}{\Sigma|A_{NB\phi}|},1\right],\alpha_o\right] \qquad (5\text{-}2.17)$$

where typically $0.1 \le S \le 0.6$ and $\alpha_{u,v,w}$ are directly incorporated in the discretized equations. Equation (5-2.17) will give $1 \ge \alpha_{u,v,w} \ge \alpha_o$ with $\alpha_{u,v,w} \to 1$ for $\Delta t \to 0$. The application of a similar expression to the turbulence equations is found unhelpful. Conjugate gradient type solvers may overcome the above mentioned solver sensitivity problem and considering these would be worthwhile future work. The application of Equation (5-2.17) is just considered for Case (E) in Chapter 11.

## 5-3 Discussion of Results

Mostly, for simpler geometry, fixed $\Delta t$, cases no form of under-relaxation is used. However, for all the adaptive approaches, because of the significant $\Delta t$ variations, to avoid divergence under-relaxation is found necessary. Generally, no effort is made here to find optimal relaxation values and what are felt to be conservative levels are used. Under-relaxation is incorporated directly through the coefficients of the discretized equations. For all cases $\varepsilon = 0.1\ \%$.

With simpler geometries/flows $C_t$ = 0.9 otherwise $C_t$ = 0.45 in Equation (5-2.6). A range of convective term treatments is used. These include first order upwinding, HYBRID, CONDIF and Central Differencing. Nominal set up parameters for adaptive scheme solutions are given in *Table 5-3.1*. The results for Cases (D)-(F) are presented in the later Case Study chapters. Despite this, it is considered convenient to maintain the text's flow and give adaptive case parameters globally here. Tabulated $\Delta t$ values correspond to those required for sensibly time-step independent solutions. Where comparisons are made with validation data, high grid densities and generally higher order schemes are used. However, for a few cases, that make just temporal scheme performance comparisons, more economical coarser grids are used and sometimes lower order spatial differencing schemes. At solid walls, for velocities, the usual no-slip and impermeability conditions are applied. At open boundaries, $p$ is fixed and generally the gradients of other variables are set to zero in a second order fashion.

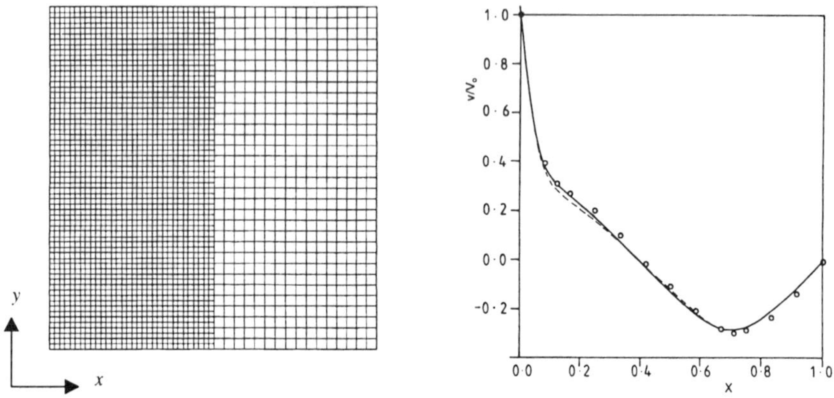


(a) (b)

*Figure 5-3.1* Spatial step refinement for transient square box flow: (a) Grid form and (b) Dimensionless $v$ component predictions (o Datum predictions, ------ Current predictions without step refinement, —— Current predictions with step refinement)

*Case (A), Transient wall driven square box flow*

The Case (A) geometry is shown in *Figure 5-1.1a*. For this case, when $t > 0$, the cavity tangential wall velocity is such that $Re = 400$. The steady flow is a popular benchmark, surprisingly for which uniform grids (see Davies and Jones (1983)) are most effective. The uniform *Figure 5-3.1a* grid, with the embedded left hand region near the moving wall (which gives rise to steep velocity gradients) is sufficient for sensibly grid independent results. Grid embedding is achieved here through the multilevel convergence scheme (see Thompson and Ferziger (1989) and Appendix C).

*Figure 5-3.1b* compares the current steady predicted variation of dimensionless velocity, $v/V_o$ ($V_o$ is the wall velocity) against dimensionless axial coordinate $X$ (where $0 \leq X \leq 1$) at the mid cavity height $Y = 0.5$ (where $0 \leq Y \leq 1$) with the datum predictions of Rayner (1993). The full and dashed lines correspond to the current predictions with and without grid embedded step refinement. Symbols represent the datum predictions. The solution verifies the spatial discretization and solution set-up. More importantly, the figure shows that additional resolution of the steep gradients for $X < 0.5$ (moving wall side of the box) improves solution accuracy. *Figure 5-3.2* shows contours of $\partial v/\partial t$ at $t = 0.25$, 0.5, 1 and 1.25 s. It is clear that, like the spatial gradients, the most significant temporal velocity gradients occur around $Y < 0.5$ (for $t > 1.5$ s the contours appear much weaker and more uniformly distributed). This information will be taken advantage of later in the grid adaptation strategy.

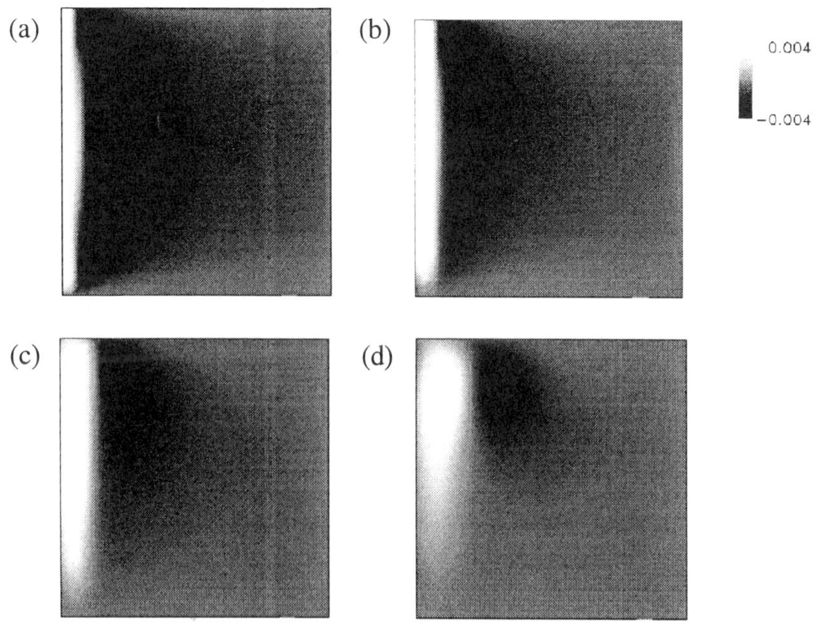


*Figure 5-3.2* Case (A) contours of $\partial v/\partial t$ (m/s$^2$) at: (a) $t = 0.25$ s, (b) $t = 0.5$ s, (c) $t = 1$ s and (d) $t = 1.25$ s

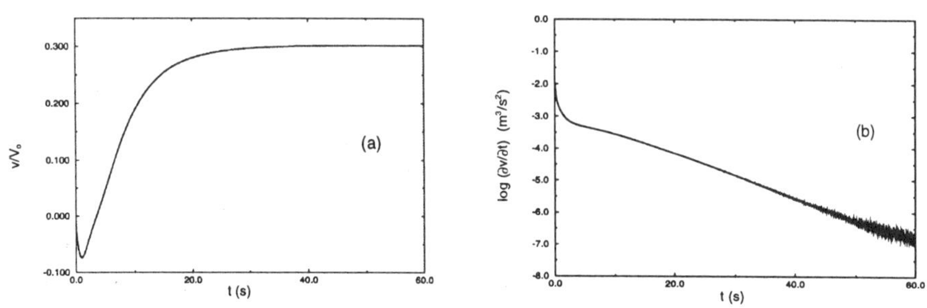


*Figure 5-3.3* Case (A) time history information: (a) variation $v/V_o$ at $X \approx 0.15$ and $Y = 0$ and (b) variation of $\partial v/\partial t$

*Figure 5-3.3* shows time history information, for predictions using the second order Central Difference (the scheme is stabilised by committing non-positive diagonal contributions to source terms – see Runchal (1987) and Appendix B) and ADI schemes with $\Delta t$ = 0.001 s. Frame (a) gives the temporal variation of $v/V_o$ at $X \approx 0.15$ and $Y = 0.5$. Severe temporal velocity gradients near the solution start can be seen. After this initial transient, the solution takes around 50 s to approach the steady state. This can be appreciated from Frame (b), giving a linear-log plot of the cell area weighted average of $\partial v/\partial t$ against $t$. As can be seen, away from the initial transient, the $\partial v/\partial t$ variation is approximately linear. This plot suggests the present flow would be amenable to the crude adaptive Approach (I). However, despite extensive efforts, no functional relationship, between previous time-step temporal gradients and new steps, that would yield computational time savings, could be found. Therefore, this approach was abandoned. It was concluded for more complex flows a relationship between temporal gradients and $\Delta t$ would generally be too difficult to find.

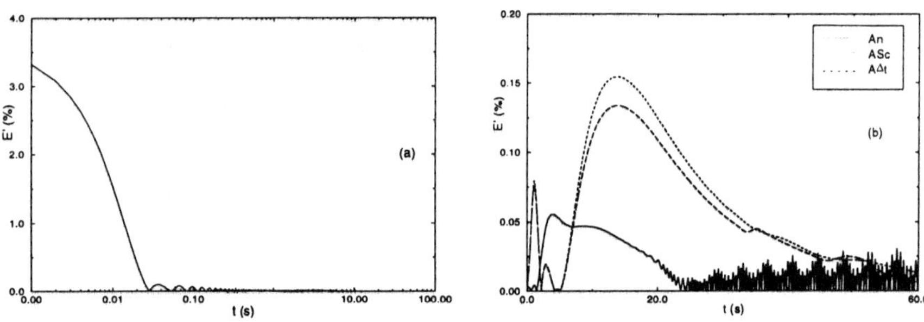


*Figure 5-3.4* Solution errors ($E$) in $v$ against $t$ at $X \approx 0.15$ and $Y = 0.5$: (a) $\Delta t = 0.03$ s. and (b) three generic adaptive schemes

*Figure 5-3.4* shows the true (relative to a $\Delta t$ = 0.001 s solution) solution errors ($E'$) in $v$ (normalised by $V_o$) against $t$ at $X \approx 0.15$ and $Y = 0.5$. Frame (a) is for a fixed $\Delta t$ = 0.03 s (please note the horizontal logarithmic axis). Frame (b) is for the three generic adaptive schemes ($n$ = 0.5 in $Vn$ method). For $\Delta t$ = 0.03 s, a significant error around the start of the transient is clear.

*Figure 5-3.5* plots $log(\Delta t)$ against $log(t)$ for the $An=0.5$ (full line), $A\Delta t$ (short dashed line) and $ASc$ (long dashed line) schemes. Also, included is the $An$ solution with $n$ = 0.333 for $|\varepsilon/E_2| \geq 1$ (dot dash line). The figure shows that this $n$ decrease makes steps more conservative and also has the beneficial effect of reducing the wiggles at around $t$ = 30 s. Therefore, for the remaining predictions, this exponent approach is used. Around $t$ = *1 s*, the *An* approaches are least conservative with regards to step, followed by the $A\Delta t$ and then *ASc* methods. Although not easy to see (because of the log-log plot) for $t \approx 30$ s, this trend changes and the *ASc* and $A\Delta t$ approaches give the largest time-steps relative to the *An* methods. Analysis indicates *Figure 5-3.4b* errors are proportional to $\Delta t$ and hence consistent with the scheme $\Delta t$ traits noted for *Figure5-3.5*. For example, around $t$ = 30 s the smallest $\Delta t$ and hence error is for the *Vn* scheme. For the *ASc* and $A\Delta t$ approaches, $E'$ can go over the specified $\varepsilon$ = 0.1 % value. However, the average solution error ($E'_{ave}$) is well below 0.1%.

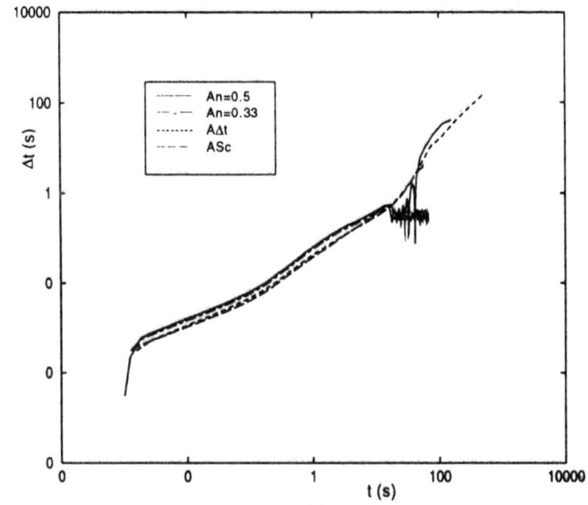


*Figure 5-3.5* Plot of *log(Δt)* against *log(t)* for the *An=0.5*, *An=0.33*, *AΔt* and *ASc* schemes

*Table 5-3.2* gives: nominal % savings in computing time ($t_s$) for $t$ < 60 s (relative to a $\Delta t$ = 0.03 s solution, giving a highly conservative estimate of adaptive $\Delta t$ performance), the % of time-steps where $E'_{ave} > \varepsilon$ (hence needing to be recomputed) and $E'_{ave}$. Also, tabulated is $E'_{ave}$ for $t$ < 5 s with $E^{old}$ evaluated for $X \leq 0.5$ (i.e. $R'/R < 1$). Errors are further considered when Equation (5-2.16) is calculated at $m$ = 2 with storage saving *pseudo-injection* (data thinning) and a 6 point restriction.

The tabulated *An* results are for both $n$ = 0.33 and $n$ = 0.5 and can be seen to be quite similar. Results show the storage saving measure of evaluating $E^{old}$ for $X$ < 0.5 has slightly improved accuracy. As might be expected, the additional storage saving measure of pseudo injection reduces accuracy. None-the-less the solution accuracy is more than acceptable. Surprisingly, relative to pseudo-injection, the use of full restriction, that absorbs more information from the finest grid level, generally reduces accuracy further (since the forward Euler component does not use multilevel convergence restriction no *ASc* result is given). However, again, overall, the solution error is low and both approaches are naturally compatible with multilevel convergence and hence computationally economical. The storage saving is more apparent when step selection based on several variables (see Equation (5-2.15)). Importantly, use of just $v$ or both $u$ and $v$ in Equation (5-2.16), on average only influences computational performance by 1.2 %, the numbers of iterations to convergence and solution error values being surprisingly similar.

| Scheme | $t_s$ (%) | %$\Delta t$ Fails | % $E'_{ave}$ t<60s | % $E'_{ave}$, t<5s | | | |
|---|---|---|---|---|---|---|---|
| | | | | Full region | Half region | | |
| | | | | | Std. | Pseudo Injection | Full Restriction |
| *GBM* | 0 | NA | NA | NA | NA | NA | NA |
| *An (n=0.5)* | 27 | 1.7 | 0.017 | - | | - | - |
| *An (n=0.33)* | 31 | 3.2 | 0.019 | 0.030 | 0.029 | 0.032 | 0.032 |
| *An(ADI)* | 31 | 0.07 | 0.017 | 0.018 | - | - | - |
| *ASc* | 57 | 0.80 | 0.055 | 0.023 | 0.022 | 0.024 | - |
| $A\Delta t$ | 56 | 0 | 0.062 | 0.024 | 0.023 | 0.025 | 0.028 |
| $An(p=b_{12})$ | 65 | - | 0.039 | - | - | - | - |
| $An(p=c_{12})$ | 51 | - | 0.009 | - | - | - | - |
| $A\Delta t(p=b_{12})$ | 73 | - | 0.109 | - | - | - | - |
| $A\Delta t(p=b_{21})$ | 79 | - | 0.130 | - | - | - | - |
| $A\Delta t(p=c_{12})$ | 73 | - | 0.077 | - | - | - | - |
| $A\Delta t(p=c_{21})$ | 72 | - | 0.099 | - | - | - | - |
| $An_m$ | 50 | 0 | 0.02 | - | - | - | - |
| $A\Delta t_m$ | 20 | 0 | 0.18 | - | - | - | - |

*Table 5-3.2* Case (A) adaptive scheme performances

The table shows that few time-steps need recomputing. Unlike the other adaptive approaches, for the *An* method around 1 or 2 time-steps failed the error test twice. Conveniently, bearing in mind the storage saving measure described earlier, no $A\Delta t$ steps needed recomputing

Although, at first sight, the *ASc* and $A\Delta t$ approaches appear cheaper than the *An* method, the $E'_{ave}$ for these is significantly higher. Therefore, it is not completely clear which scheme is the most efficient. Considering its computational limitations (see by Ascher and Petzold (1998)) the $A\Delta t$ scheme performance is surprisingly good. A positive $A\Delta t$ feature is that as $t \to \infty$, $\Delta t \to \infty$ monotonically. The Crank-Nicolson and forward Euler elements in the other two approaches do not exhibit this property. This is to be expected. For large time-steps, Crank-Nicolson, although unconditionally stable, gives oscillatory solutions. For the forward Euler scheme, with increasing $\Delta t$, unhelpfully, the Courant-Friedrichs-Lewy (CFL) condition must eventually be violated.

It should be noted that a solution with $\Delta t = 0.01$ removes much of the initial $\Delta t = 0.03$ temporal solution error. However, this requires around 5375 iterations, giving a computation that is vastly more expensive than any of the adaptive step methods used. Therefore, for this case all the adaptive schemes show highly impressive performances.

The table suggests pressure 'subcycling' is promising. Except for the $An(p{=}c_{12})$ method, as might be expected, 'subcycling' slightly reduces accuracy. Approaches (b), that conserve mass but discard the more accurate solution component have a lower accuracy than approaches (c). The latter retain the higher order solution component at the expense of strict mass conservation. Despite the minor accuracy loss (all solutions would certainly, for most industrial applications, be considered time-step independent) the computational savings arising from using the 'subcycling' are significant (on average 69 %). It is possible that the $An(p{=}c_{12})$ method shows *both* accuracy and speed up gains because, either spurious wiggles or minor convergence error components in the higher order solution component are suppressed. Any suppression probably arises from using the more robust lower order pressure field solution component.

Pleasingly, the $An_m$ approach does not appear to significantly reduce accuracy whilst producing around a 20 % $t_s$ improvement over the standard $An$ method. However, relative to the standard $A\Delta t$ scheme the $A\Delta t_m$ approach gives worse results. Presumably, use of the forcing term on coarse grids twice in element 2 suppresses changes relative to element 1. This results in excessively large time-steps and hence the larger $E'_{ave}$. Also, the larger time-steps are less stable and less economical to advance.

*Case (B), Transient Buoyancy Driven Flow between Cylinders*

The *Figure 2-12.1* Case (B) geometry is now considered. The central cylinder is suddenly heated to give $Gr = 9 \times 10^7$. In flow visualisation experiments, Castrejon and Spalding (1988) dye fluid in the bottom cavity half. Full test case details are given by these authors and in Section 2-12.

Here, adaptive scheme results use a grid density similar to Castrejon and Spalding's. According to them $\Delta t = 0.05$ s is sufficient. However, comparisons here with $\Delta t = 0.0001$ s predictions suggest, for this step, a 0.32 % temporal error in $u$ (radial velocity). For the $An$ approach $\Delta t_{\max} = 0.5$ s is

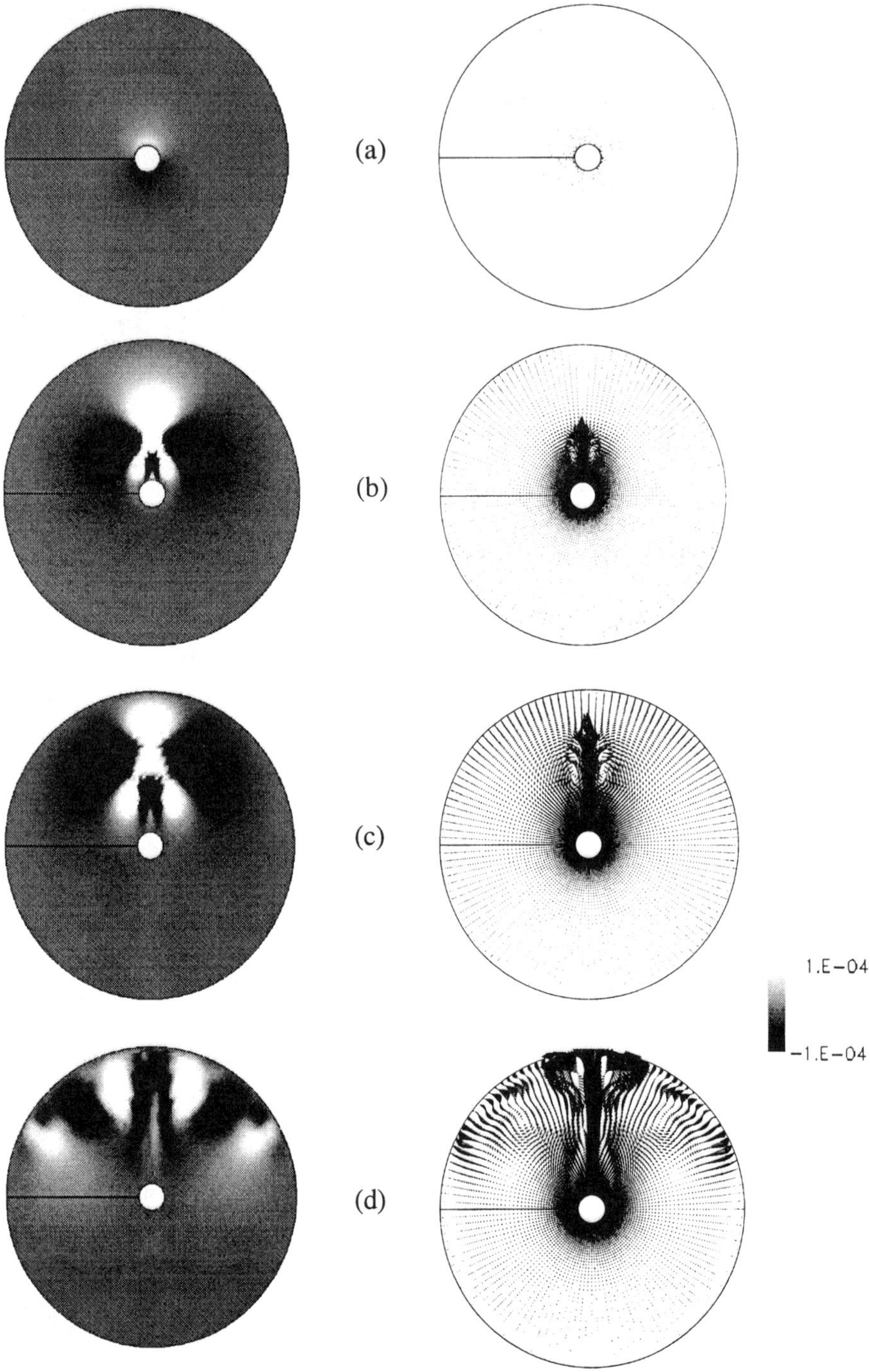


*Figure 5-3.6* Case (B) contours of $\partial u/\partial t$ (m/s$^2$) and velocity vectors at: (a) $t$ = 1, (b) $t$ = 9.5, (c) $t$ = 13.5 and (d) $t$ = 54.9 s

required to control Crank-Nicolson element wiggles. Since this higher order solution component is carried forward, it is especially important to control its error. Due to low velocities, for $t < 2.5$ s, Equation (4.5-16) gives $\Delta t \approx 1x10^{-7}$ s. Therefore, for efficiency the $\Delta t_{\min} = 0.01$ s limit, based on Castrejon and Spalding's predictions, is important. At around $t$ = 200 s, the flow becomes steady and so this period is used in later tabulated performance data. The key velocity component $u$ is used in Equation (5-2.16). The computed temporal plume growth is shown in *Figure 2-12.2*. These CONDIF predictions use a finer 65 x 120 $(r, \theta)$ grid.

Based on the particle tracking information, *Figure 2-12.3* compares predicted plume growth in terms of dimensionless ($X$ = $r/b$) plume height with measurements. *Figure 5-3-6* shows, at $t$ = 1, 9.5, 13.5 and 54.9 s contours of $\partial u/\partial t$ and velocity vectors, in the left and right-hand columns, respectively. Contours clearly show unsteadiness is focused around the top cavity half. Vectors show, for $t < 9.5$ s, the flow is slow to develop. As noted earlier, this causes Equation (5-2.16) to introduce excessively small time-steps and is remedied by clipping to $\Delta t_{min}$. The other alternative would be through a judicious choice of $v_0$. However, for the present situation, clipping is more convenient – numerical time-step data is available.

| Scheme | % Saving | %$\Delta t$ Fails | % $E'_{ave}$, t<200s | | | |
|---|---|---|---|---|---|---|
| | | | Full region | Half region | | |
| | | | | Std. | Pseudo Injection | Full Restriction |
| *An* *(n=0.33)* | 13 | 0 | 0.155 | 0.139 | 0.139 | 0.146 |
| *ASc* | 33 | 4 | 0.079 | 0.065 | 0.065 | 0.067 |
| *A$\Delta t$* | -13 | 0.05 | 0.088 | 0.085 | 0.082 | 0.085 |

*Table 5-3.3* Case (B) adaptive scheme performances

Scheme performances are summarised in *Table 5-3.3*. The average error $E'_{ave}$ in $u$ (normalised by the maximum plume velocity) at $X \approx 0.5$, on the plume centre line, is considered. Computational savings ($t_s$) are compared against the $\Delta t$ = 0.05 s solution where $E'_{ave}$ =0.32 %. Therefore, $t_s$ values are again conservative. The $A\Delta t$ has some step fails and so is correctly implemented with full storage. The table shows the *ASc* approach, with its fast explicit element, is most economical, giving $t_s$ = 33 %. As indicated by Ascher and Petzold (1998), the $A\Delta t$ approach is worst, being slower than the fixed step case. The table shows that, as well as saving storage, focusing the error estimate on the top cavity half improves accuracy. For this case, pleasingly, storage saving pseudo injection maintains accuracy. However, again, unexpectedly, the perhaps theoretically more promising full restriction gives lower accuracy.

*Case (C), Transient Wall Driven Free Disc Flow*

*Figure 5-1.1b* gives a schematic of the Case (C) system. Adair and Tucker (1998) present full case details. The axisymmetric flow induced by the suddenly rotated or stopped disc in an infinite environment is considered for the Reynolds number $\mathrm{Re}_\phi = \rho b^2\Omega/\mu \approx 1 \times 10^3$, where $b$ is the disc radius and $\Omega$ its angular velocity. Adaptive time-stepping is just considered for the suddenly started disc. However, for interest, results for the final steady flow and that originating from this when the disc is suddenly stopped are considered. *Figure 5-3.7* gives a steady state flow schematic.

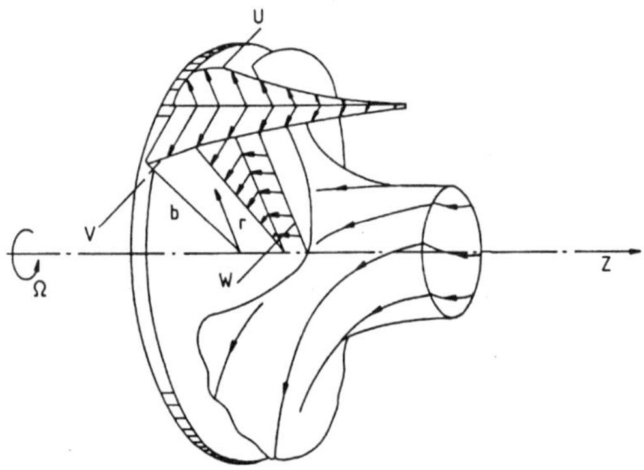


*Figure 5-3.7* Schematic of flow due to a free disc (Schlichting (1979))

As a result of the disc surface no-slip condition, the flow is entrained in the axial direction and pumped radially outwards through the boundary layers on the disc surface. von Karman (1921) showed that a similarity solution of the following form is possible

$$
\begin{aligned}
u &= \Omega r C_0(\xi) \\
v &= \Omega r C_1(\xi) \\
w &= \left(\Omega \frac{\mu}{\rho}\right)^{1/2} C_2(\xi) \\
p &= -\mu\Omega C_3(\xi)
\end{aligned}
\qquad (5\text{-}3.1)
$$

where $\xi = (\rho\Omega/\mu)^{1/2} z$ and $u$, $v$ and $w$ the radial, tangential and axial fluid velocity components, respectively.

For flow when the disc is suddenly started or stopped, Thiriot's (1940) analytical solution indicates that axial velocity distributions are similar with time, $t$, when transformed to

$$f^* = \frac{w}{-4\Omega^2 t\sqrt{\mu t/\rho}} \tag{5-3.2}$$

and plotted against

$$\eta = \frac{z}{2\sqrt{\mu t/\rho}} \tag{5-3.3}$$

It is convenient to define the following dimensionless velocities: $F = u/\Omega r$, $G = v/\Omega r$ and $H = w/\sqrt{\Omega\mu/\rho}$. For a suddenly started disc, von Karman gives the analytical slope coefficient $m$ expression below

$$m = \lim_{\eta\to 0}\frac{F}{G-1} = \frac{-0.72676\Omega t}{1+0.23(\Omega t)^2} \tag{5-3.4}$$

Similarly for a sudden disc deceleration

$$m = \lim_{\eta\to 0}\frac{F}{G} = \frac{-1.2372\Omega t}{1-0.045836(\Omega t)^2} \tag{5-3.5}$$

Cochran (1934) gives slope coefficient measurements and the following boundary layer thickness estimate

$$\delta * = bRe_\phi^{-1/2} \tag{5-3.6}$$

Unlike the other cases above, the transient solutions are unstable. Therefore, even for fixed step computations, significant under-relaxation is required and adaptive solutions require the inclusion of high false diffusion levels (despite this, temporal accuracy comparisons can be made). Also, partly for stability reasons $C_t$ = 0.45. The tangential velocity component $v$ is used in Equation

(5-2.16). To avoid excessively small steps, at the solution start, where error values are normalised by small velocities, $\Delta t_{min} = 0.001$ s. Towards the steady state, to limit less stable scheme component oscillations $\Delta t_{max} = 0.5$ s.

*Figure 5-3.8* Comparison of steady prediction with analytical

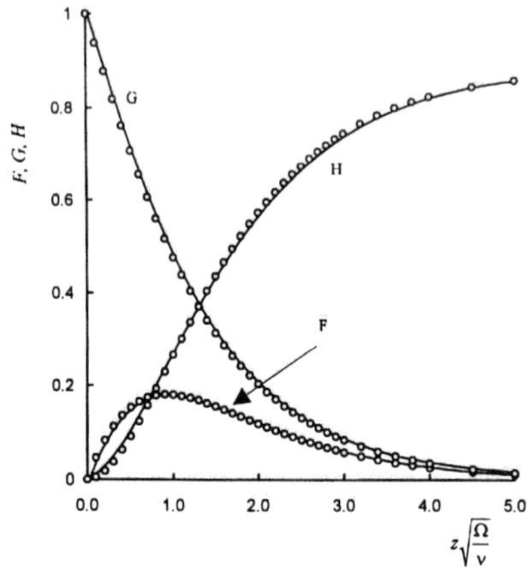


solution (— analytical solutions, o predictions)

First, it is instructive to consider accuracies for steady flow. *Figure 5-3.8* compares dimensionless $w$, $u$ and $v$ variations against dimensionless axial co-ordinate with analytical information (see Cochran (1934)). Predictions (as with other data comparisons) use a non-uniform grid with around 61 x 61 $(r,z)$ nodes.

*Figure 5-3.9* Suddenly started disc Case (C) temporal development predictions: (a)

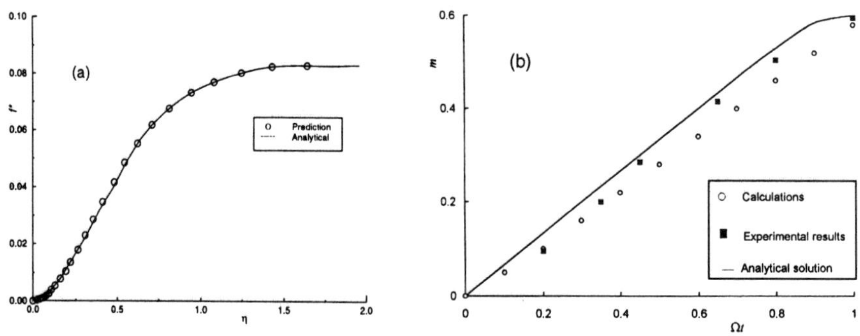


comparison for $w$ with analytical solution and (b) comparison between experimental, analytical and predicted slope coefficients

The agreement is reasonable. *Figure 5-3.9a* compares a first order backwards Euler scheme, $w$ component prediction with Thiriot's analytical solution for an acceleration. The analytical solution is represented by the full line and the prediction by symbols. As can be seen, there is agreement. *Figure 5-3.9b* compares analytical, experimental and predicted slope coefficients $m$ (see Equation (5-3.4)). The correspondence between the three sets of data is not that good. This is possibly because of experimental errors and analytical solution simplifications. *Figure 5-3.10a* compares predictions and the analytical, axial velocity, sudden disc deceleration solution of Thiriot. As with the acceleration case, there is agreement with the analytical velocity solution. *Figure 5-3.10b*, shows analytical, experimental and predicted slope coefficients for the deceleration case. Predictions agree best with the measurements. The comparisons for $w$ give confidence in the present code's temporal accuracy.

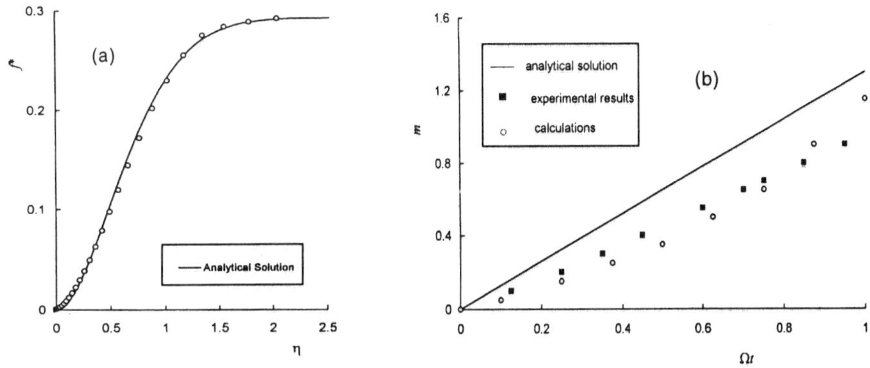


*Figure 5-3.10* Suddenly stopped disc Case (C) temporal development predictions: (a) comparison for $w$ with analytical solution and (b) comparison between experimental, analytical and predicted slope coefficients

Since the disc is in a quiescent environment, the extent of the region used for the adaptive solution error estimate is limited here by $\delta^*$ (Equation (5-3.6)) and $b$. *Figure 5-3.11* plots $\partial w/\partial t$ contours for $t = 0.25$, 0.5, 0.75 and 1 s. As might be expected, the dominant unsteadiness is adjacent to the left hand disc.

This justifies the use of Equation (5-3.6). If the temporal solution error is considered for the whole region, *more storage* is needed and *greater solution errors* should arise.

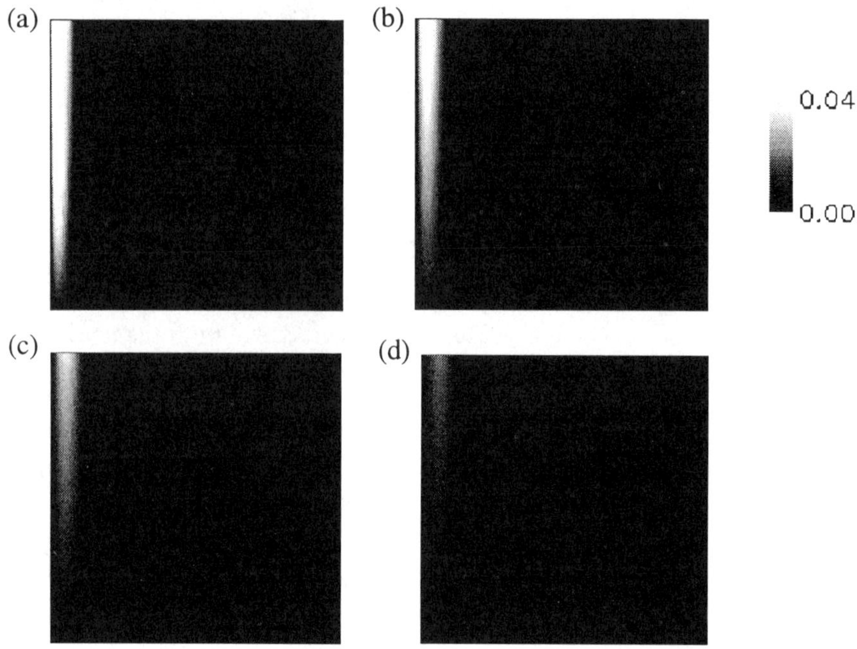


*Figure 5-3.11* Case (C) contours of $\partial w/\partial t$ (m/s$^2$) at : (a) $t = 0.25$ s, (b) $t = 0.5$ s, (c) $t = 0.75$ s and (d) $t = 1$ s.

Adaptive results are summarised in *Table 5-3.4.* Due to solution instability a limited number of schemes are tested. The % time saving $t_s$ is evaluated relative to a solution with a $\Delta t$ such that $E'_{ave} = 0.16$ %. This error level is consistent with those for the adaptive solutions. Errors are normalised by $w$ coincident with the monitoring point, at $t = \infty$.

| Scheme | $t_s$ (%) | %$\Delta t$ Fails | % $E'_{ave}$, $t$<1s |
|---|---|---|---|
| *An (n=0.33)* | 19 | 0.8 | 0.156 |
| *ASc* | 50 | 1.5 | 0.132 |
| *A$\Delta t$* | -50 | - | 0.644 |

*Table 5-3.4* Case (C) adaptive scheme performances

The table shows the *ASc* approach is most efficient followed by the *An*. As for the buoyant plume flow, the *A$\Delta t$* method is slower than the fixed $\Delta t$ solution. A key point illustrated is that the performance of adaptive stepping is highly dependent on solver robustness. Time-step variations introduce instability. The solver must be able to deal with these. Historical evidence shows solver robustness is improving. This may increase the use of adaptive time-stepping. Conjugate gradient solvers (see Section 4-2) are robust with respect to diagonal dominance and so might give improvements for this case.

The performances of the above schemes are assessed further in later Case Studies. The conclusions below assimilate findings from these. Of note, for Cases (D) (Section 8-5) and (E) (Section 11-3) adaptive scheme performance is disappointing. However, for Case (F) (Section 10-4) it is impressive.

## 5-4 Conclusions

Overall, the generic altered scheme based approaches (*ASc*) having a cheap explicit forwards Euler element show the best performances. Excluding Case (D), the *ASc* schemes yield average computing time savings ($t_s$) of around 40 %. However, the *ASc* approaches are not always best. For Case (A), the altered order ($An(p=b_{12})$) and altered step ($A\Delta t(p=c_{12})$) 'subcycling' methods are about 10 % faster. The 'subcycling' approximation generally improves $t_s$ but increases the actual solution error $E'$ relative to the prescribed value ($\varepsilon$). The exception to this is for Case (A) where the $An(p=b_{12})$ method stability appears to improve accuracy. Generally, as might be expected, the $A\Delta t$ approaches (where each step involves 3 sub-steps) give the worst performances. The standard $A\Delta t$ method shows especially poor performance for rotating flows, where it appears to introduce instability (the origins of this are difficult to explain) and, without care, divergence. However, surprisingly, for Case (F), $A\Delta t$ approaches are best, giving $t_s$ = 87%. Also, for Case (A) their performance is impressive, only being 1 % slower than the *ASc* methods. For the latter transient case, a large part of the scheme's impressive performance is attributed to its large time-step stability – each sub-step consists of just stable first order backwards Euler elements. This ensures $\Delta t \to \infty$ monotonically as $t \to \infty$. The *An* and *ASc* methods' higher order and explicit elements exhibit large $\Delta t$ numerical deficiencies. These give rise to $\Delta t$ oscillations as $t \to \infty$. For Case (F), again the generic $A\Delta t$ scheme's good performance is attributed to its stability. For the Case (F) governing equations, with quadratic sources and Neumann boundary conditions Crank-Nicolson stability is uncertain and $\Delta t$ clipping (through maximum Courant number) with damping is required for convergence. Similarly, for the complex turbulent Case (F) flow, the forwards Euler scheme stability appears problematic. A positive side to the instabilities is that generally they give conservative time-steps, but clearly this increases $t_s$ or difficulties relating $E'$ to $\varepsilon$. For all cases where storage is saved by basing $\Delta t$ on regions of greatest unsteadiness ($R'/R < 1$) accuracy improvements are found. Also, the multilevel convergence compatible, pseudo-injection storage saving introduces negligible solution error. However, surprisingly, full multilevel restriction, which absorbs more fine grid information, produces a slightly higher error (none-the-less, relative to the overall solution error the increase is minor). Tests show, saving storage by basing time-steps on a restricted range of variables can also be effective. For Case (D) – the rotating cavity with an axial throughflow – the integral of temporal derivatives over the full spatial domain $R$ is time invariant. This makes adaptive time-stepping inappropriate. Despite the extra computational overhead, correctly, only constant time-steps can be returned. For the *An* approach, advancing the subsequent comparative solution element on coarser grid levels is found effective (19% additional time saving). Higher frequency spatial components are accounted for on coarse levels through the multilevel convergence forcing function.

For certain cases, the use of $\Delta t$ clipping is found important – especially to prevent excessively small steps. These can instead be limited by the choice of appropriate reference values. Whether clipping or reference values are used partly depends on data availability. Cases reveal that solver robustness is a key issue. Time-steps and hence diagonal dominance can vary dramatically over a solution. For Case (E), with severe transients, adapting relaxation parameters to maintain diagonal dominance is found effective.

With increasing solver robustness and affordable computer storage, the use of adaptive time-stepping is likely to increase. For future work, conjugate gradient type solvers, that are less sensitive to diagonal dominance, might be worth testing. Also, with parallel architecture machines, the two comparative solution elements could be completed concurrently. This is likely to yield dramatic performance gains and so is worthy of future study. Generally, the first solution component should be a good guess to the second. Then a Newton type solver would be highly efficient. Therefore, testing the use of a Newton type solver for the second solution element also appears worthwhile future work.

Generally, overall, for the six cases considered adaptive stepping is found worthwhile and highly compatible with multilevel convergence acceleration. The performance of the different generic adaptive schemes is case dependent. Therefore, it is recommended that a code has a range of adaptive approaches. Initial tests will be required to find: the best scheme for the class of problem; perhaps the region of greatest flow unsteadiness ( $R'$ ) and also the most appropriate error normalisation scale or alternatively smallest sensible $\Delta t$.

# CHAPTER 6

# TURBULENCE MODELLING

## 6-1 Introduction

Turbulence is an unsteady three-dimensional phenomenon. Its nature is such that large eddies are generated by the mean flow. Generally, the energy from these is dissipated by successively smaller eddies (note, reverse energy cascades, i.e. transfer from smaller to larger scales, can also occur in complex turbulent flows). This is called the *turbulence energy cascade*. The larger eddies (*integral length scales*) are a strong function of the local geometry and flow system, but the smaller eddies (near the *Kolmogorov scale*) are more universal in nature. The time scales relating the integral length scales are approximately equal to the ratio of the flow system characteristic size, $L$, to a characteristic velocity scale $U_o$ (i.e. $t = L/U_o$). The Kolmogorov time scale can be evaluated as $\sqrt{\mu/\rho\varepsilon}$, where $\varepsilon$ is the viscous dissipation. Most engineering turbulence models, based on solving the RANS (Reynolds Averaged Navier-Stokes) equations, try to model both the effects of the unsteady larger and smaller turbulence scales. Since the dynamics of the larger scales are not so universal, their success is highly flow dependent. This lack of universality of such models has resulted in the availability of a wide range of turbulence models and modelling strategies. These will now be reviewed in an unsteady flow-modelling context. After this, the unsteady RANS models used in the Case Studies to be shown later will be presented. Some familiarity is assumed with basic engineering turbulence, concepts and models. For such information, see Wilcox (1998).

## 6-2 VLES/URANS Modelling

The application of RANS models to flows with unsteady vortices arising from separation is sometimes called Very Large Eddy Simulation (VLES). This involves solving the RANS equations in their Unsteady form. Hence, it is perhaps more accurate to describe VLES as URANS (see Spalart (1999)). Franke and Rodi (1993) make URANS predictions for vortex shedding behind a square cylinder. The standard $k$-$\varepsilon$, a zonal $k$-$\varepsilon$/$k$-$l$ (the $k$-$l$ model being applied near walls) and Reynolds stress equation models are used. These give significantly different predictions for temporal behaviour, the dissipative $k$-$\varepsilon$ model suggesting steady flow. Similarly, Johannson, Davidson and Ollson (1993) find the $k$-$\varepsilon$ model significantly under-predicts vortex shedding for a triangular cylinder with an after-body. Durbin (1995) proposes a $k$-$\varepsilon$-$v^2$ model ($v^2$ is a velocity scale for turbulent transport towards walls) and carries out further computations for flow over triangular cylinders. For the wake region, the domination of Reynolds stresses/velocity fluctuations from the unsteady flow component over turbulent stochastic fluctuations is noted. Przulj and Younis (1993) suggest turbulence models developed with steady flows in mind, such as the $k$-$\varepsilon$ model, fail to account for turbulence energy input from the dominant periodic wake components and so are unsuitable for vortex shedding flows. To sensitize the $k$-$\varepsilon$ model in this respect, a term involving the time rate of change of mean kinetic energy is added to the $\varepsilon$ equation production term. For flow over a square cylinder, the resulting model shows significant improvement with measured mean flow and turbulence parameters. The work Przulj and Younis also suggests the unbounded nature of the QUICK (Section 3-3) scheme is responsible for erroneous predictions in certain turbulent periodic flow studies. URANS computations for rotating cavity flows are given by Tucker (1993), Kais, Long and Tucker (1994), Long, Morse and Tucker (1997), and Tucker and Long (1996). These mixing-length model results (some with a specially bounded CONDIF (Section 3-3) scheme) show, like experiments, a three-dimensional unsteady flow. However, other work using the standard diffusive $k$-$\varepsilon$ model gives a steady axisymmetric flow. Orszag *et al.* (1997) present URANS predictions for flow past a compressor trailing edge. These are found to be in better agreement with surface pressure measurements than RANS computations. Similarly, for flow over a circular cylinder, Shur *et al.* (1996) find URANS drag coefficient predictions give improved accuracy relative to RANS solutions. Bosch and Rodi (1998) revisit URANS modelling for flow over square cylinders with the $k$-$\varepsilon$ and zonal $k$-$\varepsilon$/$k$-$l$ models. The stagnation modification of Kato and Launder (1993) is used at the cylinder front. This gives rise to a reduction in $k$ and hence, unlike with the work of Franki and Rodi, an unsteady $k$-$\varepsilon$ prediction. However, the zonal model is found to agree best with measurements, but not as well as when LES (Large Eddy Simulation) is used. Tucker (2000) makes URANS computations for flow in a complex congested internal system (an idealized electronics system). Nine turbulence models are tried including the standard $k$-$\varepsilon$, zonal (see Iacovides and Theofanopoulos (1991), and Iacovides and Chew (1993)) $k$-$\varepsilon$/$k$-$l$, $k$-$\varepsilon$/mixing length and non-linear eddy viscosity models. All models give similar time averaged velocities but significantly different values of turbulence kinetic energy and hence viscosity. This, as would be expected (see Brackenridge (1960)), gives significantly different predictions for unsteadiness amplitude. For the diffusive $k$-$\varepsilon$ model, as found by Franke and Rodi, the predicted flow is steady. Tatsumi *et al.* (1999) make URANS predictions, studying the effect of unsteadiness on heat transfer for flow over a backwards facing step. A non-linear eddy viscosity model is used. Stochastic fluctuations are found to be significantly greater than periodic fluctuations. However, despite this, the periodic component is shown to have a significant thermal effect.

Kimura and Hosoda (1999) revisit the computation of flow around square cylinders. Again, as observed by Franke and Rodi, the standard $k$-$\varepsilon$ model significantly over-predicts $k$ in the stagnation zone. Consequently, unsteadiness fluctuations are excessively damped. Furthermore, it is discovered that use of a non-linear turbulence model (see Gatski and Speziale (1993)) alone does not remedy this feature. However, when the non-linear model is used with variable coefficients (notably $C_\mu$ as used in Equation (6-5.29)), expressed as a function of strain and vorticity parameters (also used by Kato and Launder and given by Equations (6-5.35)) stagnation region $k$ values are dramatically improved (the potential of this approach was suggested by Kato and Launder (1993)). Barakos and Drikakis (1999) consider the prediction of aerofoil shock induced oscillations. For all conditions, the diffusive $k$-$\varepsilon$ model gives steady flow. Also, the temporal predictive accuracy of the one-equation Spalart and Allmaras (SA) (1994) model is found similar to that for the more computationally expensive non-linear variable $C_\mu$ models. Yao *et al.* (2000) perform unstructured grid, finite volume, low Reynolds $k$-$\varepsilon$ number URANS predictions for flow past a square trailing edge. Comparison is made with DNS (Direct Numerical Simulation – where the grid and temporal resolution is sufficient to resolve turbulence) data. Away from the near wake region satisfactory agreement is found. Strong sensitivity of temporal behaviour to the grid layout is observed. However, Alsemgeest *et al.* (2000) make $k$-$\varepsilon$ finite volume predictions for vortex shedding from an inclined plate (a throttle component). Unstructured, hexahedral and tetrahedral meshes are found to give reasonably similar results.

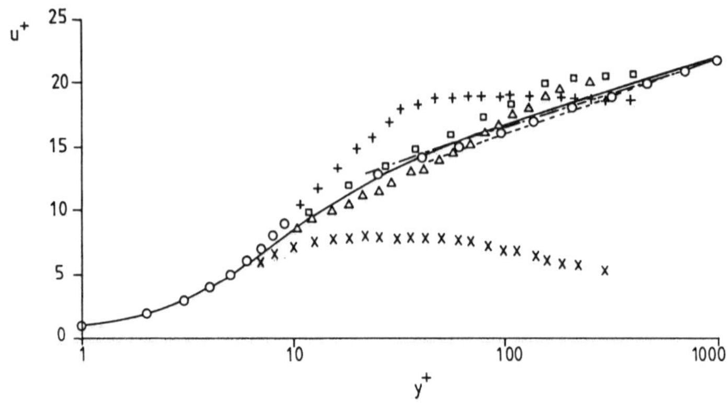


*Figure 6-2.1* Velocity distribution in channel using the universal coordinate system (o established universal profile correlated data, —— *ml* model, ······· *k-l* model, - - - - - *k-ε* model, +, □, Δ, × computations of Przulj and Younis (1993)).

Nishimura *et al.* (1999) use the *k-ε* and second moment closure models to predict the unsteady mixing between adjacent cylindrical jets. The second moment closure significantly improves the predicted jet mixing and unsteadiness characteristics. Pascal *et al.* (2000) compare DNS and URANS results for flow in an idealized cylinder with a piston (valves and other geometrical complexity that would be found in a real reciprocating compressor or engine are ignored). DNS results show that: during expansion turbulence decays; when compression effects are greater than those due to decay, turbulence increases and the turbulence field is highly anisotropic. Unlike the *k-ε* model, a low Reynolds number second moment closure gives satisfactory agreement with the DNS data. Since the production term in the former is the square of the mean strain rate, it cannot predict the turbulence decay during expansion. Due to strong non-equilibrium, wall functions are found to be totally inappropriate. Wilcox (1998) states that, for unsteady simulations, wall functions are frequently erroneously blamed for poor predictions. However, there appears to be some evidence to suggest caution is required when applying this approach to unsteady flows. *Figure 6-2.1* gives unsteady flow predictions of Przulj and Younis (1993). The +, □, Δ, × symbols, represent their computed velocity profiles at different points in time. Established universal profile correlated data is represented by the circle 'o' symbols. Steady state mixing length (*ml*), high Reynolds number *k-l* and *k-ε* predictions are also given on the graph. These, steady predictions will be discussed more later. As can be seen, the Przulj and Younis results show that for unsteady flows, wall functions are not really appropriate.

As noted above, for flows with unsteady vortices, URANS predictions can give improved accuracy over RANS solutions (see Orszag *et al.* (1997) and Shur *et al.* (1996)). However, clearly, based on the above-observed temporal sensitivity to turbulence model, other approaches which place less emphasis on the turbulence model are desirable. For the flow behind a cylinder the improved predictive accuracy of the inherently unsteady LES technique was noted and this method is outlined below.

**6-3 LES Modelling**

An attractive idea is to use temporal and spatial integration schemes of sufficient accuracy to resolve the larger more problematic turbulent scales. This leaves the turbulence model (a *sub-grid scale model*) to account for just the effects of the more universal and isotropic smaller scales. This is the basis of the Large Eddy Simulation (LES) technique. The Smagorinsky (1963) sub-grid scale model is popular, but is crude and has several drawbacks. It is basically a mixing length model, the von Karman constant being replaced by Smagorinsky's constant $C_s \simeq 0.2$. Unfortunately, as with RANS mixing length solutions, a wall damping function is required, reducing $C_s$ at walls. Also, the model does not account for effects of buoyancy, rotation and turbulence anisotropy. It is possible to follow this RANS philosophy, with more elaborate turbulence models. This approach is especially attractive for geophysical flows where grids are necessarily coarse and more turbulence length scales have to be accounted for by the subgrid scale model. For example, Kosovic (1997), use Speziale's (1987) non-linear eddy viscosity model to account for subgrid scale stresses. Deficiencies in the Smagorinsky model have given rise to the *scale similarity model* of Bardina *et al.* (1980). The basis of this, is the assumption that the smallest resolved turbulence scales are similar in many aspects to the adjacent subgrid scales. In practice, the scale similarity model does not dissipate sufficient energy from the resolved scales, but models well the transfer of energy back from the small to larger scales (*backscatter*). The problem of insufficient dissipation is overcome by combining the scaled similarity and Smagorinsky models producing a *mixed model*. A third, and perhaps the most popular model, is the *dynamic model* of Germano *et al.* (1990). This remedies the major drawback of Smagorinsky's model, which is the assumption that $C_s$ is a constant. Using the procedure of Germano *et al.*, a temporally and spatially varying value of $C_s$ can be deduced (to be used in the Smagorinsky model). A drawback of Germano *et al.*'s technique is that predicted values of $C_s$ vary dramatically with time and space making the numerical solution unstable unless damping (see Breurer and Rodi (1994)) and clipping of $C_s$ is used.

To date, the use of LES has been restricted to simple geometries and studies directed more at the physics of turbulence. For example, its application to channel and duct flows is described by Piomelli *et al.* (1988) and Breuer and Rodi (1994). Xiyun *et al.* (1997) and Rodi *et al.* (1997) describe the use of LES for oscillatory flow past round and square cylinders. Yang and Ferziger (1993) and De Matos *et al.* (1999) present computations for flow over obstacles in channels. The latter work includes the modelling of thermal behaviour. Bartello *et al.* (1994), Cambon *et al.* (1994), and Metais *et al.* (1995) describe the application of LES to examine the dynamics of rotating turbulence. The need for more reliable near wall LES treatments is noted by Rodi *et al.* (1997).

LES and also DNS (see Derksen (1996)), both require efficient (due to their computational expense) and accurate numerical schemes. Temporal and spatial fidelity is needed to ensure that the numerics do not influence the predicted turbulence behaviour. To obtain sensible turbulence statistics, simulations should be faithfully integrated over several eddy turnover periods. These demands have led to the use of some interesting temporally related schemes/strategies not generally applied to industrial computations. For example, in LES work, Armenio *et al.* (1999) use the Crank-Nicolson scheme for cross-stream diffusion terms and a low storage third order Runge-Kutta method for the convective and remaining diffusion terms. In DNS simulations, Choi and Moin (1994) use the fractional step pressure prediction scheme described in Section 4-4. These authors also demonstrate that, if Crank-Nicolson scheme inaccuracies are not to erroneously produce laminar flow, time-steps must be smaller than the Kolmogorov time scale. Despite this restriction, relative to explicit time advancement methods (with implicit treatment of viscous terms), the Crank-Nicolson scheme (when combined with the fractional step method) is shown to be efficient. For LES predictions, Mason and Cullen (1986) use the energy conserving convective term treatment of Piacsek and Williams (1970). This is accurate for long period time integrations and is commonly associated with geophysical flows (see James, Jonas and Farnell (1981), and Hignett *et al.* (1985)). Schemes of this nature are discussed in Section 3-3.

## 6-4 Zonal LES/DES

Although especially promising for unsteady separated flows, even with parallel processing ( see Strietzel (1995), Timmermans and Nieuwstadt (1996), and Jenssen (1998) ) and multilevel convergence acceleration (see Arad and Martinelli (1997)) LES is still computationally expensive. Also, in wall regions, LES presents difficulties (see Rodi *et al.* (1997)). According to Spalart (1999), due to expense, LES is unlikely to see use in aircraft design for at least a decade. Therefore, Spalart proposes the use of Detached Eddy

Simulations (DES). With this, the RANS equations are solved say in near-wall regions. In areas where there are detached unsteady eddies LES is applied. Hence DES could be viewed as a zonal technique. Shur *et al.* (1999) apply DES to flow over a wing at high incidence and hence having unsteady vortex shedding. The SA model is adapted for both the LES and RANS regions. Accuracy improvements over RANS solutions are found. However, Shur *et al.* note that the application of DES to channel or boundary layer flows is required to see how the LES and RANS solutions blend in the log-law region. Such work, of a preliminary nature, has recently been carried out by Davidson (2000). With a *k-l* subgrid scale model DES was found unsuccessful. However, the approach is found to work when a $k$-$\omega$ subgrid scale model is applied. In essence, with distance based RANS model dissipation scales, interfacing the LES and RANS regions is straightforward. The length scale in the turbulence model is just replaced by the minimum of the wall distance and grid-based scales. No other modifications are necessary (unless the sub-grid and RANS turbulence models are of a different nature). Importantly, Shur *et al.*'s DES work assumes a steady RANS region flow. This is perhaps because it is considered that the grid in this zone is too coarse to resolve unsteady turbulent structures. However, in some situations, the core flow unsteadiness (not associated with turbulence) can propagate through boundary layers. This is true for the work of Tucker (1993) and certain geophysical flows. For these, the temporal drift of cyclones and anticyclones can continue through the boundary layer with an ever-decreasing amplitude. When this is the case a URANS/LES approach would be desirable.

## 6-5 URANS Models used in Case Studies

As can be seen from the above discussion, for most engineering flows LES is too computationally expensive. Also, DES is insufficiently mature and untested to be a more economical alternative. Consequently, for the next few years, URANS/VLES modelling appears to be the most viable option.

Models of this type will now be described. The extent of the discussion is limited to those models used in the Case Studies presented later. Notable omissions are second moment closure (see Launder *et al.* (1975)) and algebraic stress models (see Gatski and Spezial (1993)). As stated by Kato and Launder (1993), generally, for unsteady three-dimensional flow solutions, the former, although promising is presently still too computationally expensive.

Unless solely used in cylindrical polar coordinates in the Case Studies, for compactness, the governing equations are presented in a Cartesian tensor form. Some discussion on the transformation between these two coordinate systems is given by Morse (1980).

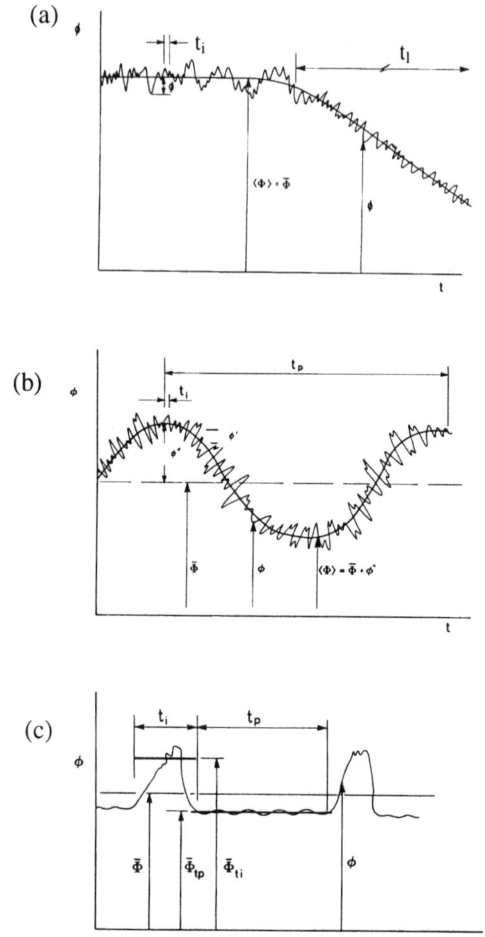


*Figure 6-5.1* Characteristic traces for the change in a general variable $\phi$ with time: (a) transient flow; (b) cyclic and (c)

*Turbulence Averaging*

It is convenient to classify the flows here as either *transient* or *periodic*. A less important classification here is *intermittent* unsteadiness. Schematics of characteristic traces for the change in a general variable $\phi$, with time, for these classifications are shown in *Figure 6-5.1*. Frame (a) gives the transient flow trace, where $t_i$ and $t_t$, represent turbulence and transient characteristic time scales, respectively. For this trace, the instantaneous velocity $\phi$ can be expressed as $\overline{\Phi}+\phi'$ where $\overline{\Phi}$ is an *ensemble* average ($\overline{\Phi}(t_n) = \left(\sum_1^N \phi(t_n)\right)\Big/N$ where $t_n$ is a fixed time for $N$ tests and $N \to \infty$)

and $\phi'$ the turbulence fluctuation. In forming the ensemble average it is necessary that $t_t >> t_i$. More formally we can write

$$\frac{\left(\int_{-t_l/2}^{t_l/2}\phi dt - \int_{t_l/2}^{3t_l/2}\phi dt\right)}{\int_0^{t_l}\phi dt} << 1 \tag{6-5.1}$$

where $t_l$ is a minimum general (not just for transient flows) period required to gain a satisfactory average.

Frame (b) shows the periodic flow trace, where $t_p$ and $\phi''$ represent the unsteadiness period and amplitude, respectively. The instantaneous velocity can now be expressed as

$$\phi = \overline{\Phi} + \phi'' + \phi' \tag{6-5.2}$$

where $\overline{\Phi}$ is the time mean. Alternatively, the following can be written

$$\phi = \langle\Phi\rangle + \phi' \tag{6-5.3}$$

where $\langle\Phi\rangle$ is the *phase* averaged $(\langle\Phi\rangle(\alpha) = \left(\sum_1^N \phi(\alpha + 2\pi(N-1)\right)\big/N$ where $\alpha$ is a fixed phase angle for $N$ cycles and $N \to \infty$) component resolved by the numerical scheme. Therefore, $\langle\Phi\rangle$ can be expressed as

$$\langle\Phi\rangle = \overline{\Phi} + \phi'' \tag{6-5.4}$$

For transient flow $\phi'' = 0$ and so

$$\langle\Phi\rangle = \overline{\Phi} \tag{6-5.5}$$

Frame (c) shows an intermittently unsteady flow trace. Such traces can be approximately observed around the outer edge of turbulent boundary layers (For a boundary layer, the peaks shown in Frame (c) would probably not be quite so distinct, but such peaks are evident in other physical processes.). At a particular point, as suggested in Frame (c), the flow may be quasi-laminar (assuming the trace applies to a turbulent boundary layer edge) for a period $t_p$ and turbulent for a period $t_i$. Different averaged quantities (*conditioned* averages) can be derived depending on whether the whole trace is considered or just specific sections. When considering such traces the intermittency $\gamma$ is an important parameter. It defines the ratio of the say turbulent flow duration to total trace duration i.e.

$$\gamma = \frac{\sum_1^N t_i}{\left(\sum_1^N t_i + \sum_1^N t_p\right)} \tag{6-5.6}$$

Using the intermittency, a conditioned average can be formed for say the Frame (c) quasi laminar period as follows

$$\overline{\Phi}_{tp} = \frac{1}{(1-\gamma)t_l}\int_{t-t_l/2}^{t+t_l/2} \phi\delta(t)dt \tag{6-5.7}$$

where

$$t_l = \sum_1^N t_i + \sum_1^N t_p \tag{6-5.8}$$

and $\delta(t)$ is a step function that assumes the value of unity for quasi-laminar periods and zero elsewhere. Conditioned averaging can in fact also be used with transient and periodic flows. When a condition is met, Equation (6-5.7) is applied; for discarded data, phase or ensemble averages are used. Bousquet *et al.* (1999) account for turbulence intermittency by solving a differential transport equation for $\gamma$ along with the usual $k$-$\varepsilon$ turbulence model equations. In this text, mostly periodic flows are considered, with, as is usual, any turbulence intermittency not being explicitly modelled. Therefore, following Johansson *et al.* (1993), Bosch and Rodi (1998) and Tatsumi *et al.* (1999) the equations are presented in a form that mainly reflects this.

The temporal derivative of $\phi$ appears in the unsteady flow governing equations (see equations (1-2.1)). On substitution of Equation (6-5.3), this derivative must be averaged in some way over a suitable time period say $t_l$ (this is the standard Reynolds averaging procedure). Therefore, we have

$$\overline{\frac{\partial\phi}{\partial t}} = \frac{1}{t_l}\int_{-t_l/2}^{t_l/2} \frac{\partial(\langle\Phi\rangle + \phi')}{\partial t}dt = \frac{\langle\Phi(t_l/2)\rangle - \langle\Phi(-t_l/2)\rangle}{t_l} + \frac{\phi'(t_l/2) - \phi'(-t_l/2)}{t_l} \tag{6-5.9}$$

The above can be re-expressed as

$$\overline{\frac{\partial \phi}{\partial t}} = \frac{\partial \langle \Phi \rangle}{\partial t} + \underbrace{\frac{\phi'(t_l/2) - \phi'(-t_l/2)}{t_l}}_{A} \tag{6-5.10}$$

With the Reynolds Averaged Navier-Stokes equations it is assumed $|\phi'| << |\langle \Phi \rangle|$. Hence term (A) can be neglected, giving

$$\overline{\frac{\partial \phi}{\partial t}} \approx \frac{\partial \langle \Phi \rangle}{\partial t} \tag{6-5.11}$$

The assumption $|\phi'| << |\langle \Phi \rangle|$ is not good close to solid surfaces and other high turbulence intensity regions (turbulence intensities of 20 % are not highly unusual in many flows). Importantly, Equation (6-5.11) also applies for spatial derivatives. Using the usual other averaging facts, i.e.

$$\begin{aligned} &\overline{\phi'} = 0, \\ &\overline{\phi_j \phi_i} = \overline{(\langle \Phi_j \rangle + \phi_j')(\langle \Phi_i \rangle + \phi_i')} = \\ &\qquad \overline{\langle \Phi_j \rangle \langle \Phi_i \rangle} + \overline{\langle \Phi_j \rangle \phi_i} + \overline{\phi_j \langle \Phi_i \rangle} + \overline{\phi_j \phi_i} = \langle \Phi_j \rangle \langle \Phi_i \rangle + \overline{\phi_j \phi_i} \end{aligned} \tag{6-5.12}$$

and assuming negligible density variations, the Navier-Stokes equations (1-2.1) can be written as the averaged unsteady *Reynolds* or, as they are referred to here, URANS equations.

*URANS Equations*

For compactness the URANS equations are expressed in the tensor form below

$$\begin{aligned} \frac{\partial \langle U_i \rangle}{\partial t} + \frac{\partial \langle U_i \rangle \langle U_j \rangle}{\partial x_j} = &- \frac{1}{\rho} \frac{\partial \langle P \rangle}{\partial x_i} \\ &+ \frac{\mu}{\rho} \frac{\partial}{\partial x_j} \left( \frac{\partial \langle U_i \rangle}{\partial x_j} + \frac{\partial \langle U_j \rangle}{\partial x_i} \right) + \frac{\partial \langle - \overline{u_i' u_j'} \rangle}{\partial x_j} \end{aligned} \tag{6-5.13}$$

where $\langle U \rangle$ is an averaged fluid velocity component, $u'$ is the averaged turbulence fluctuating component, $\rho$ the fluid density, $\mu$ viscosity, $\langle P \rangle$ averaged static pressure, $t$ time and $x$ the spatial co-ordinate. The role of the turbulence model is to approximate the *Reynolds stress* terms, $\langle -\rho \overline{u_i' u_j'} \rangle$, in Equation (6-5.13). Unless otherwise stated, to approximate these the *Boussinesq* approximation is used in the following form

$$\langle -\overline{u_i' u_j'} \rangle = -n_0 \frac{2}{3} \langle k \rangle \delta_{ij} + \frac{2\mu_t}{\rho} S_{ij} \qquad (6\text{-}5.14)$$

where $\delta_{ij}$ is the Kronecker delta ($\delta_{ij} = 1$ if $i = j$ and $\delta_{ij} = 0$ if $i \neq j$), $\mu_t$ is the turbulent viscosity, $\langle k \rangle$ turbulent kinetic energy and $n_0 = 0$ or 1 depending on the model used. The term $S_{ij}$ (the mean strain rate) has the following form

$$S_{ij} = \frac{1}{2} \left( \frac{\partial \langle U_i \rangle}{\partial x_j} + \frac{\partial \langle U_j \rangle}{\partial x_i} \right) \qquad (6\text{-}5.15)$$

*Turbulence and Scalar Transport Equations*

When $\langle k \rangle$ is required, the following standard differential transport equation is used

$$\frac{\partial \langle \Phi \rangle}{\partial t} + \frac{\partial \langle U_j \rangle \langle \Phi \rangle}{\partial x_j} = \frac{1}{\rho} \frac{\partial}{\partial x_j} \left( \Gamma_\phi \frac{\partial \langle \Phi \rangle}{\partial x_j} \right) - T_1 \langle \overline{u_i' u_j'} \rangle \frac{\partial \langle U_i \rangle}{\partial x_j} - T_2 \qquad (6\text{-}5.16)$$

where $\langle \Phi \rangle = \langle k \rangle$, $\Gamma_k = \mu + \mu_t / \sigma_k$ (where $\sigma_k$ is the diffusion Prandtl number for $k$), $T_1 = 1$ and $T_2 = \langle \varepsilon \rangle$ (the rate of dissipation of turbulence kinetic energy). When $\langle \Phi \rangle = \langle \varepsilon \rangle$, $\Gamma_\varepsilon = \mu + \mu_t / \sigma_\varepsilon$ (where $\sigma_\varepsilon$ is the diffusion Prandtl number for $\varepsilon$), $T_1 = C_{\varepsilon 1} \langle \varepsilon \rangle / \langle k \rangle$, and $T_2 = C_{\varepsilon 2} \langle \varepsilon \rangle^2 / \langle k \rangle$, Equation (6-5.16) can also be used to estimate the rate of dissipation of $\langle k \rangle$. For some modelling approaches, the rate of dissipation is related to the normal distance from walls. A novel technique for evaluating these is described later. Usefully, with $\langle \Phi \rangle = \langle C \rangle$, $T_1 = T_2 = 0$ and $\Gamma_c = \mu + \mu_t / \sigma_c$ Equation (6-

5.16) can be used to solve for the transport of a general scalar variable, $\langle C \rangle$. Alternatively, with $\langle \Phi \rangle = \langle T \rangle$, $T_1 = T_2 = 0$ and $\Gamma_T = \mu / Pr + \mu_t / Pr_t$ it serves, in the absence of fractional and compressive heating, as an energy or rather temperature, $T$, equation.

Nominally, twelve turbulence models are used. These are briefly described below in the approximate order of complexity/accuracy.

*Mixing Length Models*
For this ($ml_1$) model

$$\mu_t = \rho \langle l_\mu \rangle^2 \left[ 2 \frac{\partial \langle U_i \rangle}{\partial x_j} S_{ij} \right]^{1/2} \tag{6-5.17}$$

where $\langle l_\mu \rangle$ is a mixing length. Generally, in the near wall region the original Van Driest damping function is used, giving

$$\langle l_\mu \rangle = \kappa y \left( 1 - e^{-\langle y^+ \rangle / 26} \right) \tag{6-5.18}$$

where $\langle y^+ \rangle$ is the dimensionless wall distance equal to $y\sqrt{\rho \langle \tau_w \rangle} / \mu$, $\langle \tau_w \rangle$ is the averaged wall shear stress and $\kappa$ the von Karman constant. The modified function (see Koosinlin *et al.* (1974)) below

$$\langle l_\mu \rangle = \kappa y \left( 1 - e^{-(\langle \tau \rangle / \langle \tau_w \rangle)^{1.5} \langle y^+ \rangle / 26} \right) \tag{6-5.19}$$

is also implemented. Equation (6-5.19) (for the $ml_2$ model) is used for the zonal model to be described later and some rotating flows. Outside the near wall region

$$\langle l_\mu \rangle = C_1 \langle \delta \rangle \tag{6-5.20}$$

where $\langle \delta \rangle$ is an approximation to the boundary layer thickness and for non-rotating and rotating systems $C_1 = \kappa$ and 0.085, respectively. Following Koosinlin *et al.*, where there are cylindrical surfaces, the mixing length is modified as

$$\langle l_\mu \rangle = \langle l_\mu \rangle (1 \; - \; C_1\, Ri) \tag{6-5.21}$$

where $C_l$ = 5 (selected by Koosinlin *et al.* to give best agreement of predictions with measurements) and the Richardson number ( $Ri$ ) is defined as

$$Ri = \frac{\dfrac{2\langle V \rangle}{r^2}\dfrac{\partial (r\langle V \rangle)}{\partial r}}{\left(\dfrac{\partial \langle W \rangle}{\partial r}\right)^2 + \left(r\dfrac{\partial (\langle W \rangle / r)}{\partial r}\right)^2} \tag{6-5.22}$$

Note, in the above, $\langle V \rangle$ and $\langle W \rangle$ correspond to tangential and axial velocity components respectively. The above are used with $n_O$ = 0 in Equation (6-5.14).

*l-v Model*

The basis for this model is Spalding's (1961) single equation, Taylor series based, law of the wall

$$\langle y^+ \rangle = \langle u^+ \rangle + \frac{1}{E}\left[ e^{\kappa \langle u^+ \rangle} - 1 - \frac{\kappa \langle u^+ \rangle}{1!} - \frac{(\kappa \langle u^+ \rangle)^2}{2!} - \frac{(\kappa \langle u^+ \rangle)^3}{3!} - \frac{(\kappa \langle u^+ \rangle)^4}{4!} \right] \tag{6-5.23}$$

This can be differentiated to give the dimensionless effective viscosity. This differentiated result can then be rearranged to yield the following effective viscosity expression

$$\mu_t = \mu \frac{\kappa}{E}\left[ e^{\kappa \langle u^+ \rangle} - 1 - \frac{\kappa \langle u^+ \rangle}{1!} - \frac{(\kappa \langle u^+ \rangle)^2}{2!} - \frac{(\kappa \langle u^+ \rangle)^3}{3!} \right] \tag{6-5.24}$$

The dimensionless velocity $\langle u^+ \rangle = \langle U \rangle \sqrt{\rho / \langle \tau_w \rangle}$ in Equation (6-5.24) is calculated using a Newton-Raphson procedure needing $y$. In Equation (6-5.14), $n_O = 0$.

*k-l Models*

Equation (6-5.16) is solved for $\langle \Phi \rangle = \langle k \rangle$, and $\langle \varepsilon \rangle$ defined using

$$\langle \varepsilon \rangle = \frac{\langle k \rangle^{3/2}}{\langle l_\varepsilon \rangle} \tag{6-5.25}$$

where

$$\langle l_\varepsilon \rangle = C_{\varepsilon 0}\, y \left( 1 - n_l e^{-A_\varepsilon \langle y^+ \rangle / C_\mu^{1/4}} \right) \tag{6-5.26}$$

and $\langle y^+ \rangle = y\rho \langle k \rangle^{1/2} C_\mu^{1/4} / \mu$. The turbulent viscosity can be expressed as

$$\mu_t = \rho\, C_\mu \langle l_\mu \rangle \langle k \rangle^{1/2} \tag{6-5.27}$$

where

$$\langle l_\mu \rangle = C_{\mu 0}\, y( 1 - n_l e^{-A_\mu \langle y^+ \rangle / C_\mu^{1/4}} ). \tag{6-5.28}$$

When $n_l$ = 1, the length scales $\langle l_\varepsilon \rangle$ and $\langle l_\mu \rangle$ have near wall damping functions and the model ($k$-$l_2$) is as described by Wolfshtein (1969). Predictions are also made with $n_l$ = 0 and computationally economical logarithmic wall functions (see Launder and Spalding (1974)). When wall functions are implemented the diffusion of turbulence kinetic energy at walls is set to zero. Also, at first off-wall grid nodes $\langle \varepsilon \rangle = C_\mu \langle k \rangle^{3/2} / \kappa y$ and when $\langle y^+ \rangle$ < 11.5, $\mu_t$ = 0, otherwise $\mu_t = \mu(\kappa \langle y^+ \rangle / \ln E \langle y^+ \rangle - 1)$. This high Reynolds number model is called here the $k$-$l_l$ model.

*k-ε Model*

For this, Equation (6-5.16) is solved with $\langle \Phi \rangle = \langle k \rangle$ and $\langle \varepsilon \rangle$ and the turbulent viscosity found using

$$\mu_t = \rho\, C_\mu \frac{\langle k \rangle^2}{\langle \varepsilon \rangle} \tag{6-5.29}$$

For the Case Studies, the standard model with logarithmic wall functions is used ($k$-$\varepsilon_l$).

*Figure 6-2.1* gives law of the wall predictions for the $ml_l$, $k$-$\varepsilon_l$, and $k$-$l_l$ models described above. For the $ml_l$ predictions a full line is used. The $k$-$\varepsilon_l$ and $k$-$l_l$ predictions are shown using dashed lines. There is agreement between these models and the established correlation data represented by the circle 'o' symbols.

*Zonal models*

In this approach, for $\langle y^+ \rangle \geq 60$ the $k$-$\varepsilon$ method is used and for $\langle y^+ \rangle < 60$ the $k$-$l_2$, $ml_l$ or $ml_2$ models are applied. To connect these differing models at the $\langle y^+ \rangle = 60$ interface, patching conditions are needed. For the $k$-$\varepsilon$/$k$-$l_2$ method Equation (6-5.30) below

$$\langle k \rangle^{3/2} = \langle l_\mu \rangle \langle \varepsilon \rangle \tag{6-5.30}$$

is used (rearranged for $\langle \varepsilon \rangle$). For the $k$-$\varepsilon$/$ml_l$ and $k$-$\varepsilon$/$ml_2$ methods, the following additional condition is needed.

$$\langle \varepsilon \rangle = \left\langle -\overline{u_i' u_j'} \right\rangle \frac{\partial \langle U_i \rangle}{\partial x_j} \tag{6-5.31}$$

*Non-linear Eddy Viscosity Models*

These models ($nl$-$k$-$l_1$, $nl$-$k$-$l_2$, $nl$-$k$-$\varepsilon$, $nl$-$k$-$\varepsilon$/$k$-$l_2$ and $nl$-$k$-$\varepsilon$/$nl$-$k$-$l_2$) are based on Speziale's (1987) model. They essentially involve using the quadratic 'Boussinesq' approximation below

$$\left\langle -\overline{u_i' u_j'} \right\rangle = -\frac{2}{3}\langle k \rangle \delta_{ij} + 2\frac{\mu_t}{\rho} S_{ij} + \tag{6-5.32}$$

$$4C_D C_\mu \frac{\mu_t}{\rho}\frac{\langle k \rangle}{\langle \varepsilon \rangle}\left( S_{ik} S_{kj} - \frac{1}{3} S_{mn} S_{nm} \delta_{ij} \right) + 4C_E C_\mu \frac{\mu_t}{\rho}\frac{\langle k \rangle}{\langle \varepsilon \rangle}\left( \overset{\circ}{S}_{ij} - \frac{1}{3} \overset{\circ}{S}_{mm} \delta_{ij} \right)$$

The $\overset{\circ}{S}_{ij}$ term, is called the Oldroyd derivative, which can be expressed as

$$\overset{\circ}{S}_{ij} = \frac{\partial S_{ij}}{\partial t} + \langle U_k \rangle \frac{\partial S_{ij}}{\partial x_k} - \frac{\partial \langle U_i \rangle}{\partial x_k} S_{kj} - \frac{\partial \langle U_j \rangle}{\partial x_k} S_{ki} \tag{6-5.33}$$

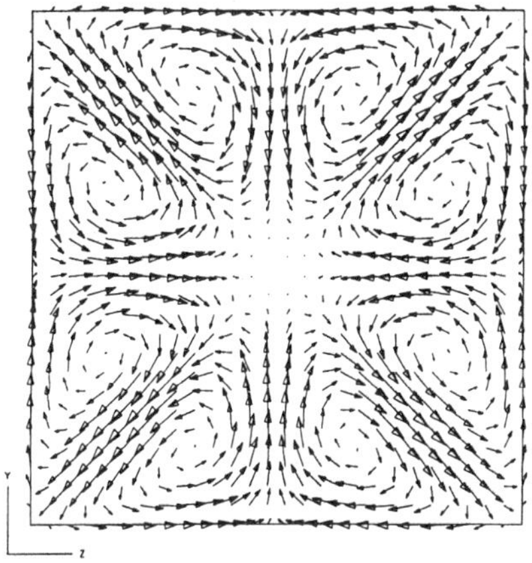

*Figure 6-5.2* Non-linear model predicted flow structure for a three-dimensional channel

Here the quadratic 'Boussinesq' approximation is incorporated in various $k$-$l$, $k$-$\varepsilon$ or zonal (as its descriptor suggests, the $nl$-$k$-$\varepsilon$/$k$-$l_2$ model uses a linear near wall $k$-$l$ model) frameworks, Equation (6-5.32) replacing Equation (6-5.14) in the linear models. Non-linear contributions from Equation (6-5.32) are differentiated and then directly incorporated into the momentum equations. Following Speziale, velocities in the non-linear part of the 'Boussinesq' approximation are smoothed. Here, for convenience this is done using the restriction operators (essentially consisting of an 18-point distance weighted averaging) from the multilevel algorithm described in Section 4-2. Both, high ($nl$-$k$-$l_1$, $nl$-$k$-$\varepsilon_1$) and low Reynolds ($nl$-$k$-$l_2$, $nl$-$k$-$\varepsilon$/$nl$-$k$-$l_2$, $nl$-$k$-$\varepsilon$/$k$-$l_2$) number non-linear model forms are used. *Figure 6-5.2*, gives three-dimensional channel flow predictions for the $nl$-$k$-$l_2$ model. A Van Driest type damping function based on Equation (6-5.18) is implemented. The significant cross-stream vortices, that linear models fail to predict, are clear.

*Stagnation Modification*

As noted earlier (see Franke and Rodi (1991)), with the $k$-$\varepsilon$ model, excessive $k$ production in stagnation regions can erroneously give rise to the prediction of steady flow. For $\langle\Phi\rangle=\langle k\rangle$, the term $\langle\overline{u_i'u_j'}\rangle\partial\langle U_i\rangle/\partial x_j$, in Equation (6-5.16), is an exact representation of turbulence energy production. However, except for simple shear flows, the $\langle\overline{u_i'u_j'}\rangle$ terms are not well modelled through Equation (6-5.15). Indeed, in the stagnation region, $S_{ij}$ can become high, giving rise to the overprediction of $\langle k\rangle$. A partial remedy to this is the stagnation region modification given by Kato and Launder (1993). With this, in the stagnation region, the production is re-expressed as

$$\left\langle \overline{u_i' u_j'} \right\rangle \frac{\partial \langle U_i \rangle}{\partial x_j} = \frac{\mu_t}{\rho} S\,\Omega \qquad (6\text{-}5.34)$$

where $S$ and $\Omega$ are strain and vorticity related parameters

$$S = \sqrt{\frac{1}{2}\left(\frac{\partial \langle U_i \rangle}{\partial x_j} + \frac{\partial \langle U_j \rangle}{\partial x_i}\right)^2} \;;\; \Omega = \sqrt{\frac{1}{2}\left(\frac{\partial \langle U_i \rangle}{\partial x_j} - \frac{\partial \langle U_j \rangle}{\partial x_i}\right)^2} \qquad (6\text{-}5.35)$$

Because in stagnation regions, the flow is very nearly irrotational, i.e. $\Omega \approx 0$, Equation (6-5.34) leads to the reduction of turbulence generation. For simple shear flows, the equation gives the original production term. A Case Study, involving the application of Equation (6-5.34) is given in Chapter 11.

The following standard constants are used in the above models: $\sigma_k$ = 1, $\sigma_\varepsilon = 1.3$, $\sigma_c \approx 1.$, $A_\varepsilon = 0.263$, $A_\mu = 0.016$, $C_E = C_D = 1.68$, $C_{\varepsilon 0} = 2.4$, $C_{\varepsilon l}$ = 1.44, $C_{\varepsilon 2} = 1.92$, $C_{\mu 0} = 2.4$, $C_\mu = 0.09$, $\kappa = 0.41$, $E \approx 9.0$, $Pr_t = 0.9$.

*Calculation of Normal Wall Distances*

Some of the above turbulence models require distances for nearest walls, $y$. For complex, cluttered geometries $y$ can be calculated using a method derived by Spalding (1994). This involves the solution of a steady Poisson equation of the following form,

$$\frac{\partial}{\partial x_j}\left(\frac{\partial L}{\partial x_j}\right) = C \qquad (6\text{-}5.36)$$

where $C = -1$, and at solid walls $L = 0$ otherwise the condition $\partial L/\partial y = 0$ is used. Unlike when nearest wall search procedures are used, at corners, this method produces smooth variations in $y$ and hence $\mu_t$. This can help convergence. Here, $y$ is related to $L$ using the slightly modified (see Tucker (1998)) expression to Spalding's given below

$$y=-\sum_{j=1,3}\left|\frac{\partial L}{\partial x_j}\right|\pm\sqrt{\left(\left(\sum_{j=1,3}\left|\frac{\partial L}{\partial x_j}\right|\right)^2+2L\right)} \qquad (6\text{-}5.37)$$

Derivation of the above requires the assumption that the surface perpendicular to $y$ is infinite. This approximation is reasonable in the regions of importance, which are close to surfaces. Equation (6-5.37) has negative and positive roots corresponding to nearest ($y_{min}$) and furthest ($y_{max}$) normal wall distances. Therefore, the approximate distance between two surfaces is equal to $y_{\min} + y_{\max}$. Solution of the Poisson and fluid flow equations is accelerated using linear multilevel convergence.

An alternative to the Poisson equation based algorithm above would be the fast marching methods of Sethian (1999). For wall distances, these involve solution of the generic equation $|\nabla L| = 1$. However, since, for this equation, $L$ is not in corners (where there can be two nearest wall distances) continuous, a small diffusion term is added to the right hand side (if not small this constitutes an error). This gives an essentially convection/diffusion type equation which can be solved by marching (treating the spatial coordinate in a time like fashion) from boundaries. The technique has the advantage that it can also accurately find distances between surfaces. This is useful when evaluating things such as radiative exchange factors in heat transfer problems. However, for standard Navier-Stokes equation solvers, relative to the Poisson equation technique, its implementation does not appear to be as natural.

*Particle Trajectories in Turbulent Flow*

As noted in Chapters 2 and 3, the governing equations for fluid or other particle trajectories in a Lagrangian frame are time dependent. Therefore, for this reason alone it appears reasonable to discuss their solution in this frame. However, a further reason is that, the influence of flow unsteadiness on particle trajectories in a turbulent flow is assessed in Chapter 10. Therefore, the stochastic particle transport technique used is outlined here. For the case considered, the presence of the particles does not influence the flow field. Such flows are called *dilute*. For dilute problems, particle movements can be calculated by solving the time dependent particle trajectory and momentum equations presented in Chapter 2. In these, $\Sigma$ **F** represents the sum of external forces acting on a particle. Here, these are the drag

$$\mathbf{F}_D = \frac{1}{2}\rho C_D \pi \frac{d^2}{4}|\Delta\mathbf{U}|\Delta\mathbf{U} \tag{6-5.38}$$

(where $\Delta\mathbf{U} = \langle\mathbf{U}\rangle - \mathbf{U}_p$ and $\langle\mathbf{U}\rangle$ is the averaged fluid velocity, $d$ the particle diameter and $C_D$ a drag coefficient) and gravitational $\mathbf{F}_g = (\rho_p - \rho)\pi d^3\mathbf{g}/6$ forces. Therefore, $\Sigma\mathbf{F} = \mathbf{F}_D + \mathbf{F}_g$. Other forces, of which there can be many (see Hinze (1975)), are ignored. The particle Reynolds number is defined as $Re_p = \rho d|\Delta U|/\mu$. As a particle drag estimate, for $Re_p < 1$, $C_D = 24/\,Re_p$. If $1000 \geq Re_p > 1$,

$$C_D = \frac{24}{Re_p}(\,1{+}0.15Re_p^{\,0.687}\,) \tag{6-5.39}$$

and when $Re_p > 1000$, $C_D = 0.44$. Equation (2-11.2) is integrated using a 4th order Runge-Kutta method.

The instantaneous velocity field, with which the particle interacts, is expressed as the sum of the phase averaged velocities and a turbulence fluctuation magnitude. Therefore, we have

$$u = \langle U\rangle + N\sigma_1, \quad v = \langle V\rangle + N\sigma_2, \quad w = \langle W\rangle + N\sigma_3 \tag{6-5.40}$$

For isotropic turbulence predictions

$$\sigma_1 = \sigma_2 = \sigma_3 \approx \sqrt{2\langle k\rangle/3} \tag{6-5.41}$$

For anisotropic turbulence it is assumed

$$\sigma_1 = \sqrt{\langle\overline{u_1 u_1}\rangle}, \quad \sigma_2 = \sqrt{\langle\overline{u_2 u_2}\rangle}, \quad \sigma_3 = \sqrt{\langle\overline{u_3 u_3}\rangle} \tag{6-5.42}$$

where $\langle\overline{u_1u_1}\rangle$, $\langle\overline{u_2u_2}\rangle$ and $\langle\overline{u_3u_3}\rangle$ are given by Equation (6-5.32). $N$ is a Gaussian random number with a standard deviation of unity. It is updated either when an eddy life ($t_l$) is exceeded or the particle traverses an eddy (this taking a time $t_t$, the eddy transit time). To evaluate the above two time scales an assumption about the eddy size $\langle l_e\rangle$ is required. Following Gosman and Ioannides (1983) (this model on really applies to isotropic turbulence, however, for comparative purposes its use with Equations (6-5.42) is perhaps not unreasonable) it is assumed,

$$l_e = C_\mu^a \frac{\langle k \rangle^{3/2}}{\langle \varepsilon \rangle} \tag{6-5.43}$$

In the work of others (see Milojevic (1990)), to fit experimental data, the exponent $a$ is varied between 0.32 and 0.9. Here $a$ = 0.58, which is consistent with the predictions of Milojevic (1990). The eddy life time can be approximated as

$$t_l = \frac{\langle l_e \rangle}{\sigma}, \tag{6-5.44}$$

where $\sigma = \sqrt{2\langle k \rangle / 3}$. The transit time $t_t$ can be expressed as

$$t_t = \frac{\langle l_e \rangle}{v_d} \tag{6-5.45}$$

where $v_d$ is the drift velocity $|\langle \mathbf{U} \rangle - \mathbf{U}_p|$. However, within an eddy $v_d$ is not constant. Therefore, small time-steps are implemented and $v_d$ is computed each time-step. Values of $v_d$ are used to calculate relative displacements within an eddy. When the total of these is greater than $\langle l_e \rangle$ it is known the particle has left the eddy and $N$ is recomputed.

For unsteady flows, the stochastic particle trajectory algorithm described above can be solved sequentially with the flow field, i.e. once the flow field has been advanced one time-step the particle trajectory equation is also advanced this period. This procedure is then repeated. In certain circumstances, for stability, accuracy or computational economy reasons the time-steps used in the fluid simulation may not be the same as used in the particle integration. This is fairly common in coupled problems. When this is the case, for each stage many to one time-step ratios must be used. Alternatively, the order of one of the schemes can be altered to give time-step compatibility. It would seem ideal if the fluid velocities used in the particle integration are consistent with the underlying fluid flow temporal discretization. Therefore, the fluid related variables used in the particle transport equations could be expressed using Equation (2-2.1) ($\phi = W_f^{new}\phi^{new} + W_f^{old}\phi^{old} + W_f^{older}\phi^{older}$).

### 6-6 Conclusions

URANS computations offer the potential for greater accuracy over RANS predictions. However, the computed temporal behaviour is highly sensitive to turbulence model choice. For unsteady flows, second moment closure models mostly show some predictive accuracy improvements relative to eddy viscosity models. However, with presently available computing resources they are still, in general, excessively expensive for routine complex geometry industrial computations. LES, which places less emphasis on the turbulence model, shows promise, but at present, for realistic engineering flows is also too computationally expensive. DES represents an interesting zonal compromise, but is insufficiently mature and untested to be an alternative to URANS. Consequently, for the next few years, URANS/VLES modelling appears to be the most viable option. For periodic flows, phase averaged properties, $\langle \Phi \rangle$, can be solved for in both fluid flow and particle transport equations.

CHAPTER

7

# CYCLIC ANNULAR CAVITY FLOWS

## 7-1 Introduction

The flow in a heated rotating annulus, as shown in *Figure 7-1.1*, has similarities to that in the earth's atmosphere, sometimes consisting of cyclones and anticyclones between which relatively fast moving jets of fluid, called 'jet streams', flow. The *Figure 7-1.1* annulus consists of two discs connected by concentric cylinders of length $s$, having a temperature difference $\Delta T$, and rotates at an angular velocity $\Omega$. The inner and outer cylinders have radii of $a$ and $b$, respectively. The earth's acceleration due to gravity, $g$, acts along the axis of rotation. Hide (1965), experimentally investigated the flow for a water filled annular cavity with a top surface that was free (open).

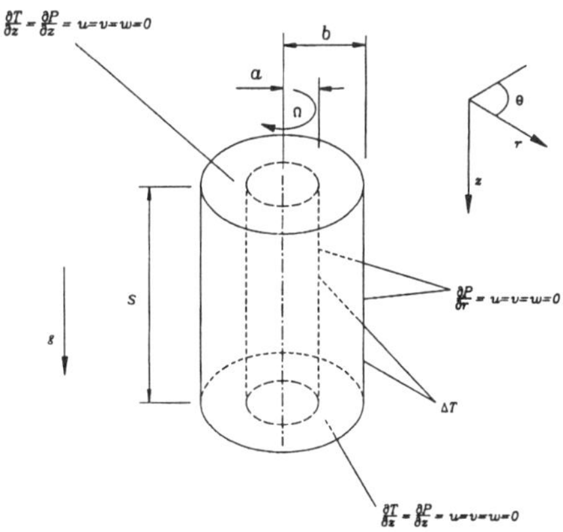


*Figure 7-1.1* Schematic diagram of geometry and numerical boundary conditions

*Figure 7-1.2* Less coherent flow structure for $\Phi < 0.5$

Hide observed that the flow structure was a function of the discriminant

$$\Phi = \frac{g s \Delta \rho}{\Omega^2 (b-a)^2 \rho_{ave}} \tag{7-1.1}$$

where $\rho_{ave}$ is the average density of the fluid at the surface of the inner and outer cylinders and $\Delta\rho$ the difference in density of fluid at these two surfaces.

*Figure 7-1.2* shows a typical flow structure for $\Phi < 0.5$ and *Figure 7-3.2b* a more ordered regime for $\Phi > 0.5$. An interesting feature of this annular cavity flow is that when the outer cylinder is colder than the inner ($\Delta T < 0$) the flow structure angular velocity $\omega < \Omega$. For $\Delta T > 0$ the reverse is true. This latter effect can be observed by comparing frames (b) and (c) of *Figure 7-3.2*. At the top edge of the inner cylinder there is a white reference marker and the change in the flow structure relative to this is clear. It is important to note that for $\Phi = 0.53$, in a sense, the flow *structure* is quasi steady, with the main unsteadiness arising from the angular drift of the flow relative to the cavity. Therefore, for an observer fixed in a frame of reference rotating at $\omega$, the flow would appear steady. For $\Phi < 0.5$, the nature of the flow is significantly different. From whatever frame of reference it is viewed, it will not appear steady.

Traditionally, special convective term treatments (see *Aliasing* in Section 3-4), combined with explicit temporal finite difference techniques have been used to predict this flow, which requires long time period integrations. For such integrations, accurate implicit schemes offer the advantage of larger and hence fewer time-steps. Here, the predictive performance of fully implicit, various semi-implicit and explicit, finite volume/difference techniques are examined and comparisons made with the data of Hignett *et al.* (1985) and others. As well as studying the behaviour of various temporal schemes, the performances of the HYBRID, CONDIF and central difference (CD) (Section 3-3) convective term treatments are explored.

## 7-2 Numerical Details

The momentum equations (1-2.5)-(1-2.7) are solved in a cylindrical polar coordinate system rotating at $\Omega_g$, with SIMPLEC (Section 4-4) and the reduced pressure ($p$) defined by Equation (4-4-34). The hydrostatically reduced $p$ is essential if numerical precision errors are not to result in steady axisymmetric predictions.

*Substantial Derivative*

Also, the choice of coordinate system angular velocity is important. This can be appreciated by considering the substantial derivative Equation (3-2.6), which is restated below

$$\left(\frac{\partial \phi}{\partial t}\right)_{fp} = \left(\frac{\partial \phi}{\partial t}\right)_{g1} + \left(\frac{\partial \phi}{\partial t}\right)_{f} + \left(v - \Omega_g r\right)\frac{1}{r}\frac{\partial \phi}{\partial \theta} + \frac{u}{r}\frac{\partial r\phi}{\partial r} + w\frac{\partial \phi}{\partial z} \tag{7-2.1}$$

The first point to note is that if $v \approx \Omega_g r$, the magnitude of the tangential derivative convective term is reduced in the above. Convective terms are always problematic (see Section 3-3). Therefore, this is desirable. In fact, as observed by Tucker (1997), if steady, axisymmetric predictions are avoided, to sufficiently reduce false diffusion, it is essential that $\Omega_g \approx \omega$ (i.e. $v \approx \Omega_g r$). Secondly, (see Section 3-2) for certain classes of unsteady flows, as embodied in Equation (7-2.1), the time derivative can be split into the following two components.

$$\frac{\partial \phi}{\partial t} = \left(\frac{\partial \phi}{\partial t}\right)_{g1} + \left(\frac{\partial \phi}{\partial t}\right)_{f} \tag{7-2.2}$$

The first, right hand side component, $(\partial \phi / \partial t)_{g1}$ is the unsteadiness arising just from the flow drift (i.e. because $\omega \neq \Omega_g$). The $\Phi > 0.5$ flow structure(*Figure 7-3.2*a) does not change significantly with time but merely drifts relative to the cavity. Therefore, minimizing $|\Omega_g - \omega|$ will reduce $(\partial \phi / \partial t)_{g1}$ and hence $\partial \phi / \partial t$ (clearly, if the overall flow structure is time invariant, setting $\Omega_g = \omega$ will give steady flow, i.e. $(\partial \phi / \partial t)_f = 0$). This offers the potential for increased temporal accuracy and computing time savings.

The above property could perhaps be taken advantage of for a series of studies exploring the variation of flow structure with just one independent variable. If from experimental data or preliminary computations a monotonic $\omega$ trend could be observed, conservative $\Omega_g$ estimates could be made. Alternatively, for individual solutions the condition $\Omega_g - \omega = 0$ could possibly be achieved through iterative means, thus giving maximum temporal accuracy. However, the flow shown in *Figure 7-2.1*, where $\Phi <$

0.5, is far more complex and the solution cannot be made steady ($(\partial\phi / \partial t)_f = 0$) by a simple coordinate system change.

*Solution Initialization*

The numerical solution is initialized in the same way as James *et al.* (1981). An axisymmetric steady state solution is produced and represented on the three-dimensional grid. A non-axisymmetric sinusoidal temperature perturbation is then introduced. The perturbation is made at $r = (a + b) / 2$ for $0 < z/s < 1$ and $0 < \theta \leq 2\pi$. The initialization is of the form, $T^n|_{r=\frac{a+b}{2}} = T^o|_{r=\frac{a+b}{2}} + \delta T \sin(m\,\theta)$ where $T^o$ and $T^n$ are previous and modified grid point temperatures, respectively. Since the comparison is to be made with the experimental data of Hignett *et al.*, which is for flow with 3 cyclone/anticyclone pairs, a value of $m = 3$ is chosen with $\delta T = 0.6$ K. No tests have been carried out to explore the sensitivity of predictions to initial conditions. However, this would perhaps be interesting.

*Grid Structure and Numerical Parameter Settings*

Uniform grids, intended to resolve the vortices which flank the 'jet stream', and non-uniform, to resolve boundary layers at the cavity surfaces, are used. Boundary layer estimates (see Hignett *et al.*) show that the cylindrical surface (Stewartson) and disc surface (Ekman) layers are 2.5 mm and 1.25 mm thick, respectively. To resolve these layers a heavily stretched (using geometric expansions) 22 x 60 x 28 ($r$, $\theta$, $z$) grid is required. The grid density for the current predictions is comparable to that used by Hignett *et al.* (16 x 16 x 64) and was the maximum possible for the computing resources available at the time. To model temporal variations the following schemes are tried: (I) Fully Implicit 1st order backwards Euler; (II) Galerkin; (III) Crank-Nicolson with fully implicit coefficients; (IV) Crank-Nicolson scheme and (V) Fully explicit 1st order forwards Euler. Time-steps of $\Delta t$ = 0.1 s are used. For Scheme (IV), time-step independent solutions are found for $\Delta t < 0.25$ s.

*Convergence Criteria*

All real variables are solved to 16 significant figures. To test for convergence, the momentum equation residuals are normalized by the inertial force $[\rho_o (u,v,w)^2 \pi (b^2 - a^2) s] / 2L$, where $L = b - a$ and $u$, $v$ and w are the computed maximum velocity components ($u \approx v/2 \approx$ w $\approx 0.003$ m/s). The pressure correction equation is normalised by the net mass flow ($\dot{m}$) across the $z$ - $\theta$ plane at $(r - a) / (b - a) = 0.5$ ($\dot{m} \approx 0.03$ kg/s) and the energy equation is normalised by $\dot{m}\ \Delta T$, where $\Delta T$ is the temperature difference between the outer and inner cylinders. Convergence is also checked by monitoring the root mean square ( *rms* ) changes. At convergence, the *rms* (see Equation (4-2.12)) change and maximum residual are less than 5 x $10^{-5}$ and 5 x $10^{-4}$, respectively. Convergence of each time-step requires approximately 4 minutes of central processor unit time on a *SPARC Center 2000* with four 40 MHz processors.

## 7-3 Discussion of Results

*Experimental Conditions and Fluid Properties*

Two sets of experimental data are used. The first set is experimental flow visualization evidence obtained from apparatus described by Tucker (1993) and the second measurements of Hignett *et al.* Experimental conditions for the data used are given in *Table 7-3.1*. Fluid property expressions are given by Hignett *et al.* and stated here for completeness. The fluid density is given by, $\rho = \rho_o(1 - c_1(T - T_0) - c_2(T - T_0)^2)$ where $c_1$ = 3.07 x $10^{-4}$ $K^{-1}$ and $c_2$ = 7.83 x $10^{-6}$ $K^{-2}$.

| Parameter | Symbols | Current Data | Experiments of Hignett et al. | |
|---|---|---|---|---|
| Inner radius of annulus | $a$ | 29.0 | 25.0 | mm |
| Outer radius of annulus | $b$ | 96.9 | 80.0 | mm |
| Depth of annulus | $s$ | 48.5 | 140.0 | mm |
| Rotational Speed | $\Omega$ | 0.707 | 1.0 | rad / sec |
| Inner wall temperature | $T$ | 282.06 | 295.0 | K |
| Outer wall temperature | $T$ | 297.60 | 299.0 | K |
| Reference temperature | $T_o$ | 282.06 | 295.0 | K |
| Temperature difference | $\Delta T$ | 15.54 | 4.0 | K |
| Fluid density | $\rho_o$ | 999.8 | 1043.0 | kg / $m^3$ |
| Kinematic viscosity | $\nu_o$ | $1.35 \times 10^{-6}$ | $1.62 \times 10^{-6}$ | $m^2$ / s |
| Diffusion coefficient | $k_o/c_{po}$ | $1.38 \times 10^{-4}$ | $1.345 \times 10^{-4}$ | kg / m s |
| Flow discriminant | $\Phi$ | 0.532 | 0.6 | |
| Taylor number | $Ta$ | $3.25 \times 10^{7}$ | $5.48 \times 10^{6}$ | |

*Table 7-3.1* Experimental conditions

The diffusion coefficient in the energy equation is modelled using $k/c_p = k_o/c_{po}(1 + c_1(T - T_0))$ where $c_1$ = 2.33 x $10^{-3}$ $K^{-1}$. The kinematic viscosity is given by, $\nu = \nu_o(1 - c_1(T - T_0) - c_2(T - T_0)^2)$ where $c_1$ = 2.79 x $10^{-2}$ $K^{-1}$, $c_2$ = 6.73 x $10^{-4}$ $K^{-2}$.

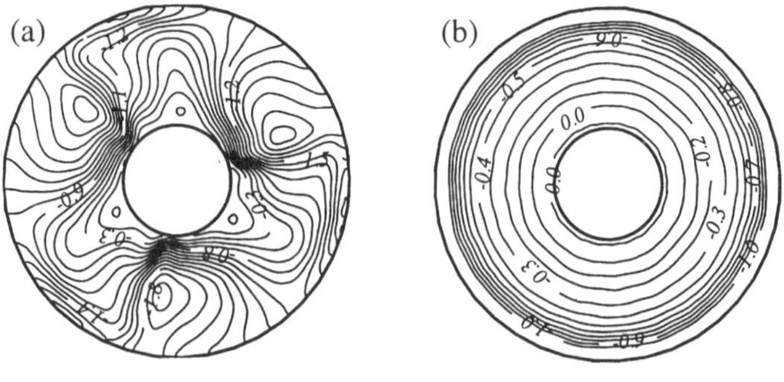


*Figure 7-3.1* Predicted dimensionless stream function field for: $\Omega_g = \Omega$ and $\Omega_g = 0$.

*Comparisons with Experimental Data*

Little difference is found between predicted velocities for the two grids, but, computations made using the more refined are presented, since these slightly more accurately predict Fourier components. Solutions are shown when the flow reaches a cyclic steady state ($t > 1000$ s). *Figure 7-3.1* shows predicted streamfunction plots made with $\Omega_g = \Omega$ (frame (a)) and $\Omega_g = 0$ (frame (b)). Due to false diffusion the frame (b) prediction is steady and axisymmetric, and the importance of coordinate system choice is clearly illustrated.

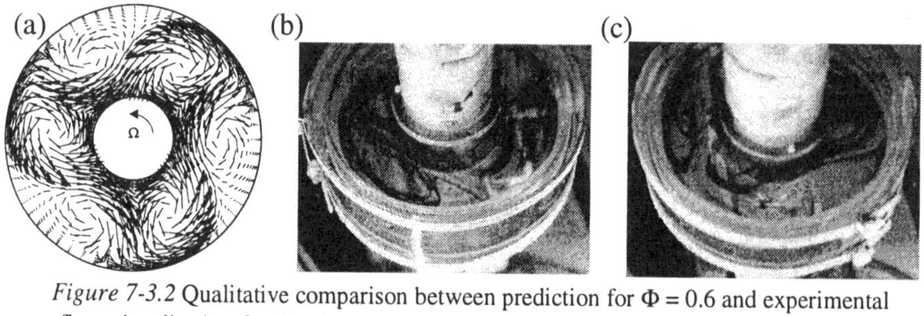


*Figure 7-3.2* Qualitative comparison between prediction for $\Phi = 0.6$ and experimental flow visualisation for $\Phi = 0.53$: (a) Predicted velocity vectors; (b)Experimental flow visualisation and (c) Experiment visualisation at a later time.

The HYBRID, CONDIF and CD discretizations are found to be stable for $t > 1000$ s. A qualitative comparison between predictions at $z/s = 0.5$ and current experimental flow visualization evidence at $z/s = 0$ is shown in *Figure 7-3.2*. There is reasonable agreement between the experimentally observed and numerically predicted flow structures. *Figure 7-3.3* shows comparisons of predictions, for radial variation of circumferentially averaged tangential velocity, with measurements of Hignett *et al.*, and velocity vector plots in the $r$ - $\theta$ plane. The comparisons are made at the following axial locations: $z/s$ = 0.3; $z/s$ = 0.5; $z/s$ = 0.7. Computations made using the CD and HYBRID schemes are represented by the full and dashed lines, respectively.

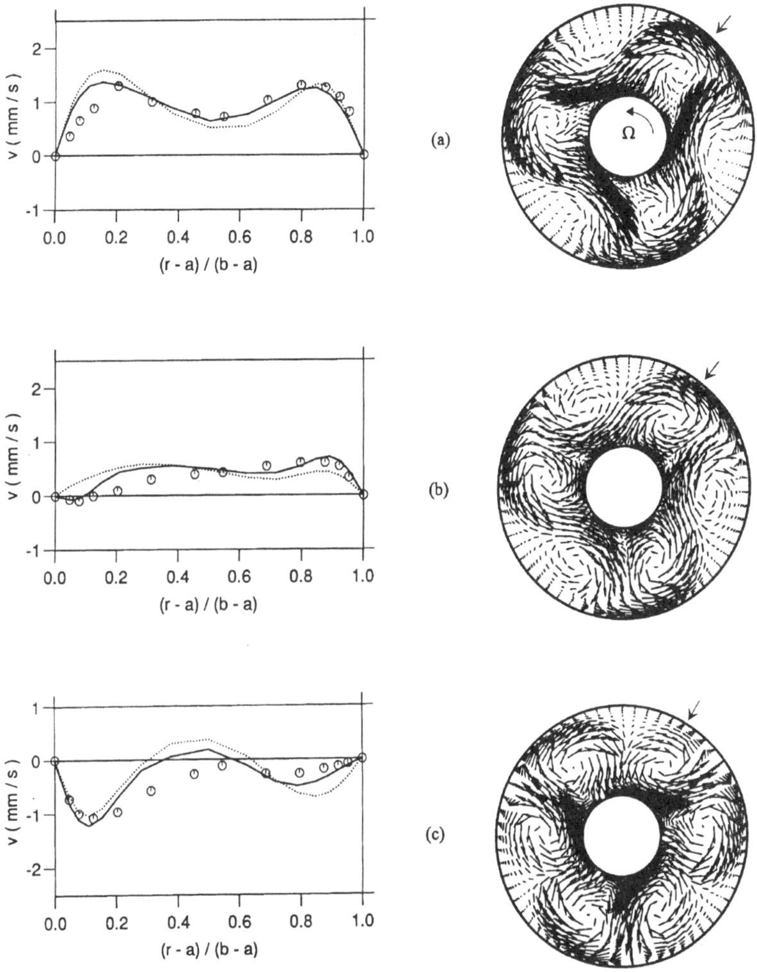


*Figure 7-3.3* Comparisons of predictions, for radial variation of tangential velocity, with measurements of Hignett *et al.* and velocity vector plots in the $r$-$\theta$ plane at: (a) $z/s$ = 0.3; (b) $z/s$ = 0.5 and (c) $z/s$ = 0.7. (o Measurements, CD ——, HYBRID ------)

The symbols represent the measurements. The jet stream structure can be observed most clearly towards the top of the cavity. As would be expected, the second order CD scheme gives the best agreement with the measurements. The CONDIF accuracy is found to be worse than that for the CD scheme but better than HYBRID.

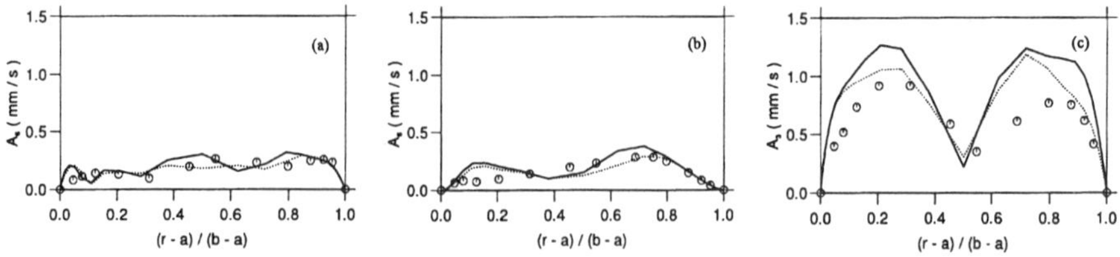


*Figure 7-3.4* Comparison of different wavenumber Fourier components with experimental measurements: (a) $m = 6$, tangential velocity; (b) $m = 6$ radial velocity and (c) $m = 3$, tangential velocity (o Measurements, CD ——, HYBRID ------)

The radial variation of different wavenumber Fourier components at $z/s = 0.5$ is compared with experimental measurements in *Figure 7-3.4*. The variation of tangential and radial velocity wavenumber six ($m = 6$) Fourier components is shown in *Figures 7-3.4a* and *b*, respectively. The variation of the tangential velocity $m = 3$ component is shown in *Figure 7-3.4c*. Surprisingly, *Figure 7-3.4c* suggests the CD scheme is the least accurate.

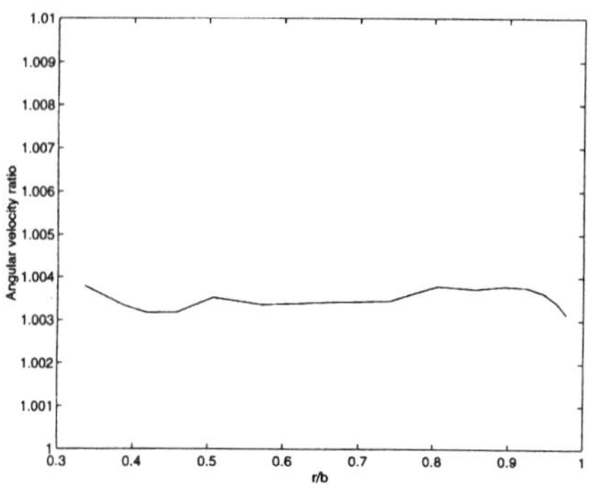


*Figure 7-3.5* Mid-axial variation of $\omega / \Omega$ with radius

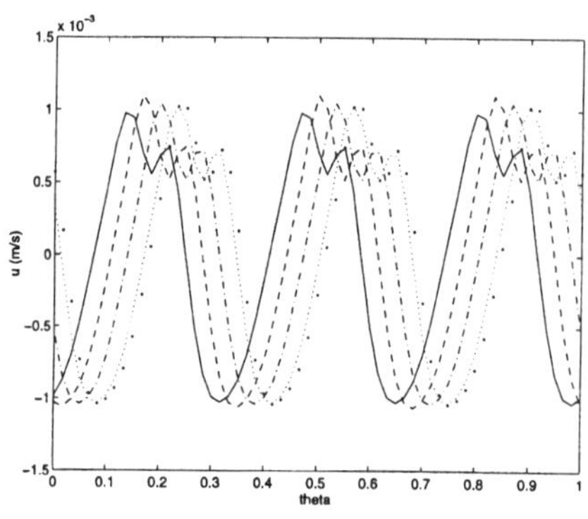


*Figure 7-3.6* gives the circumferential $u$ variation at $z/s = 0.5$ and $(r - a)/(b - a) = 0.5$

*Figure 7-3.5* shows the mid-axial plane variation of $\omega / \Omega$ with radius. Non-uniformities in $\omega / \Omega$ are attributed to flow vacillations distorting the Fourier analysis used to estimate $\omega$. In reality it is expected that $\omega / \Omega$ is nearly constant.

*Figure 7-3.6* gives the circumferential $u$ variation at $z/s = 0.5$ and $(r - a)/(b - a) = 0.5$. The lines show solutions at 50 s intervals for $\Omega_g = 1$ rad/s. Their movement from left to right represents the temporal drift (note the flow structure is actually rotating *faster* than the cavity). The symbols correspond to a solution made with $\Omega_g$ = 1.0035 rad/s and integrating for 150 s from the solution represented by the far right hand dotted line. As can be seen for $\Omega_g$ = 1.0035 rad/s there is little temporal drift indicating $(\partial\phi/\partial t)_{g1} \approx 0$. As a consequence of this reduced unsteadiness just 1 rather than 7 iterations is required for time-step convergence. This represents a dramatic computing time saving.

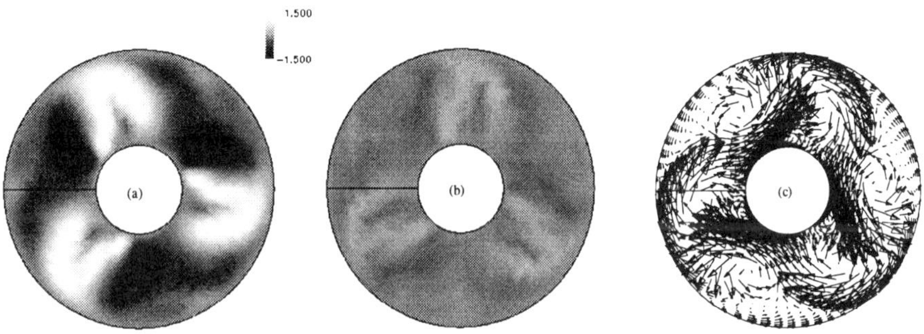


*Figure 7-3.7* Mid-axial contours of (1 x $10^5$)($\partial u/\partial t$) and a velocity vector plot: (a) Contours for $\Omega_g$ = 1.0, (b) Contours for $\Omega_g$ =1.0035 rad/s and (c) Velocity vectors

*Figure 7-3.7*, frames (a) and (b) give mid-axial contours of (1 x $10^5$)($\partial u/\partial t$) for $\Omega_g$ = 1.0 and 1.0035 rad/s, respectively. The more uniform $\Omega_g$ = 1.0035 rad/s contours signify greater flow steadiness. Frame (c) is the corresponding velocity vector plot.

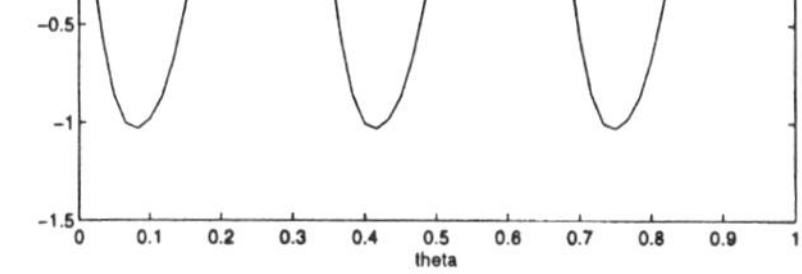


*Figure 7-3.8* Circumferential *u* variation for $\Omega_g$ = 1 rad/s and temporal schemes (I) to (V)

*Figure 7-3.8* gives the circumferential *u* variation at $z/s$ = 0.5 and $(r-a)/(b-a)$ = 0.5 for $\Omega_g$ = 1 rad/s and temporal schemes (I) to (V). A line is also included for Scheme (IV) with $\Delta t$ halved confirming sensible time-step independence. The lines are produced after a 150 s integration period from the same initial conditions. As can be seen, for all schemes, the curves are indistinguishable.

*Table 7-3.2* shows the number of iterations to convergence per time-step for the different schemes.

| Scheme | No. iterations | Comments |
|---|---|---|
| (I) | 5 | - |
| (II) | 5 | - |
| (III) | 5 | - |
| (IV) | 5 | - |
| (V) | * | Converges for $\Delta t \leq 0.05$s |
| (IV) with $\Delta t=0.05$s | 4 | |

*Table 7-3.2* Iterations to convergence per time-step for the different temporal schemes

For the above implicit schemes, the time-step to number of iterations to convergence ratio is about optimal. This suggests that surprisingly, for this case (unlike all others in this text) the explicit forward Euler scheme is most economical.

## 7-4 Conclusions

'Jet stream' flows are predicted, in a differentially heated rotating annulus, by solving the time dependent equations describing the conservation of momentum and energy in a three-dimensional cylindrical polar coordinate system. The performances of the following temporal schemes are addressed: (I) Fully implicit backwards Euler; (II) Galerkin; (III) Crank-Nicolson with fully implicit coefficients; (IV) Crank-Nicolson and (III) Fully explicit forwards Euler. Also, comparison is made between the HYBRID, CONDIF and central difference (CD) schemes. To accurately model flow driving pressure differences a gravitationally reduced pressure is required.

In general, there is agreement between predictions and measurements. Overall, the CD predictions are the most accurate followed by CONDIF and then HYBRID. All are found stable for integrations over long time periods ( $t > 1000$ s ). For certain non-vacillating flow regimes it appears that increases in temporal accuracy and computational economy can be made by minimizing $|\Omega_g - \omega|$, where $\Omega_g$ and $\omega$ are the coordinate system and flow angular velocities, respectively.

CHAPTER

# 8

# CYCLIC AERO ENGINE MOTIVATED CAVITY FLOWS

## 8-1 Introduction

In certain modern aero-engine gas turbines, cooling air is ducted axially between the bores of adjacent discs in the high-pressure compressor. A schematic diagram showing a cross-sectional view through a high-pressure compressor drum is shown in *Figure 8-1.1*. Some of the central axial throughflow enters the relatively warm cavities formed between the adjacent discs and a parasitic temperature rise $\Delta T$ occurs. Knowledge of the heat transfer in these cavities is important for designers.

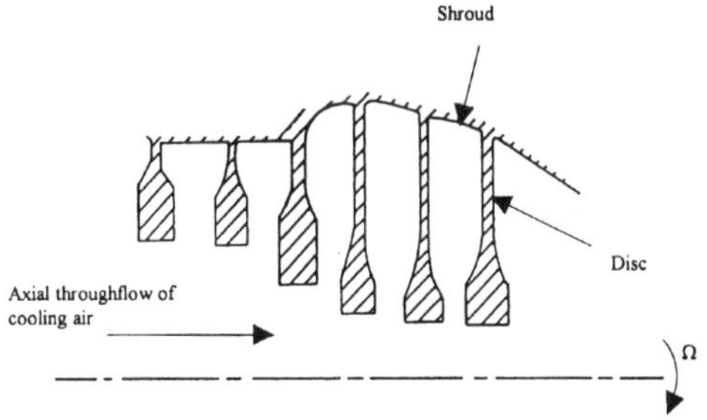


*Figure 8-1.1* Simplified arrangement of a high pressure compressor drum

An idealised version of the above is shown in *Figure 8-1.2*. This comprises a cylindrical cavity (formed by two discs of outer radius, $b$, and a peripheral shroud of axial width, $s$) which rotates with an angular velocity, $\Omega$. The

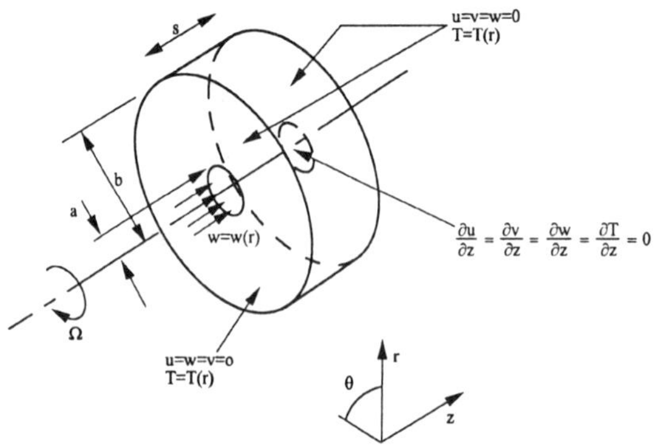


*Figure 8-1.2* Idealised high pressure compressor cavity

central throughflow enters the cavity, with an average velocity $W$, through a hole of radius $a$, in the upstream disc and leaves via an identical hole in the downstream disc. Farthing *et al.* (1992a,b) experimentally studied the flow in a rotating cavity with an axial throughflow. The essential features of the observed flow are illustrated in *Figure 8-1.3*. For unheated discs, the structure is similar to that shown in *Figure 8-1.3a*. In this case, the central jet passing through the cavity (the throughflow) generates a toroidal vortex. In the absence of various modes of isothermal vortex breakdown (see later), the flow is steady and two-dimensional axisymmetric.

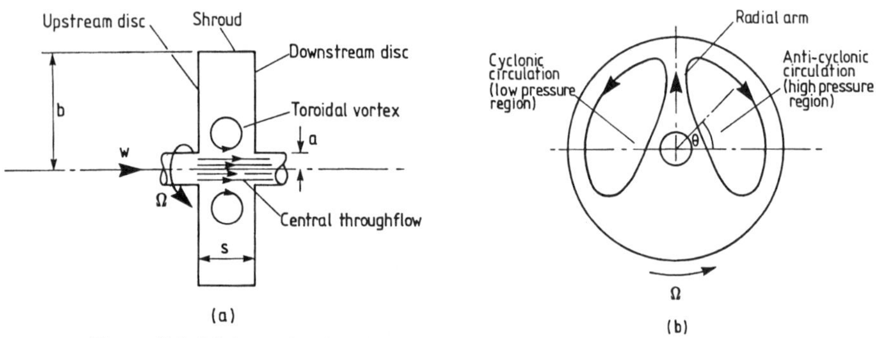


*Figure 8-1.3* Schematic diagrams of flow in a rotating cavity with an axial throughflow: (a) Isothermal flow, $T_s = T_i$ ($r$-$z$ plane) and (b) heated flow, $T_s > T_i$ ($r$-$\theta$ plane)

When the discs are heated, there is a dramatic change in the flow structure. Centripetal acceleration forces the denser axial throughflow into the less dense cavity fluid. The flow situation is similar to the classic Rayleigh-Bernard instability problem, but this time in a rotational sense. The resulting flow is time-dependent and three-dimensional.

A schematic of the typical heated flow structure, for a disc surface temperature that decreases with radius, given in the $r$-$\theta$ plane as this best illustrates the relevant features observed by Farthing *et al.*, is shown in *Figure* 8-1.3b. The flow now comprises: a radial arm and a pair of circulation regions. Multiples of this basic flow structure have also been observed, i.e. two radial arms and two pairs of circulations etc. As with the cyclonic/anticyclonic flow described in the previous chapter, this entire flow rotates/drifts at an average angular velocity $\omega \neq \Omega$. This drift constitutes a key flow unsteadiness element. The radial arm(s) are flanked by cyclones and anticyclones. When the disc temperature increases with radius or the shroud is heated the flow structure is found by Farthing *et al.* to be significantly different to that described above, other less coherent unsteadiness elements being introduced.

For isothermal flow, when $Ro \approx 1$, ($Ro = W/\Omega a$), Owen and Pincombe (1979) observed a so-called bubble vortex breakdown, where there is stagnation and occasional axial flow reversal near the cavity outlet at the rotation axis. This again presents a key unsteadiness component for predictions. Experiments have shown that the isothermal vortex breakdown can, in a complex fashion, be superimposed on the non-isothermal flow structure (see Long and Tucker (1994) and Iacovides and Chew (1993)).

## 8-2 Numerical Details

For the flow considered, the Eckert number, $E = \Omega^2 r^2/2C_p\Delta T$ is small, so compressive heating is neglected. Relative to the discs, the flow structure of the cavity is slow moving, because $\omega \cong \Omega$, therefore frictional heating can also justifiably be ignored. This allows the formal stagnation enthalpy equation to be substituted by one in which the temperature is the dependent variable: no source terms arise with consequent savings in computing time. All real variables are computed to 16 significant figures. Also, the reduced pressure Equation (4-4.34) is used. Either the HYBRID (Section 3-3) or CONDIF (in a bounded form) (Section 3-3) convective term treatments are implemented. The following temporal schemes are tested: (I) Implicit backwards Euler; (II) Implicit Coefficient Crank-Nicolson; (III) Crank-Nicolson; (IV) Alternating Direction Implicit (ADI); (V) Galerkin and (VI) Explicit forwards Euler.

The SIMPLEC (Section 4-4) method is generally used. However, a novel hybrid AVPI/SIMPLEC algorithm is also tested (see also Section 4-4). To improve momentum equation coupling Equation (4-2.8) is implemented with $0.5 < \alpha_0 < 2$ and $\alpha_l = 2$. To further reduce solution times, multilevel convergence (Section 4-2) is employed with three grid levels in all coordinate directions.

*Solution Initialization and Boundary Conditions*

Various tests are carried out to verify that the unsteady solutions are independent of the chosen initial conditions (notably different initial tangential velocity distributions). The results presented are independent of these conditions. However it is interesting to consider the anatomy of a steady computation for an unsteady flow. When the flow is treated as steady and large $\alpha_0$ values used in Equation (4-2.8), results can be obtained with different numbers of $r$-$\theta$ plane vortices. The number of these depends on initial fluid tangential velocity distributions. However, since the residuals cannot be driven to machine round-off these results can certainly not be regarded as solutions. If $\alpha_0$ is small, a particular flow pattern will not be settled upon. This is indicative of the classic symptoms when an unsteady flow is treated as steady – the residual dropping to a plateau and with increasing number of iterations exhibiting oscillations about this. However, as noted in Chapter 1, there is the possibility that, depending on initial conditions, different final steady solutions (*pitchfork* bifurcations) can be produced and the observed patterns may be weakly related to such bifurcations.

For unsteady predictions, at $t = 0$, the initial conditions for variables $u$, $v$, $w$, $p$, $T$ and $\rho$ inside the cavity are: $u = v = w = 0$, $p = 1.03 \times 10^5$ N/m$^2$, $T = T_i$, $\rho = p/RT_i$. For $t > 0$, the disc and shroud surface temperatures are specified. At the cavity inlet $T = T_i$, $u = v = 0$ and $w = (Re_z\rho/\mu a) = W$.

Both discs are heated with the same axisymmetric surface temperature distribution, and, for initial predictions, four different nominal distributions (profiles) are investigated. These are illustrated in *Figure 8-2.1*. Frames (a), (b), (c) and (d) will be referred to as profiles A, B, C and D, respectively. General boundary conditions are outlined in Section 4-6.

*Convergence Criteria*

Mostly, the parameters used to monitor convergence are the normalized residuals and also the root-mean square (*rms*) (Equation (4-2.12)) change in all the variables. Generally, at the end of a time-step, *rms* values are below $5 \times 10^{-5}$ (but, mostly an order of magnitude smaller than this). Typically, the residuals for the axial, radial and tangential momentum equations are normalized using $\rho u^2_{ref}\pi bs/2$, where $u_{ref}$ is the maximum radial velocity at $r = b/2$ and $z = s/2$. The residual of the pressure correction is made dimensionless by dividing by the net mass flow rate of fluid actually entering the cavity (typically 1 to 2% of the throughflow flow rate), $\dot{m}_{cav}$. For the energy equation, the product of flow rate entering the cavity and the maximum temperature difference between the cavity walls and the incoming air is used to normalize the residuals. These ($u_{\rm ref}$ and $\dot{m}_{cav}$) normalising parameters are chosen in preference to the perhaps more obvious choice of bulk axial velocity, $W$, and total throughflow rate, $\dot{m}_{cav}$, since they are considered to be more appropriate. Use of the latter can bring about a relaxation in the convergence criteria with the result that insufficiently converged solutions can be obtained. Generally, the normalized residuals for all the equations are less than 2 x $10^{-2}$. A typical average value is 0.75 x $10^{-2}$.

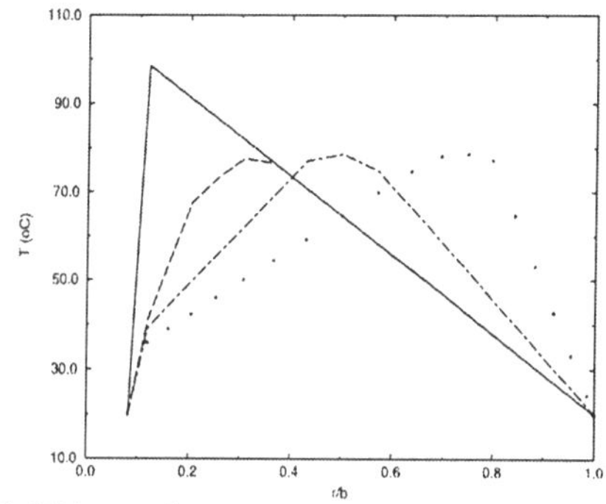


*Figure 8-2.1* Disc surface temperature distributions investigated
( — Profile A; ------ Profile B; -·-·-· Profile C; ······ Profile D)

*Numerical Parameter Settings*

Generally, results are presented for extremely modest 21 x 40 x 17 ($r$, $\theta$, $z$) grids with $\Delta t$ = 0.05 s. Although by present day standards coarse, at the time the grid was chosen due to limitations in computing speed and available storage. In more recent tests, for Profiles A-D, involving a factor of 25 grid density increase, the basic flow structure is found not to greatly alter. Hence the mostly illustrative points to be made here are considered valid. Also, the coarse grids give the advantage that, if desired, solutions can be readily repeated by the reader on a PC. Solutions are found time-step independent for $\Delta t < 0.2$ s. However, $\Delta t$ = 0.05 s corresponds to about the optimum time-step to number of iterations ratio.

## 8-3 Lower Reynolds Number Results

In this section, predictions focus on a cavity with $a/b$ = 0.1 and $G$ = 0.267, for the lower Reynolds numbers of $Re_z$ = 2180 and $Re_\phi = 1.3 \times 10^4$.

*Flow Structure Features*

The essential *cyclic* Profile A flow features are illustrated in *Figure 8-3.1*. On the left-side of the figure are the computed velocity vectors at $z/s$ = 0.5 and on the right-side are schematic diagrams of the flow. The vector plot at $t \approx 30$ s, shows flow entering the cavity in a radial arm flanked by cyclones and anticyclones. At $t \approx 40$ s (Frame (b)), there are now two radial arms, two regions of cyclonic flow and two smaller regions of anticyclonic flow, i.e. the flow has a *limit* cycle.

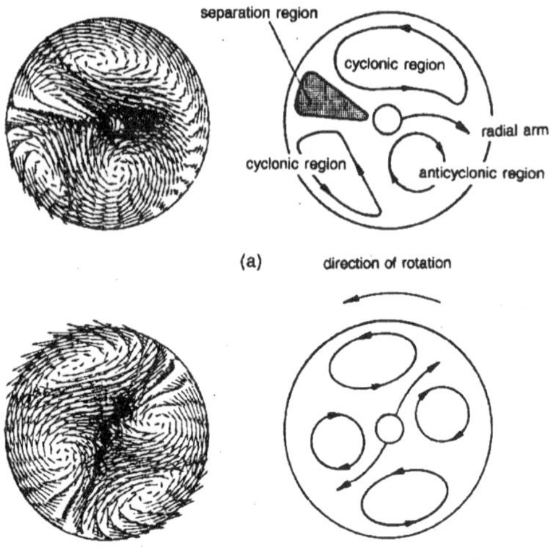


(b)
*Figure 8-3.1* Essential features of the basic flow structure for: (a) $t \approx 30$ and (b) $t \approx 40$ s

The dramatic effect of the disc temperature profile on flow structure is illustrated in *Figure 8-3.2*. Frames (a), (b), (c) and (d) show the velocity vectors corresponding to profiles A, B, C and D, respectively. In all these cases, except for Frame (e), the cavity shroud is at $T = T_i$ (= 20°C). In Frame (e) the shroud is also heated with $T_s$ = 80°C. Analysis of the Profile A limit cycle suggests the predominant flow structure is that with the single radial arm; this is shown in *Figure 8-3.2a*. For temperature distribution B, the flow structure (see Frame (b)) always has two radial arms. A similar (two-armed) flow structure occurs with temperature distribution C. Shifting the location of the maximum disc temperature towards the outer radius, as in distribution D, creates a significant difference in the flow structure (see Frame (d)). It is now difficult to distinguish radial arms and the flow mildly vacillates. In Frame (e), where the discs are heated with distribution D and the shroud is also heated, the flow inside the cavity appears to have broken down into a number of less orderly vacillating streams. The predicted flow structure is not inconsistent with that found by Farthing *et al.* who observed multiple radial arms.

As previously noted, the flow visualization work of Farthing *et al.* (1992b) focussed on disc surface temperature distributions A and B. The numerical results presented here have shown that under these conditions the

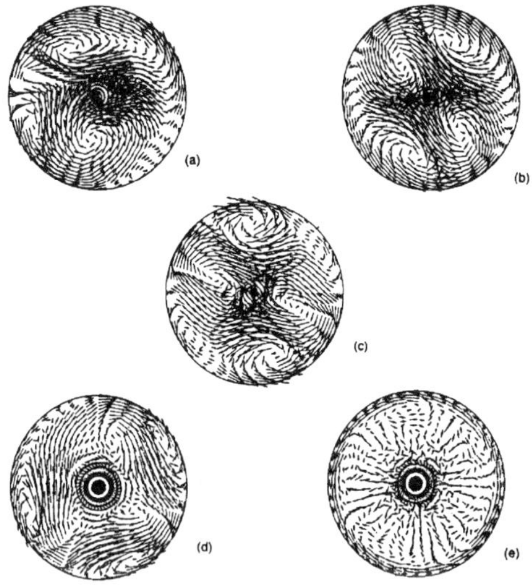


*Figure 8-3.2* Velocity vectors for the following profiles: (a) A; (b) B; (c) C; (d) D and (e) D with a heated shroud

flow may commute from having a single radial arm etc., to a pair of radial arms etc. This sensitivity of the flow structure to the imposed temperature profile may serve to explain differences in the observed experimental flows, where in some cases there would be one radial arm and in others two.

*Comparison between Numerical Predictions and Experimental Flow Visualization*

A comparison between the numerical (with temperature distribution A and $Sc = \mu/\Gamma\rho, \ = \infty$) and experimentally observed flow structures is given in *Figure 8-3.3* at $t$ = 1, 3 and 7 s. The grey fixed scale chosen attempts to match that in the experimental results. In Frame (a) both the numerical and experimental results show the radial arm. In Frame (b) this bifurcates and begins to form the cyclonic and anticyclonic regions, which are evident in Frame (c).

*The Flow Drift*

The above unheated shroud predictions give for the bulk of the flow $\omega < \Omega$. This is consistent with the observations of Farthing *et al. Figure 8-3.4* shows the Profile B, mid-axial $\omega/\Omega$ radial variation for $\Omega_g = 17.0168$ (the coordinate system angular velocity). Away from the axial throughflow ($r/b > 0.1$) $\omega$ is relatively constant and any variations are attributed to small flow vacillations distorting the Fourier analysis used to estimate $\omega$. However, in the vicinity of the throughflow $\omega$ increases significantly. The above predictions and those later suggest that when the *shroud is heated* (giving the stongest positive radial temperature gradient) $\omega > \Omega$, which is consistent with the positive radial temperature gradient 'jet stream' flow in Chapter 7.

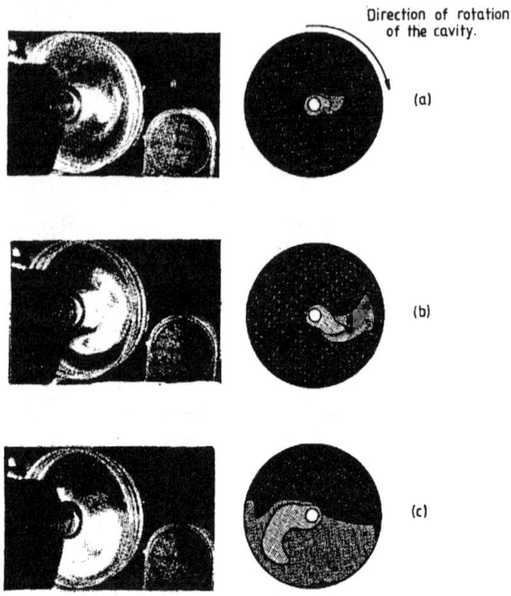


*Figure 8-3.3* Comparison between flow visualization results and predictions with: (a) $t = 1$; (b) $t = 3$ and (c) $t = 7$ s

*The Substantial Derivative*

As in Chapter 7, the choice of coordinate system angular velocity ($\Omega_g$) in the above predictions is important. Therefore, it is recommended that the subsection on the Substantial Derivative in Section 7-2 is revisited before reading on. The points made in this Chapter 7 section again apply to this flow. This can be appreciated by considering the substantial derivative Equation (3-2.6), which again is restated below

$$\left(\frac{\partial\phi}{\partial t}\right)_{fp} = \left(\frac{\partial\phi}{\partial t}\right)_{g1} + \left(\frac{\partial\phi}{\partial t}\right)_{f} + \left(v - \Omega_g r\right)\frac{1}{r}\frac{\partial\phi}{\partial\theta} + \frac{u}{r}\frac{\partial r\phi}{\partial r} + w\frac{\partial\phi}{\partial z} \qquad (8\text{-}3.1)$$

As noted before, for certain classes of unsteady flows (see Section 3-2), as embodied in Equation (8-3.1), the time derivative can be split into the following two components.

$$\frac{\partial\phi}{\partial t} = \left(\frac{\partial\phi}{\partial t}\right)_{g1} + \left(\frac{\partial\phi}{\partial t}\right)_{f} \qquad (8\text{-}3.2)$$

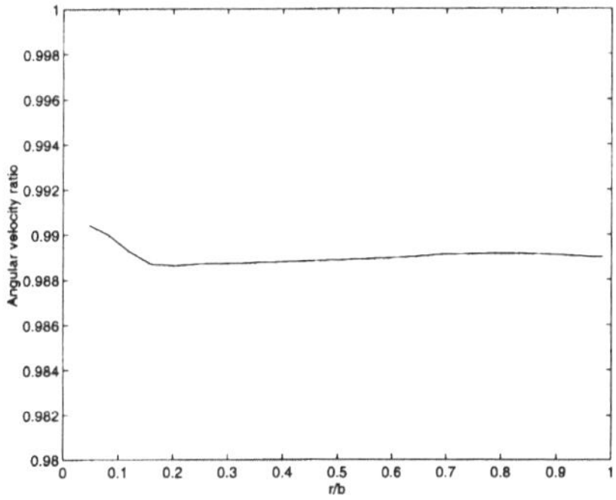


*Figure 8-3.4* Profile B, mid-axial $\omega/\Omega$ radial variation for $\Omega_g = 17.0168$

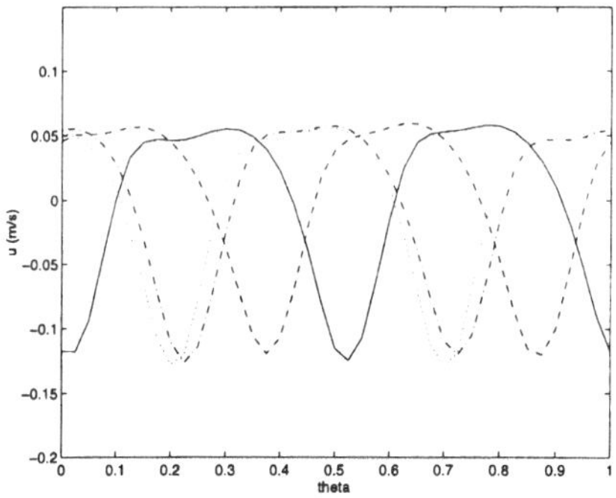


*Figure 8-3.5* Circumferential variation of $u$ at $r/b = z/s = 0.5$ (— $t = 0$, ---- $t = 5$ s, -·-·-·-$t = 10$ s, ····· $t = 15$ s ($\Omega_g/\Omega = 0.989$))

The first, right hand side component, $(\partial\phi/\partial t)_{g1}$ is the unsteadiness arising just from the flow drift (i.e. because $\omega \neq \Omega_g$). *Figure 8-3.5* shows the $\theta$ variation of $u$ at $r/b = z/s = 0.5$. The full, dashed and dash-dot lines correspond to solutions at $t = 0$, 5 and 10 s, where $t = 0$ corresponds to a point in time relative to a datum solution at around 100 s. As a result of the flow structure drift ($\omega < \Omega$) the curves move with equal spacing to the left. The solution for the dotted line is produced by integrating for 5 s from the $t$ = 10 s solution with $\Omega_g$ reduced to 16.8379 rad/s. This corresponds to $\Omega_g/\Omega = 0.989$. The smaller displacement of the line signifies the temporal drift relative to the new frame of reference has decreased.

*Figure 8-3.6*, frames (a) and (b) give mid-axial contours of $\partial u/\partial t$ for $\Omega_g$ = 17.0168 and 16.8379 rad/s, respectively. The more uniform $\Omega_g$ = 16.8379 rad/s contours signify greater flow steadiness. Frame (c) is the velocity vector plot corresponding to Frame (a).

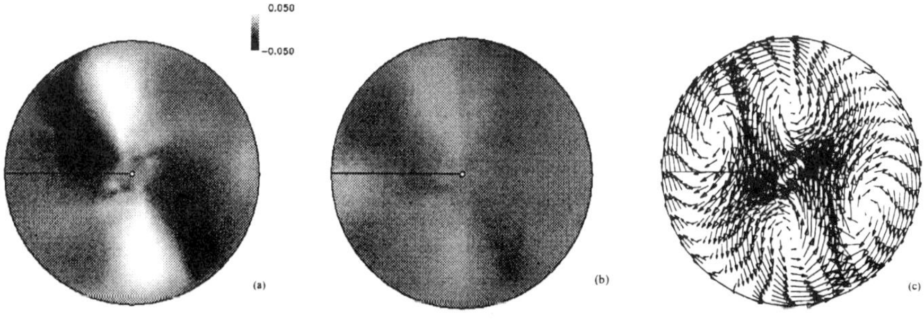


*Figure 8-3.6* Mid-axial contours of $\partial u/\partial t$ and a velocity vector plot: (a) Contours for $\Omega_g$ = 17.0168, (b) Contours for $\Omega_g$ =16.8379 rad/s and (c) Velocity vectors

An important point to make for the above plots is that when $\Omega_g$ is altered to 16.8379 rad/s (unlike when a similar procedure is carried out in Chapter 7) after about 15 s, presumably due to increased accuracy, the flow begins to exhibit strong vacillations similar to those for Profile A. However, this time, the flow appears to commute between flows having two and then three radial arms.

Much of the above suggests there is potential for computational efficiency gains through minimizing $\left|\Omega_g - \omega\right|$. This could perhaps be achieved as outlined in Section 7-2.

Potential computing time savings are illustrated *Figure 8-3.7*. This shows the drop in averaged momentum residual against number of iterations through a $\Delta t$ = 0.05 s time-step. The full and dashed lines correspond to $\Omega_g$ = 17.0168 and 16.8379 rad/s. As can be seen for $\Omega_g$ = 16.8379 rad/s, because of increased flow steadiness, less computational work is required. However, for profiles A and D (with the heated shroud) the significant vacillations, contributing to $(\partial\phi/\partial t)_f$, will to an extent negate any potential computational savings. Further more, as suggested by *Figure 8-3.4*, for a rotating cavity with an axial throughflow, theoretically the $\Omega_g$ choice will never make solutions completely steady. If $\Omega_g$ is chosen to match the $r/b > 0.1$ flow drift, the throughflow unsteadiness will increase.

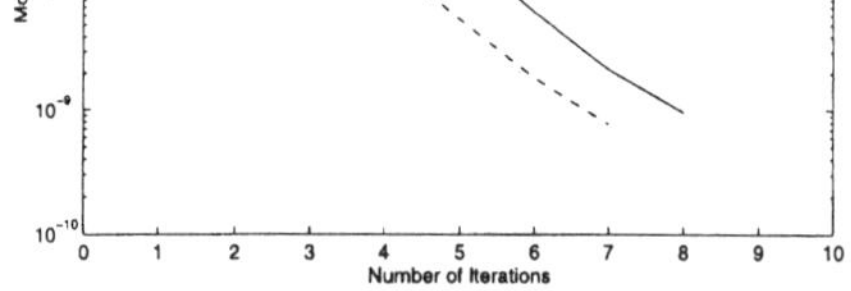


*Figure 8-3.7* Drop in momentum residual against number of iterations for $\Omega_g$ = 17.0168 and 16.8379 rad/s ($\Omega_g$ = 17.0168 ——, $\Omega_g$ = 16.8379 -----)

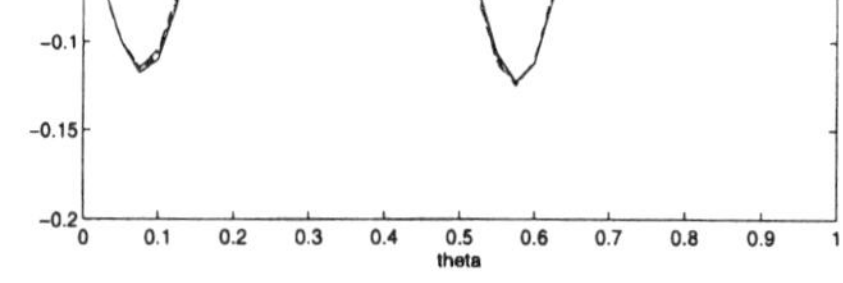


*Figure 8-3.8* Curves contrasting performances with multilevel convergence of the schemes given in *Table 8-3.1*

*Performances for Different Numerical Schemes*

*Figure 8-3.8* shows five curves contrasting performances with multilevel convergence of the schemes given in *Table 8-3.1*. For the explicit forward

Euler and ADI methods, even with $\Delta t << 0.01$ s the solution diverges and so no curves are shown.

| Case | Scheme Description | Δt (s) | Its | Comments |
|---|---|---|---|---|
| (I) | Implicit backwards Euler | 0.05 | 13 | - |
| (I) | Implicit backwards Euler | 0.025 | 8 | - |
| (II) | Implicit Coefficient Crank-Nicolson | 0.0125 | 4 | $\Delta t$>0.0125s diverges |
| (III) | Crank-Nicolson | 0.0125 | 4 | $\Delta t > 0.0125$ s diverges |
| (IV) | Alternating Direction Implicit (ADI) | 0.0125 | 4 | $\Delta t > 0.0125$ s diverges |
| (V) | Galerkin | 0.0125 | 4 | - |
| (VI) | Explicit forwards Euler | - | 4 | $\Delta t << 0.01$ s diverges |

*Table 8-3.1* Performance summary for different temporal schemes

As can be seen, from *Figure 8-3.8*, predicted velocities are similar. Notably, the implicit backwards Euler solutions for $\Delta t$ = 0.025 and 0.05 s confirm time-step independence. The 'higher order' schemes (II-IV) require $\Delta t \leq$ 0.0125 s to converge with 4 iterations per time-step. Therefore, to advance 1 s in time nearly 25 % more iterations/computing time is needed relative to Scheme (I). Clearly, the performances of the explicit and ADI schemes are poor. Since, for this case, the 'higher order' schemes (II) and (IV) diverge when $\Delta t$ = 0.2 s, where first order scheme time-step independence is already reached, it is not possible to demonstrate their second order accuracy.

For this axial throughflow case, the hybrid AVPI/SIMPLEC (AVPI for axial momentum terms and SIMPLEC for others) pressure correction scheme described in Section 4-4 is tried. For $\Delta t$ = 0.01 s the convergence rate is similar to that for a purely SIMPLEC solution.

## 8-4 Higher Reynolds Number Results

In Section 8-3 the flow structure is considered in the absence of vortex breakdown. Here a prediction suggestive of vortex breakdown is presented along with some URANS heat transfer computations. For these, higher Reynolds number predictions, viscous dissipation influences are re–examined. Comparisons between numerical predictions with and without viscous dissipation, for the highest $Re_\phi$ considered here, show variations in local Nusselt number ($Nu$) which are less than 5%.

*Computation of Bubble Vortex Breakdown*

Predictions are now considered for the Section 8-3 geometry with $Ro$ = 1.1, where for isothermal flow bubble vortex breakdown is expected to occur.

Profile B (see Section 8-3) is used with $Re_\phi = 1 \times 10^5$ and $Re_z = 2180$. For this higher Reynolds number, to ensure time-step independence $\Delta t = 0.005$ s. Also, to reduce computing time $\Omega_g$ is set at 97.5% of $\Omega$. The predicted form of the $r$-$\theta$ plane flow structure is as shown in *Figure 8-4.1*. As can be seen (cf. *Figure 8-3.2b*) the general flow structure is the same as that for $Re_\phi = 1.3 \times 10^4$, but superimposed on this is a time dependent vortex breakdown with behaviour similar to that observed by Pincombe (1983).

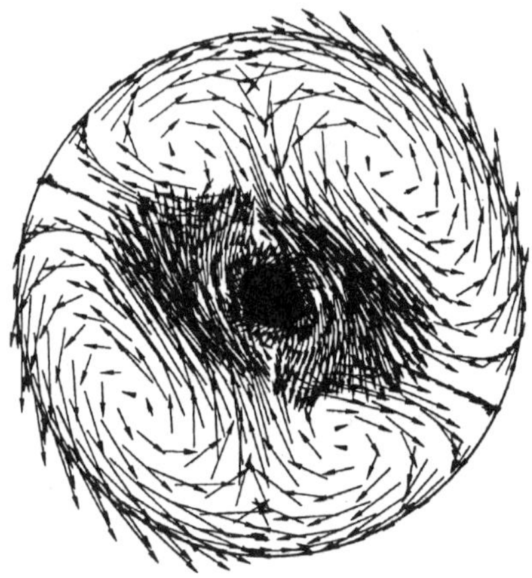

*Figure 8-4.1* Predicted $r$-$\theta$ plane flow structure for $Re_\phi = 1 \times 10^5$, $Re_z = 2180$ and Profile B

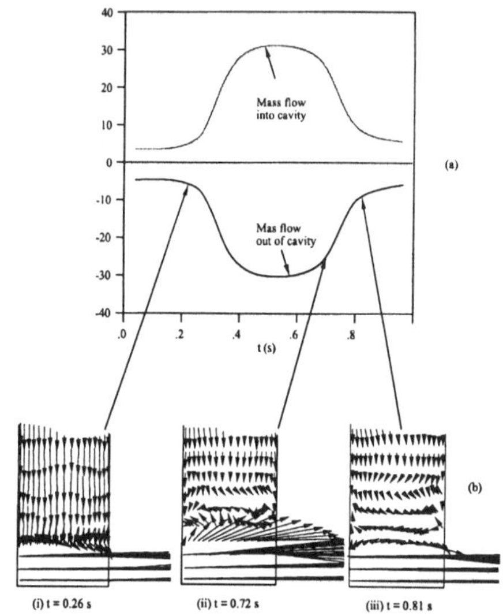


*Figure 8-4.2* Change in the mass flow rate entering the cavity and fluid flow structure with time: (a) variation of mass flow rate entering the cavity; (b) velocity vector plots in the $r$-$z$ plane

The predicted change in mass flowrate entering the cavity and fluid flow structure, with time is shown in *Figure 8-4.2*. The variation of the mass flow rate entering the cavity, expressed as a percentage of the axial throughflow rate, over a one second period is shown in *Figure 8-4.2a*. Note, cycles presented are for $t > 25$ s, but to simplify discussion cycle start times are redefined as $t = 0$ *s*. Pincombe's smoke flow visualisation observations for $Ro \approx 1$, show that vortex breakdown is accompanied by an increase in smoke penetration into the cavity, which is certainly consistent with the numerical prediction presented here. Velocity vector plots in the $r$-$z$ plane at $\theta = 0$ and $t = 0.26$, 0.72 and 0.81 *s* are shown in *Figure 8-4.2b*. These also suggest a modulation in the mass flow rate entering the cavity.

Monitoring the variation with time of axial velocity adjacent to the cavity axis (see *Figure 8-4.3*) suggests signs of the stagnation observed by Pincombe.

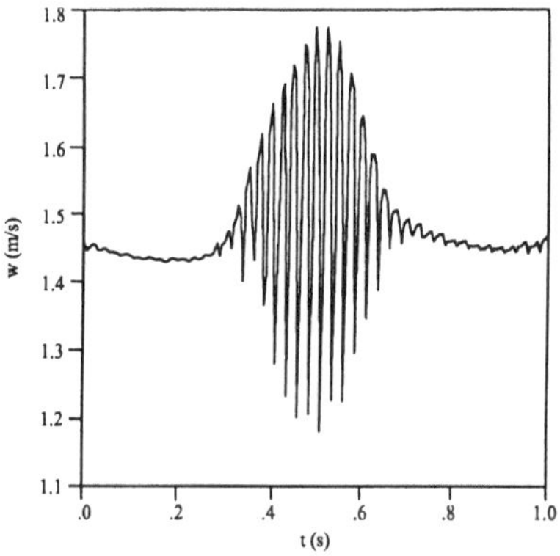


*Figure 8-4.3* Variation of axial velocity with time adjacent to axis at fluid outlet

*URANS/VLES Predictions*

The prediction of shroud heat transfer is now considered. The data is for a relatively high *Re* and so the flow assumed turbulent. However, it should be noted that the conditions under which laminar turbulent transition occurs are not, as yet, known. Farthing *et al.* (1992a) showed heat transfer (for $2 \times 10^4 \leq Re_z \leq 1.6 \times 10^5$ and $2 \times 10^5 \leq Re_\phi \leq 5 \times 10^6$) to be correlated by $Gr^{0.25}$. The exponent of the Grashof number suggests laminar free convection. It is quite likely that both the axial Reynolds number and the Grashof number affect transition. It is also considered likely that laminar and turbulent conditions could co-exist, as for example in a rotating cavity with superimposed radial flow (Owen, Pincombe and Rogers (1985)) where there can be a turbulent region outside laminar non-entraining disc boundary layers.

Transitional flows in rotating disc geometries can perhaps best be handled by the use of low Reynolds number $k$-$\varepsilon$ models. Indeed, Morse (1991a, 1991b) and Morse and Ong (1992) have obtained predictive agreement for the flow and heat transfer in various geometries over a wide range of operating conditions. However, such models require fairly dense clustering of grid nodes in near-wall regions to obtain sufficient resolution of the rapidly-changing turbulence quantities for the predictions to be sensibly grid-independent. This approach is viable for two-dimensional flow computations, but not generally for three-dimensional flows. Here, to reduce computing time and core memory requirements, the Richardson corrected ($ml_2$) mixing length model described in Section 6-5 is used. The diffusive standard $k$-$\varepsilon_l$ model is tried but erroneously gives the flow as steady and axisymmetric. This highlights the strong dependence of results on the turbulence model when making URANS/VLES predictions. All heat transfer data presented here is averaged circumferentially and in time.

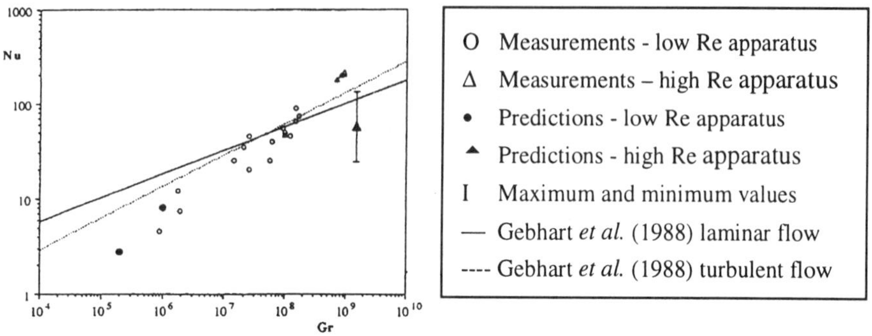


*Figure 8-4.4* Variation of shroud Nusselt number with Grashof number

The variation of measured and predicted values of average shroud Nusselt number ($Nu = qL/k\Delta T$) with Grashof number ($Gr = \Omega^2 b\beta\Delta TL^3/\nu^2$) is shown in *Figure 8-4.4*. The choice of characteristic length, $L = s/2$, and temperature difference, $\Delta T$, are as discussed by Long and Tucker (1994). The measured values are taken from Long Morse and Tucker (1997) and the predictions are made using the HYBRID scheme. Also shown on the figure are laminar and turbulent correlations for free convection from a horizontal surface, heated with a constant surface temperature, facing upwards and in a gravitational field (Gebhart *et al.* (1988)). These correlations are found by Long and Tucker (1994) to be in reasonable accord with measurements. For the highest $Gr$ number prediction, to resolve the shroud boundary layer the number of radial grid nodes is increased by over 75% and a high grid expansion factor used. This result is shown on the graph, using the symbol ▲ for the shroud-average value and a vertical line denoting the variation from maximum to minimum values. As can be seen, the numerical results for $Gr < 1 \times 10^7$, are in reasonable accord with the measurements. These two points are obtained for a decreasing disc temperature distribution (similar to Profile B) and nominal shroud temperatures of 75°C and 100°C. The conditions of the highest $Gr$ produce steeper velocity gradients, which are less well resolved by the grid. It is quite likely that agreement will be improved with further refinement of the grid. However, at the time when these predictions were carried out, due to computing time limitations, this was not possible. Also, a higher level of turbulence is expected which will highlight any deficiencies in the mixing length model used for the current predictions. Use of more sophisticated turbulence models, such as the one-equation energy model of Iacovides and Chew (1993) or a low Reynolds number $k$-$\varepsilon$ model (e.g. Morse and Ong (1992)) might improve agreement, although these would necessitate refinement of the finite-volume mesh, which appears to underpin the errors found here. Zonal models may be a sensible compromise.

## 8-5 Adaptive Time-stepping

Adaptive time-stepping is considered for the low *Re* case shown in *Figure 8-3.2b*. *Figure 8-5.1* gives the % deviation of temporal velocity gradients against $t$ for this case (System (a)) and also for the cyclic flow system considered in Chapter 10 (System (b)). The gradients are volume-weighted averages over the complete systems. Frame (a) corresponds to System (a), giving the variation of $\partial(u+v+w)/3\partial t$. For this, unlike System (b) in Frame (b), the average temporal change is quasi constant. System (a) traces are constant because the key unsteadiness is caused by a quasi-constant flow structure rotating at a velocity less than the cavity. Clearly, for such a flow (at the cyclic steady state) adaptive time schemes will give constant step values and hence be in a sense ineffective. However, it is worth considering if the steps returned by the various adaptive schemes are similar to the computationally optimal value of $\Delta t$ =0.05 s. Also, the number of iterations required to advance 1 s in time, for different schemes, is worth examining. This data is given in *Table 8-5.1* for the Altered Order (*An*), Altered Scheme (*ASc*) and Altered Step (*A$\Delta t$*) methods outlined in Chapter 5.

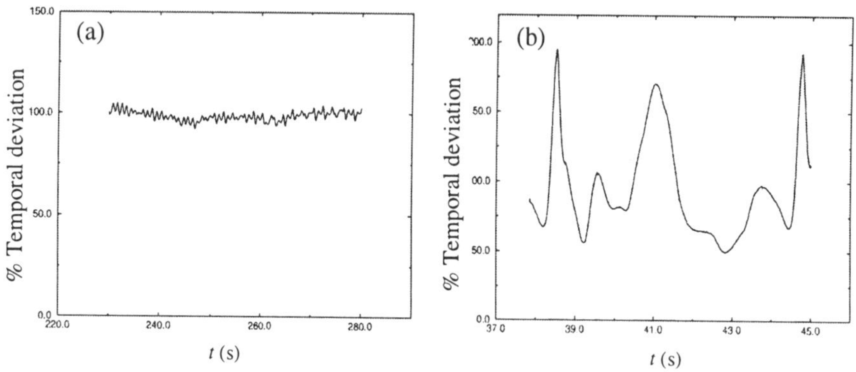


*Figure 8-5.1* Plot of the % deviation of temporal velocity gradients against time for: (a) System (a) and (b) System (b)

As for the rotating flow case in Chapter 5, the *A$\Delta t$* approach is unstable and here diverges. However, encouragingly, the time-steps predicted by the *An* and *ASc* schemes are about a similar magnitude to the optimal $\Delta t$ =0.05 s value. Also, as observed with other cases presented in this text, their values are conservative. However, the penalty of each adapted $\Delta t$ having two solution components is clear (it is worth bearing in mind that the adaptive solutions have greater accuracy and that increasing the specified solution error would improve matters). Despite this, the former observation means that a preliminary design calculation could be carried out with greater confidence than when a fixed step solution is used (to establish any degree of confidence at least two fixed step solutions would be required). In this sense, perhaps real time savings could be made, especially if the flow is evolved using adapted steps.

| Scheme | $\Delta t$ (s) | Iters/$\Delta t$ |
|---|---|---|
| Fixed $\Delta t$ | 0.0500 | 80 |
| An | 0.0225 | 267 |
| ASc | 0.0120 | 250 |
| $A\Delta t$ | NA | $\infty$ |

*Table 8-5.1* Adaptive scheme performances

## 8-6 Conclusions

For economy, preliminary computations to ascertain the optimum time-step (in the time-step independence range) to number of convergence iterations ratio is worthwhile.

The dramatic effect of surface temperature distributions on the computed flow structure was demonstrated. Consistent with experimental observations, unheated shroud predictions gave, for the bulk of the flow, $\omega < \Omega$, where $\omega$ and $\Omega$ are the flow structure and cavity angular velocities, respectively. Computations suggested that when the *shroud is heated* (giving the stongest positive radial temperature gradient) $\omega > \Omega$. This is consistent with the positive radial temperature gradient heated annulus 'jet stream' flow explored in the previous chapter. It was noted that, for the current unsteady flow class the time derivative can be split into two components - the unsteadiness arising just from the flow drift and that due to flow structure vacillations. Potential computational efficiency gains through minimising $|\Omega_g - \omega|$ were discussed.

Higher order temporal schemes generally required nearly 25 % more computing time relative to the fully implicit first order scheme. Due to instability, it was not possible to demonstrate potential improved accuracies for the higher order schemes. For small time-steps the hybrid AVPI/SIMPLEC pressure correction scheme convergence rate was found similar to that for a purely SIMPLEC solution

A prediction suggestive of vortex breakdown was presented along with some URANS heat transfer computations. To reduce computing time, whilst exploring the complex vortex breakdown phenomenon, the grid angular velocity was set at 97.5% of the cavity angular velocity.

Shroud heat transfer predictions for $Gr < 1 \times 10^7$, were in reasonable accord with measurements. For the highest *Gr*, it is quite likely that agreement will be improved with further grid refinement. For these, a higher level of turbulence was expected, which would highlight deficiencies in the mixing length model used. Use of more sophisticated turbulence models, such as the one-equation energy model or a low Reynolds number $k$-$\varepsilon$ model might improve agreement, although these would necessitate refinement of the finite-volume mesh, which appears to underpin the errors found here. Zonal models may be a sensible compromise and the use of these is explored in Chapters 10 and 11.

Adaptive time-stepping was not found to be entirely compatible with the cyclic case considered here.

CHAPTER

# 9

# CYCLIC MOVING BOUNDARY FLOWS

## 9-1 Introduction

In this chapter the prediction of unsteady flow and heat transfer in bearings will be addressed. Mostly hydrodynamic journal bearings are considered. But a final example, illustrating the modelling of pressure distributions in magnetic bearings will also be presented.

Essentially a journal bearing consists of a shaft of radius $a$ and length $L_s$ inside a cylindrical sleeve of inner radius $b$, outer radius $b_0$ and length $L$. The area of shaft inside the sleeve is called the *journal* and the sleeve itself called the *bearing*. Fluid is fed into the space between the bearing inner surface and journal via one or more inlets at a pressure $p_i$ and temperature $T_i$. Here, the velocity components $u$, $v$ and $w$ correspond to the $r$, $\theta$, $z$ directions, respectively. Also, the subscripts $s$, $b$, $l$, $v$ and $a$ are used to refer to shaft, bearing, lubricant, vapour (see later) and ambient conditions, respectively.

The bearing and shaft are not concentric, the distance, $h$, between the journal and inner bearing radius varying with the angular co-ordinate $\theta$. In the area where the shaft is nearest to the bearing the eccentricity creates a wedge shaped fluid volume (see *Figure 9-2.1*). The shaft rotation, $\Omega$, continuously forces this fluid wedge between the journal and bearing, thus creating a high pressure region which prevents mechanical contact. Obviously, for a magnetic bearing, the field induced by poles is used to prevent this contact.

Due to balancing imperfections, in both the above types of bearings, the shaft will always exhibit some kind of orbit. Clearly the resulting boundary movement will give rise to flow unsteadiness. For magnetic bearings, the interaction of aerodynamic forces arising from this unsteadiness, with the field forces is a thing of importance. For journal bearings, the interest here is the thermal consequences of unsteadiness. Generally, shaft orbits will be elliptic, with the ellipse centre displaced some distance $e_0$ from the bearing centre. However, under certain conditions, this may tend to be more circular, with an orbit radius $e_1$ and an orbit angular velocity $\omega$. Here, attention is focused on the circular centred orbit case $(e_o = 0, e_1 > 0)$ where $\Omega$ and $\omega$ have the same direction. This is economical to compute and illustrates a worst design case.

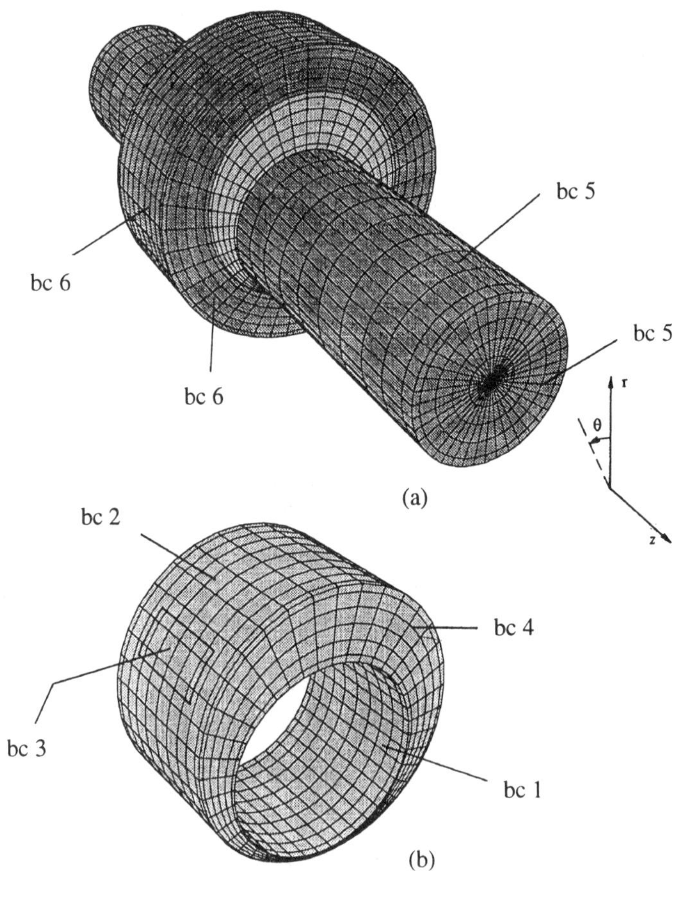


*Figure 9-2.1* Grids used for journal bearing predictions

## 9-2 Numerical Details

Generally, an implicit time scheme is implemented with the HYBRID (Section 3-3) convective term treatment and SIMPLEC (Section 4-4) algorithm. The space conservation law is not explicitly enforced (see Section 3-5). *Figure 9-2.1*, shows typical grids used for the journal bearing predictions in the solid and lubricant regions along with boundary conditions.

*Boundary Conditions*

Basically, heat transfer coefficient boundary conditions are used on all outer solid surfaces. At fluid inlets (bc3) the pressure ($p_i$) and temperature ($T_i$) are specified. At flow outlets (bc4) the normal gradients of all variables, except pressure, are set to zero. Also, the pressure is set to the atmospheric value $p_a$.

*Dynamic Cavitation Modelling*

For journal bearings, significant sub-ambient pressures can occur. These produce a cavitation region. When modelling this, it is reasonable to use averaged cavitated fluid properties, based on the fraction of lubricant and gas/vapour. The fraction of fluid $\Phi(\theta, z, t)$ can be obtained from conservation of mass considerations using the following equation.

$$\frac{\partial}{\partial t}(h\Phi)+\frac{1}{a}\frac{\partial}{\partial \theta}\left(\int_a^{a+h} v\,dr\,\Phi\right)+\frac{\partial}{\partial z}\left(\int_a^{a+h} w\,dr\,\Phi\right)=0 \qquad (9\text{-}2.1)$$

Equation (9-2.1) is solved here for $p < p_c$ (the fluid's cavitation pressure) using the explicit 1$^{st}$ order forward difference Euler scheme. More cavitation modelling details are given in Appendix D.

*General Case Details*

For the journal bearing predictions, to achieve sensible grid independent solutions, $5 \times 40 \times 9$ $(\eta, \theta, z)$ grids are used in the bearing and lubricant regions. Since the radial velocity profile is either nearly linear or parabolic just 5 radial grid nodes are needed. The corresponding shaft grid is $45 \times 40 \times 29$. The $\eta$ spacings are decreased in a geometric manner towards the shaft and inner bearing surfaces, but other grid spacings are uniform. For time-step independence $0.008 \geq \Delta t \geq 0.0004$ s. Typically, 5 iterations are required each time-step to achieve velocity and temperature *rms* (see Section 4-2) errors of below $1 \times 10^{-7}$.

*The Substantial Derivative*

For this case, as with those in Chapters 7 and 8, significant computational effort savings can be made by careful choice of the coordinate system angular velocity $\Omega_g$. It is helpful to consider the substantial derivative, $(\partial\phi/\partial t)_{fp}$, in the form given by Equation (3-2.7). This is restated below

$$\left(\frac{\partial\phi}{\partial t}\right)_{fp}=\left(\frac{\partial\phi}{\partial t}\right)_{g1}+\left(\frac{\partial\phi}{\partial t}\right)_{f}+\left(\mathbf{U}-\mathbf{U}_g\right)\nabla\phi+G_D \tag{9-2.2}$$

where $\mathbf{U}_g=\Omega_g r$ and $G_D$ is a temporal grid distortion term. In the above, the usual time derivative is split into the following two components

$$\frac{\partial\phi}{\partial t}=\left(\frac{\partial\phi}{\partial t}\right)_{g1}+\left(\frac{\partial\phi}{\partial t}\right)_{f} \tag{9-2.3}$$

The first, $(\partial\phi/\partial t)_{g1}$, comes from temporal *grid distortions* arising from the journal orbit (for a circular centered orbit, from the viewpoint of an observer on the shaft this term will be zero) and the second, $(\partial\phi/\partial t)_f$, relates to the remaining sources of unsteadiness. Equation (3-5.4), for the grid distortion term, $G_D$, reduces, for the simple co-ordinate transformation used here ($\eta$ = $(r-a)/h$), to

$$G_D=-\left(\frac{\eta}{h}\right)\left(\frac{\partial h}{\partial t}\right)\frac{\partial\phi}{\partial t} \tag{9-2.4}$$

For a circular centered orbit, minimising $|\Omega_g-\omega|$ will reduce $(\partial\phi/\partial t)_{g1}$. If $\Omega_g=\omega$ there will be no temporal grid distortions and $(\partial\phi/\partial t)_{g1}=0$ (i.e. from the viewpoint of an observer moving with the shaft the grid appears stationary). However, for bearings with fluid inlets as shown in *Figure 9-2.1*, when $\Omega_g\neq 0$ the inlets will spin relative to the grid at $-\Omega_g$ and hence $(\partial\phi/\partial t)_f\neq 0$ (i.e. relative to the observer moving with the journal, the inlet grooves will be moving). However, in the absence of grooves, and hydrodynamic instabilities for $\Omega_g=\omega$ and a circular centered orbit the solution will become completely steady. Therefore, in the solutions presented here this is the preferred coordinate system.

For non-circular orbits there is still the possibility of minimising grid distortion by careful selection of $\Omega_g$. The shaft orbit can be considered as a superposition of harmonics and $\Omega_g$ can be chosen to minimise the dominant harmonic and hence give the potential for improvements in accuracy and computational expense.

## 9-3 Discussion of Results for Journal Bearings

*Analytical Isothermal Case*

First, comparison is made with the Sommerfeld isothermal solution (see Pinkus and Sternlicht (1961)) for the time dependent pressure in an infinitely long bearing with a circular orbiting journal:

$$p = \frac{6\mu_l a^2(\Omega - 2\omega)\varepsilon_1(2+\varepsilon_1 cos\theta)sin\theta}{c^2(2+\varepsilon_1^2)(1+\varepsilon_1 cos\theta)^2} \tag{9-3.1}$$

In the above $c = b - a$, $\varepsilon_o = e_o/c$ and $\varepsilon_1 = e_1/c$. *Figure 9-3.1* compares the analytical and numerical predictions, without cavitation, obtained in the stationary $\Omega_g = 0$ frame for: $a$ = 0.0425 m, $c$ = 0.05 mm, $\rho_l$ = 840 kg/m$^3$, $\mu_l$= 0.066 kg/ms, $\Omega$ = 19.5 rad/s, $\omega$ = 92.7 rad/s, $e_0$ = 0 mm and $e_1$ = 0.025 mm. The position of the maximum in predicted pressure moves with the minimum film thickness, $h_{min}$, position and there is reasonable agreement between the analytical and numerical solutions. With this stationary distorting grid solution, $\Delta t$ should be sufficiently small to allow accurate mapping of the journal orbit. In the rotating $\Omega_g = \omega$ system, the grid will not distort. Therefore, the solution effectively becomes steady, and hence is more economical to compute and potentially more accurate.

*Experimental Isothermal Comparison*

Comparison is now made with the pressure measurements of Sun *et al.* (1993) for dynamic cavitation, $\Omega$ = 0 and an effective (see Sun *et al.*) dynamic eccentricity ratio $\varepsilon_1 \approx 0.26$. *Figure 9-3.2a* shows comparison between predictions, obtained in the stationary $\Omega_g = 0$ system, and measurements for the circumferential centre line pressure variation. The data for this prediction is as follows: $a$ = 0.0425 m, $L$ = 0.0423 m, $c$ = 0.21 mm, $p_c$ = - 1.01 bar, $\rho_l$ = 840 kg/m$^3$, $\rho_v$ = 1.2 kg/m$^3$, $\mu_l$ = 0.049 kg/ms$^2$, $\mu_v$ = 2 × 10$^{-5}$ Ns/m$^2$, $\Omega$ = 0 rad/s, $\omega$ = 250 rad/s, $e_0$ = 0 mm and $e_1$ = 0.055 mm. Considering the large experimental errors, there is reasonable agreement between the measurements and prediction. For completeness, *Figure 9-3.2b*

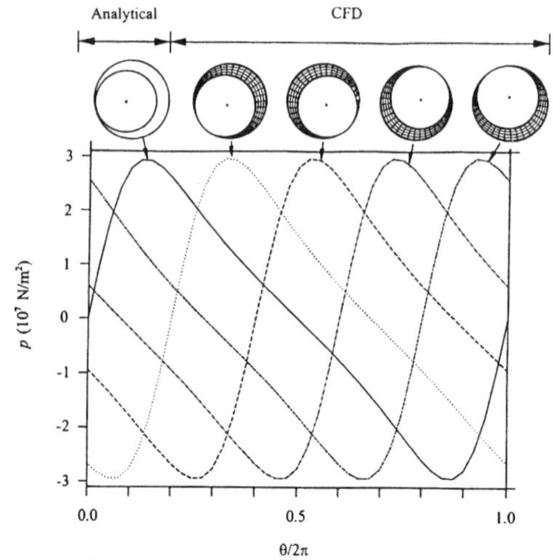


*Figure 9-3.1* Comparison between isothermal predictions and analytical solution

shows a map of the predicted instantaneous pressure distributions adjacent to the bearing surface in the $\theta - z$ plane. Again this prediction would become steady and hence more economical to computed if $\Omega_g = \omega$.

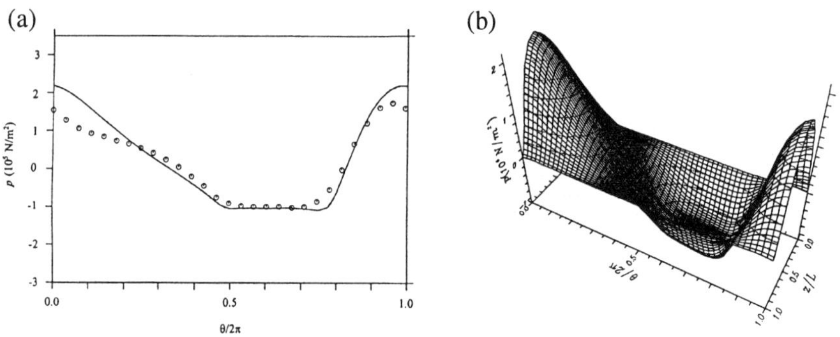


*Figure 9-3.2* Isothermal dynamic bearing pressure including cavitation: (a) comparison with measurements ( o measurements, — predictions) and (b) $\theta$–$z$ plane pressure map

The thermal effects of a non-orbiting and orbiting journal are now considered for the geometry of Lund and Tonnesen (1984). Data for this $\Omega = 367\, rad/s$ case is given at the end of Appendix D (*Table D.1*). The bearing has two diametrically opposed fluid inlets of length $L_i$ = 0.0275 m and angular extent $\theta_i$ = 9°. Considering *Figure 9-3.3*, it can be seen that if

$\omega = \Omega$ shaft point H will always be nearest to the bearing and hence be exposed to the greatest frictional heating. Point C will experience the least and so a temperature differential will develop across the shaft. This will cause bending of the shaft and bearing failure (see de Jongh and Morton (1994)).

When $\omega = 0$ all points on the shaft experience the maximum heating effect for the same short time period. Therefore, shaft temporal temperature variations are highly concentrated at the shaft surface. Consequently, the $\omega = 0$ solution can be considered steady.

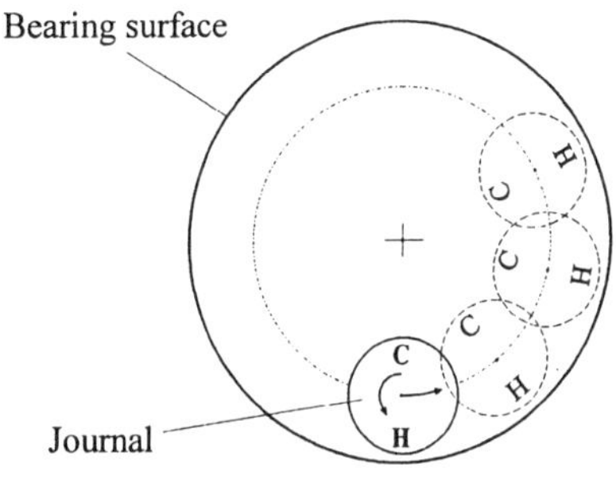


*Figure 9-3.3* Schematic of mechanism for journal differential heating

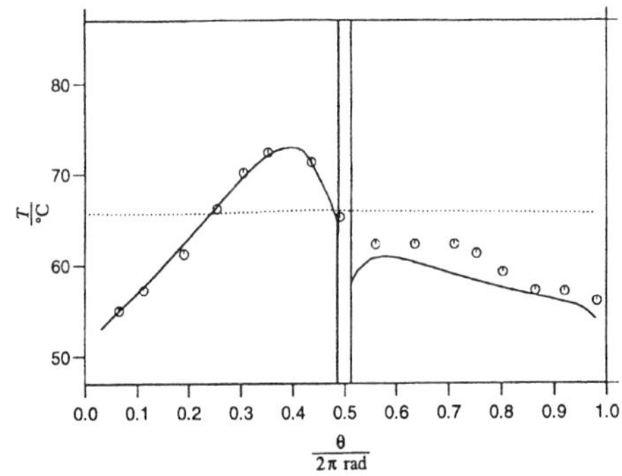


*Figure 9-3.4* Comparison of predicted and measured temperatures for $\omega = 0$

*Figure 9-3.4*, compares predictions with the data of Lund & Tonnesen for $\omega = 0$. This particular computation is made by applying the exponential convective term treatment of Patankar (1980) in the rotating shaft region. However, the solution is most accurate (numerical precision and dissipation errors are minimised) when sliding grids are used with the $\Omega_g = \omega$ co-ordinate system being used for the shaft (the convective term in Equation (9-2.2) then vanishes) and the $\Omega_g = 0$ system used elsewhere. A computation made using this technique is shown later.

The thermal time constants associated with the lubricant, shaft and bearing (see Tucker and Keogh (1996)) are $\rho_l c_{vl} h^2/k_l$, $\rho_s c_{vs} h^2/k_s$ and $\rho_b c_{vb} a^2/k_b$, respectively. Clearly, the time constant for the fluid is orders of magnitude lower than that for the bearing a shaft. Hence, for the orbiting shaft case, the problem is numerically stiff and would require many of thousands of time-steps to integrate to the steady state. However, when solved in the $\Omega_g = \omega$ frame the solution becomes quasi steady, the only unsteadiness being due to the fluid inlets which appear to rotate at an angular velocity of - ω. This inlet movement would cause minor temperature fluctuations. These are removed here by linearly interpolating the heat flow between the leading and trailing edges of each inlet.

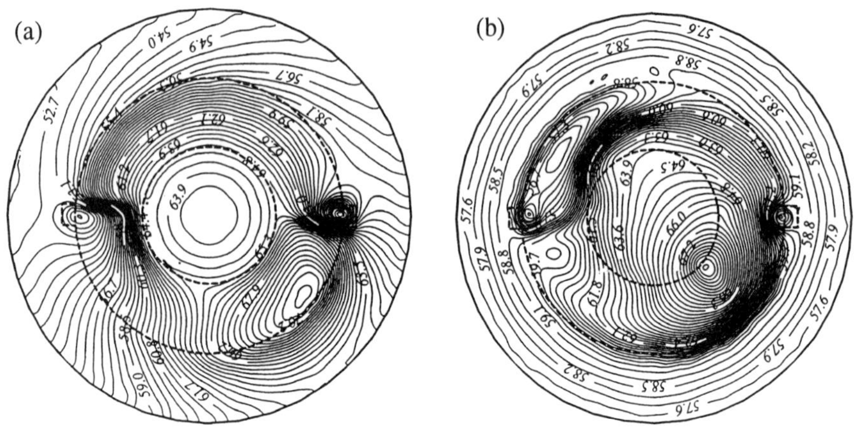


*Figure 9-3.5* Temperature contours through the half mid-axial $\eta$ - $\theta$ plane for (a) no orbit, (b) circular centred orbit.

Therefore, it is possible to use large time-steps to predict the shaft and bearing temperature distributions and small steps for the lubricant, thus removing the numerical stiffness. Note, that to model the inlets accurately (avoiding interpolation errors) time-steps must be chosen so that at each step the inlet can be correctly mapped to the grid. *Figure 9-3.5a*, shows temperature contours for the conditions studied by Lund & Tonnesen (note, for clarity the fluid region has been greatly scaled and displaced). For this, sliding grids are used, the shaft being solved in a rotating frame (thus minimising numerical precision and dissipation errors) of reference and the lubricant and bearing with $\Omega_g = 0$. Correctly, the shaft temperature does not vary circumferentially. In *Figure 9-3.5b* the shaft is allowed to move off from its static position, in a circular centred orbit. As can be seen, a significant quasi-steady shaft temperature differential has now developed. It should be noted that, in *Figure 9-3.5b*, $\varepsilon_o = 0$, $\varepsilon_1 = 0.58$, $\omega = \Omega$ and the shaft external surface is made adiabatic.

## 9-4 Discussion of Results for Magnetic Bearing

As a final example, flow in a magnetic bearing with $L$ = 0.055 m, shown in *Figure 9-4.1*, having four equally spaced pole faces at a radius $b$ = 0.055 m, supporting a shaft with $a$ = 0.051 m and $\omega = \Omega$ = 100 rad/s, is considered. The shaft has a circular centred orbit with $e_o = 0$ and $e_I = (b - a)/2$.

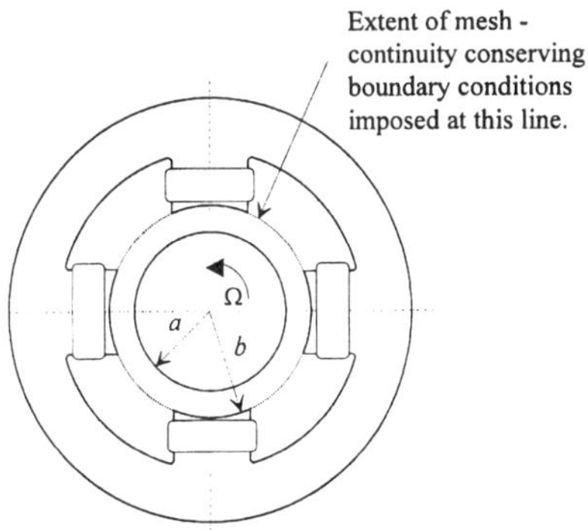


*Figure 9-4.1* Schematic diagram of magnetic bearing assembly

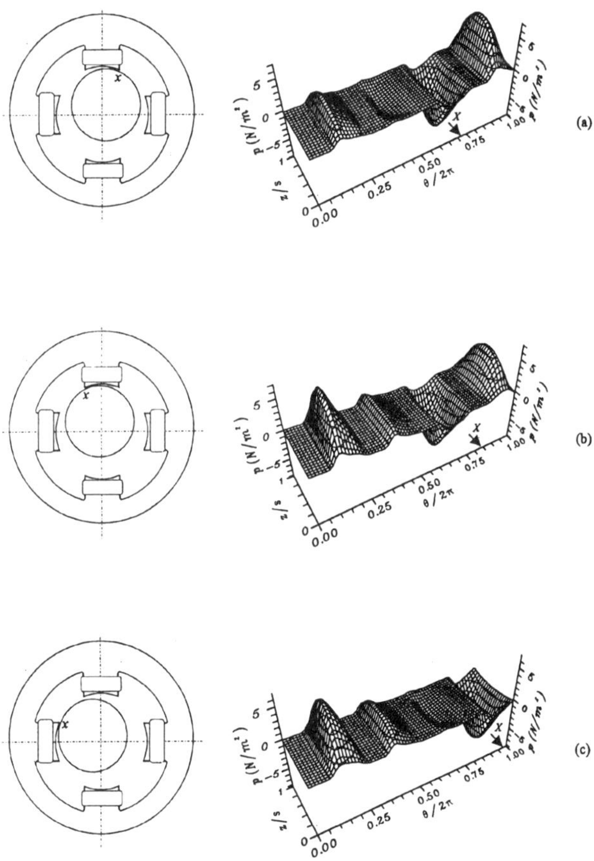


*Figure 9-4.2* Variation with time of air pressure between a magnetic bearing and shaft

To reduce computing time the flow away from the poles, outside the dashed circle $(r > b)$ shown in *Figure 9-4.1*, is not calculated. Sensible boundary conditions must be applied along the boundary at $r = b$. At this, the pressure is assumed to be atmospheric and continuity conserving boundary conditions imposed. Similar boundary conditions are used at the sides of the bearing. The solution is made on a $11 \times 40 \times 21$ $(\eta, \theta, z)$ grid with $\Delta t$ = 0.001 s and $\Omega_g = 0$.

*Figure 9-4.2* shows the variation of pressure in the $z$ - $\theta$ plane at $r = b$ ($\eta=1$, the plane coincident with the poles of the magnetic bearing) when the shaft is in three different positions. The point $x$ is used to indicate where the shaft is closest to the surface of the plane. In frames (a) and (c) point $x$ is at the leading edge of a pole face and so the pressure distributions are the same but shifted through 90

The time period between frames is approximately 0.008 s and so the pressure variations can be seen to be sudden but, due to the low angular velocity, relatively small (< 7 N/m$^2$). Again, choosing $\Omega_g = \omega$ would prevent temporal distortion of the grid and so is to be the preferred choice. The poles faces would then again effectively rotate at $-\omega$.

## 9-5 Conclusions

Several orbiting bearing cases have been presented. These give rise to moving boundary problems. For cases where the shaft orbits, solutions can be made steadier (or for some configurations completely steady) by careful choice of the coordinate system. This approach should increase predictive accuracy by removing temporally related grid distortion terms along with other time derivative terms. Coordinate system selection can also dramatically reduce numerical stiffness enabling solutions that would otherwise be impractical.

# CHAPTER 10

# CYCLIC COMPLEX SYSTEM FLOW

## 10-1 Introduction

In this chapter the prediction of cyclic turbulent flow for the complex electronic system shown in *Figure 10-1.1* is considered. The geometry consists of three main hollow block like components defined as regions 1, 2 and 3. Regions 1 and 2 have fans attached to them. A key feature of this chapter is that the predictive behaviour of a wide range of turbulence models is compared when computing a complex system cyclic flow. Some of the adaptive time-step approaches described in Chapter 5 are also further tested.

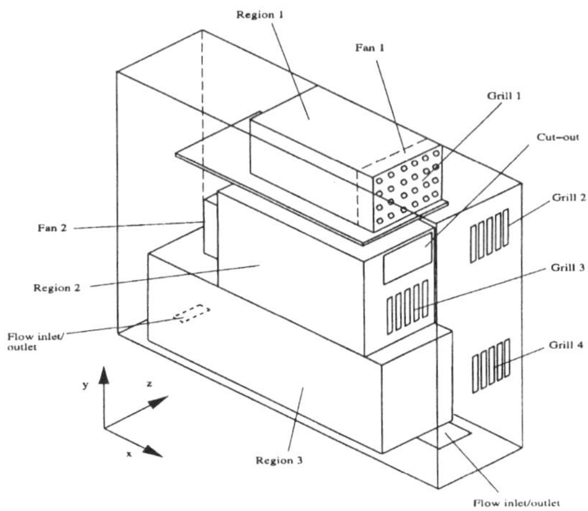


*Figure 10-1.1* Schematic of modelled geometry

Some cyclic, laminar flow, electronics cooling related CFD work is reviewed in Chapter 1. Cyclic flow, URANS predictions, for relatively simple geometries (mostly flow over rectangular cylinders), are reviewed in Section 6-2.

To assess the accuracy of numerical models, comparison is made with Pitot static tube, hot wire anemometry and LDA data. Although the flow is unsteady, this component has not been removed from the turbulence intensity data. Instead, predictions are corrected by adding the effect of the unsteadiness component, thus enabling direct comparisons.

The models tested are summarised below:

(I) standard Van Driest damping function based mixing length model ($ml_1$);
(II) the single equation law of the wall based ($l$-$v$) model;
(III) a high Reynolds number $k$-$l$ model ($k$-$l_1$);
(IV) the low Reynolds number $k$-$l$ model ($k$-$l_2$);
(V) the standard $k$-$\varepsilon$ model ($k$-$\varepsilon_1$);
(VI) a zonal turbulence technique with the $k$-$l_2$ model ($k$-$\varepsilon$/$k$-$l_2$);
(VII) a zonal turbulence technique with the $ml_1$ model ($k$-$\varepsilon$/$ml_1$);
(VIII) the above with a more refined damping function in the $ml_1$ model ($k$-$\varepsilon$/$ml_2$);
(IX) a non-linear high Reynolds number $k$-$l$ model ($nl$-$k$-$l_1$);
(X) a non-linear low Reynolds number $k$-$l$ model ($nl$-$k$-$l_2$);
(XI) a zonal model linear at walls but non-linear away from walls ($nl$-$k$-$\varepsilon$/$k$-$l_2$) and
(XII) a fully non-linear variant of the above ($nl$-$k$-$\varepsilon$/$nl$-$k$-$l_2$).

The phase-averaged form of the above models is outlined in Chapter 6, Section 6-5. The influence of unsteadiness on particle transport is considered at the end of the present chapter, where a transient scalar transport comparison is also made with flow visualisation data.

## 10-2 Numerical Details

Numerical features of the program used are given in Chapter 4, Section 4-6. Generally, the second order central difference based CONDIF (Section 3-3) convective term treatment is used. The pressure field is computed using the SIMPLE (Section 4-4) method. For unsteady predictions the fully implicit backwards Euler time scheme is mostly used. For $k$-$l_1$ solutions, the Crank-Nicolson scheme (this is used to give confidence that solutions are time-step independent) and three adaptive time-stepping strategies discussed in Chapter 5 are tried. The three adaptive strategies are: (I) Adaptation with a forward/backward Euler error estimate ($ASc$); (II) Adaptation with an altered step error estimate ($A\Delta t$) and (III) Adaptation with an altered order ($An$) error estimate.

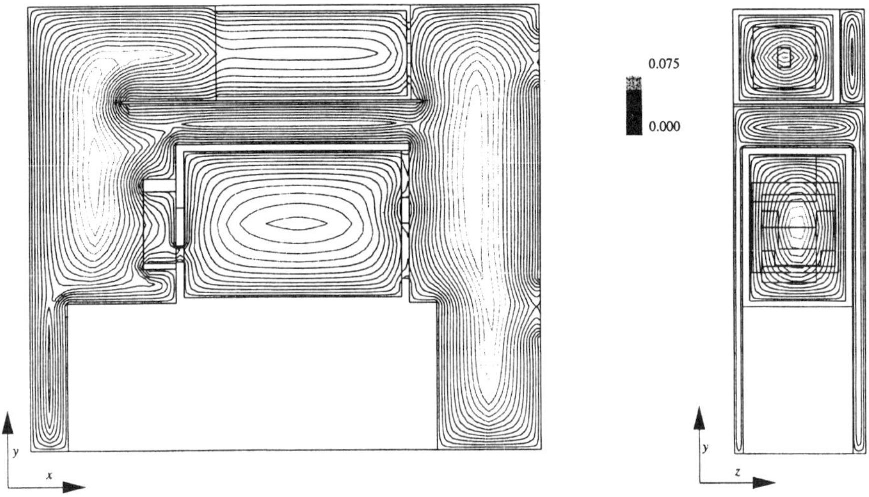


*Figure 10-2.1* Wall distance contours around mid $x$-$y$ and $z$-$y$ planes

*Wall Distance Algorithm*
Due to the complicated geometry, the elliptic wall distance algorithm, outlined in Section 6-5, is used here. *Figure 10-2.1*, shows wall distance contours around the mid $x$-$y$ and $z$-$y$ planes. These are predicted using equations (6-5.36) (subject to the boundary conditions described in the next section) and (6-5.37). Correctly, contours are parallel to solid walls. Also, at corners, distance variations are smooth. This can be beneficial to convergence. In the important regions, close to surfaces and away from corners, distances are accurate to within 1 %.

*Boundary Conditions and Solution Initialization*
At inflow boundaries the total pressure is fixed, the normal velocity set to conserve mass and the remaining velocity components made zero. Flow outlet and most other boundary conditions are as indicated in Section 4-6. Prior to the solution starting it is not known which are inflow and outflow boundaries. Therefore, these boundary conditions are set automatically depending on the flow direction at each iteration. The slotted grills 1 to 4 are modelled using loss coefficients and fans 1 and 2 are modelled using quadratic momentum sources. For the wall distance algorithm, in Equation (6-5.36) at solid walls the dependent variable is set to zero and at flow

inlets/outlets its gradient set to zero. Tucker (2000) gives fuller boundary condition details.

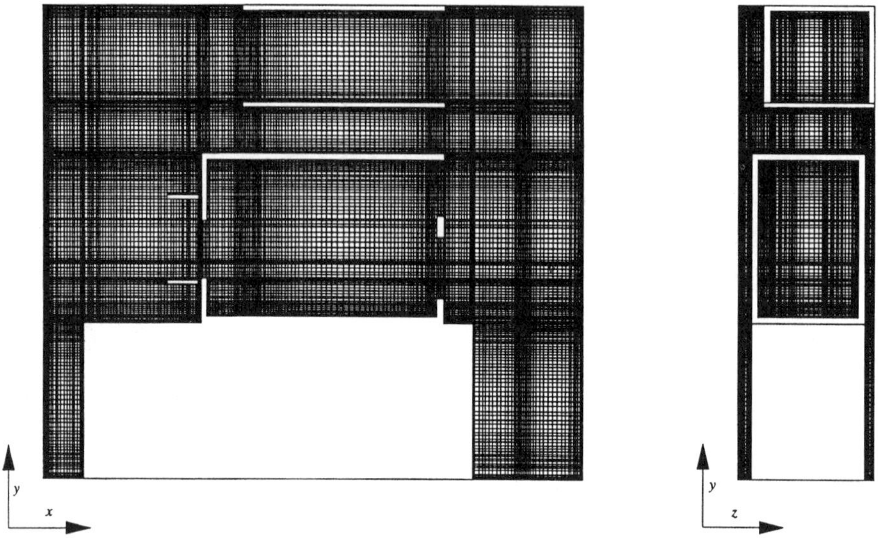


*Figure 10-2.2* Typical form of grid used for low Reynolds number predictions

For initial conditions, at $t$ = 0, in fluid regions, solutions from previous predictions are sometimes used. When a previous prediction is not used, interior velocities and the hydrostatically reduced gauge pressure are set to zero. However, $k$, $\varepsilon$ and $\mu_t > 0$, values are set based on a high (for stability reasons) uniform initial turbulence intensity distribution. For particle transport integrations, initial particle velocities are assumed zero. For scalar transport solutions, when $t$ = 0, except at the scalar source, $C$ = 0, everywhere.

*Structure of Grids*

Low Reynolds number turbulence model predictions are made using a base grid with at least 105 x 97 x 51 non-uniform control volumes in the $x$, $y$ and $z$ directions, respectively. For high Reynolds number model predictions a mostly uniform 101 x 89 x 45 grid is used. To study grid independence a 209 x 193 x 101 grid is implemented. The typical form of the grids used for low Reynolds number predictions is illustrated, for the finest 209 x 193 x 101 grid, in *Figure 10-2.2*.

The views shown are at $Z$ = 0.5 and $X$ = 0.5, $Z$ and $X$ being dimensionless distances between 0 and 1 in the $z$ and $x$ directions, respectively. For the zonal techniques, the $k$-$\varepsilon$ model is only solved when $y^+ \geq 60$. *Figure 10-2.3*, illustrates, using a coarser grid, node points with $y^+ \geq 60$

at $Z = 0.5$ and $X = 0.5$. For the grids described, at first off-wall nodes $y^{+}_{ave}$ is around 2 and 15 for the low and high Reynolds number turbulence models, respectively.

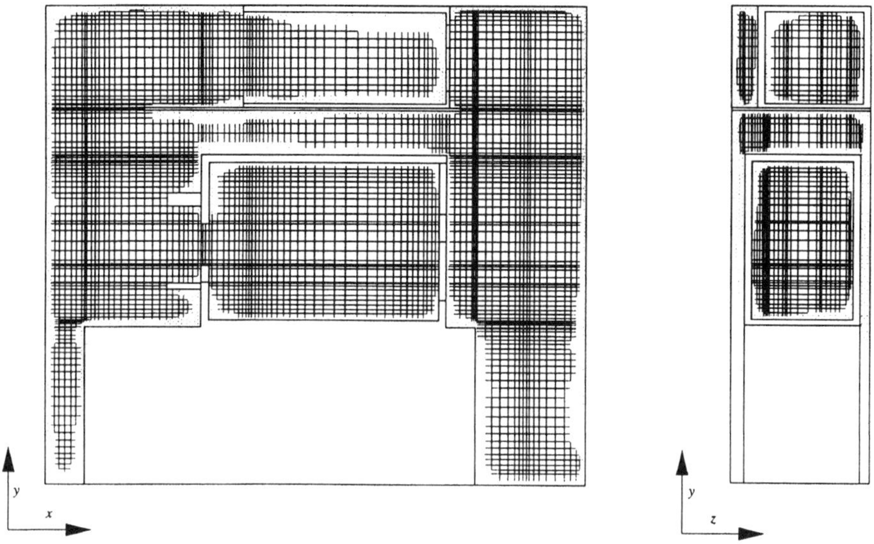


*Figure 10.2-3* Diagram indicating nodes where two equation model is used in zonal solutions

*General Case Details*

For unsteady predictions (including those for scalar and particle transport) sensibly time-step independent predictions are gained when $\Delta t = 0.001$ s.

Measurements of turbulence intensity to be compared with have not been corrected for the effect of flow unsteadiness. These total intensities are denoted here as $T_i'$ values. Comparisons are made with predictions of $T_i$ (where the periodic fluctuation is filtered out) and $T_i'$. In this way the contribution of the predicted unsteadiness component to the total intensity $T_i'$ is explored.

For profiles, where predictions and measurements are compared, percentage root mean square errors are given. For each profile the following summation is made

$$Error = \sqrt{\frac{\sum_{\exp} (\phi_{\exp} - \phi_{num})^2}{\sum_{\exp} \phi_{\exp}^2}} \qquad (10\text{-}2.1)$$

where $\phi_{exp}$ corresponds to an experimental data point and $\phi_{num}$ a numerical value. At points where numerical data points do not coincide exactly with measurements (this is mostly the case) a Bessel (stiff quadratic spline) interpolation is used.

The influence of unsteadiness on particle transport is also examined. The particle tracking techniques, described in Sections 2-11 and 6-5, are implemented.

## 10-3 Discussion of Results for Fluid Flow

*Spatial Velocity Variations*

Comparisons are made with measurements along the lines/profiles shown in *Figure 10-3.1*. The exact locations of profiles 1-6 are: $X$=0.53, $Y$=0.73; $X$=0.37, $Y$=0.73; $X$=0.41, $Z$=0.06; $X$=0.41, $Z$=0.10; $X$=0.37, $Z$=0.57 and $X$=0.41, $Z$=0.96, respectively.

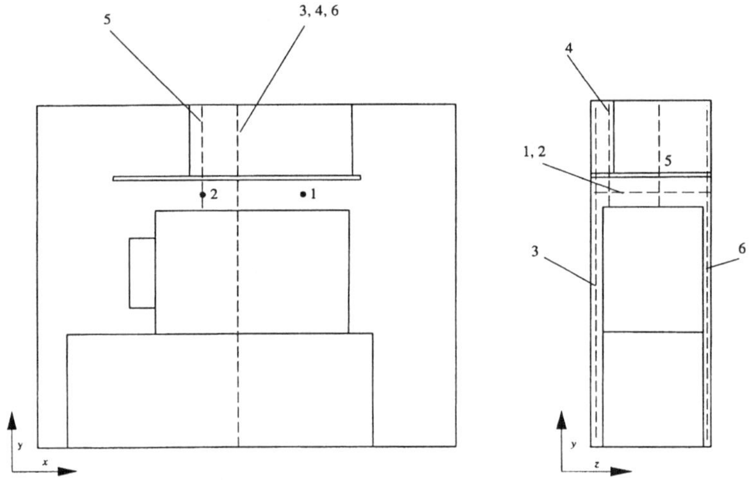


*Figure 10-3.1* Schematic diagram showing positions for velocity and turbulence intensity profiles

*Figure 10-3.2*, compares the zonal $k$-$\varepsilon$/$k$-$l_2$ model velocity predictions with time averaged measurements. The LDA measurements, represented by the circle symbols, have an estimated accuracy of ± 5%. For interest Pitot-static tube measurements are also shown using the + symbols. Velocities are normalized by the approximate average axial velocity at the two fans ($U_o \approx$ 4.5 m/s). The line represents the predictions. The percentage error for each profile and model is summarized in *Table 10-3.1*. The table shows the average error for all models is similar, ranging between 14 and 22%. For velocities, the $k$-$\varepsilon$ model is the least accurate and the linear zonal models the most accurate. Of note is the fact that the non-linear models give no accuracy improvement.

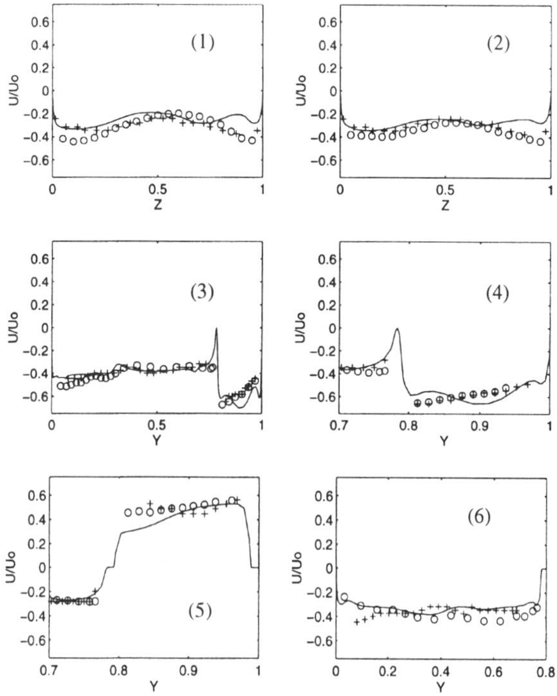


*Figure 10-3.2* Comparison of the zonal $k\text{-}\varepsilon/k\text{-}l_2$ model velocity predictions with measurements (o LDA, + Pitot-static tube, —— Predictions)

| Model | 1 | 2 | 3 | 4 | 5 | 6 | Average |
|---|---|---|---|---|---|---|---|
| *(I) $ml_1$* | 32 | 25 | 13 | 15 | 6 | 18 | 18 |
| *(II) l-v* | 27 | 18 | 13 | 18 | 12 | 21 | 18 |
| *(III) $k\text{-}l_1$* | 27 | 23 | 13 | 14 | 20 | 19 | 19 |
| *(IV) $k\text{-}l_2$* | 26 | 22 | 14 | 15 | 7 | 15 | 16 |
| *(V) $k\text{-}\varepsilon$* | 24 | 20 | 16 | 14 | 22 | 38 | 22 |
| *(VI) $k\text{-}\varepsilon/k\text{-}l_2$* | 24 | 19 | 10 | 12 | 9 | 13 | 14 |
| *(VII) $k\text{-}\varepsilon/ml_1$* | 20 | 20 | 12 | 25 | 12 | 12 | 16 |
| *(VIII) $k\text{-}\varepsilon/ml_2$* | 17 | 14 | 11 | 23 | 8 | 11 | 14 |
| *(IX) $nl\text{-}k\text{-}l_1$* | 28 | 19 | 24 | 17 | 14 | 18 | 20 |
| *(X) $nl\text{-}k\text{-}l_2$* | 29 | 23 | 16 | 20 | 16 | 19 | 20 |
| *(XI)$nl\text{-}k\text{-}\varepsilon/k\text{-}l_2$* | 20 | 19 | 13 | 17 | 11 | 15 | 16 |
| *(XII)$nl\text{-}k\text{-}\varepsilon/nl\text{-}k\text{-}l_2$* | 19 | 18 | 11 | 20 | 15 | 11 | 16 |

*Table 10-3.1* Summary of percentage errors in velocity

*Figure 10-3.3*, shows mid-axis *x-y* plane, *k-*$l_l$ model, instantaneous streamline and axial velocity ($\partial U/\partial t$) contour plots at approximately $t^* \cong$ 0.125, 0.25, 0.5, 0.625 and 0.875 corresponding to frames (a), (b), (c), (d) and (e), respectively (note the dimensionless time is expressed as a fraction of one complete unsteadiness cycle). The contour plots clearly show the major flow unsteadiness if focused around $Y > 0.75$ and $t^* \cong 0.25$. The latter suggests the present computation would benefit from adaptive time-stepping, where away from the $t^* \cong 0.25$ temporal region, the time-step is automatically increased. The application of adaptive time-stepping is discussed later. The streamline plots suggest the major unsteadiness source is separation arising from Fan 1 rapidly turning the flow through 180°. This, at $t^* \cong 0.125$ (Frame (a)) causes massive *x-y* plane separation. At $t^* \cong 0.25$ (Frame (b)) the vortex arising from separation collapses. By Frame (e), the vortex is re-established. The collapse and regeneration of this vortex repeats exactly in time. A similar but less intense behaviour can also be viewed in *x-z* plane at around $Y \approx 0.9$ (this *y* location corresponds to the centre-line of Fan 1). The dominant *x-y* plane separation and reattachment cycle causes a temporal modulation of the system mass throughflow (see Tucker and Pan (2000)).

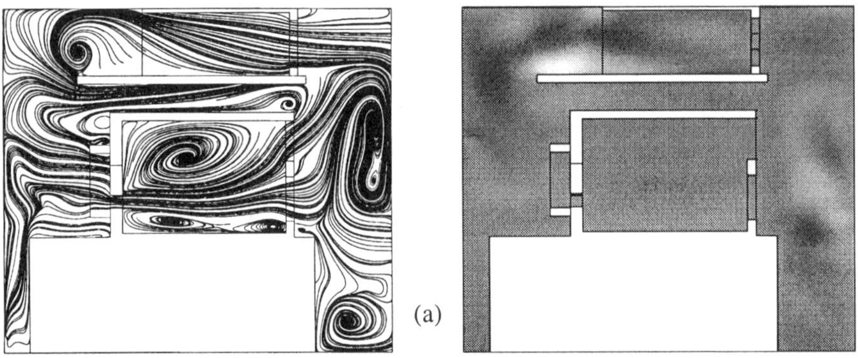

(a)

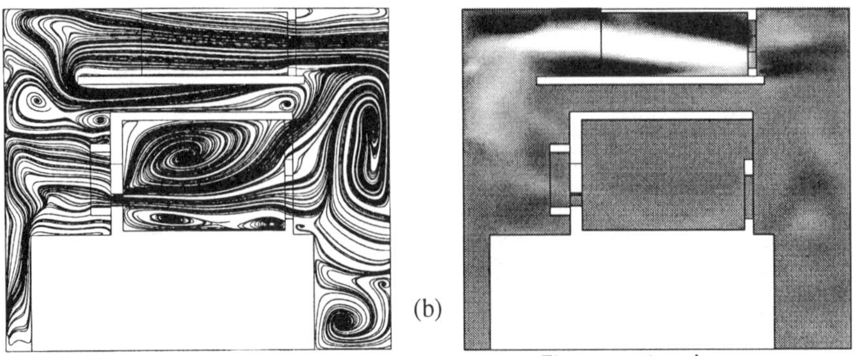

(b)

*Figure continued on next page...*

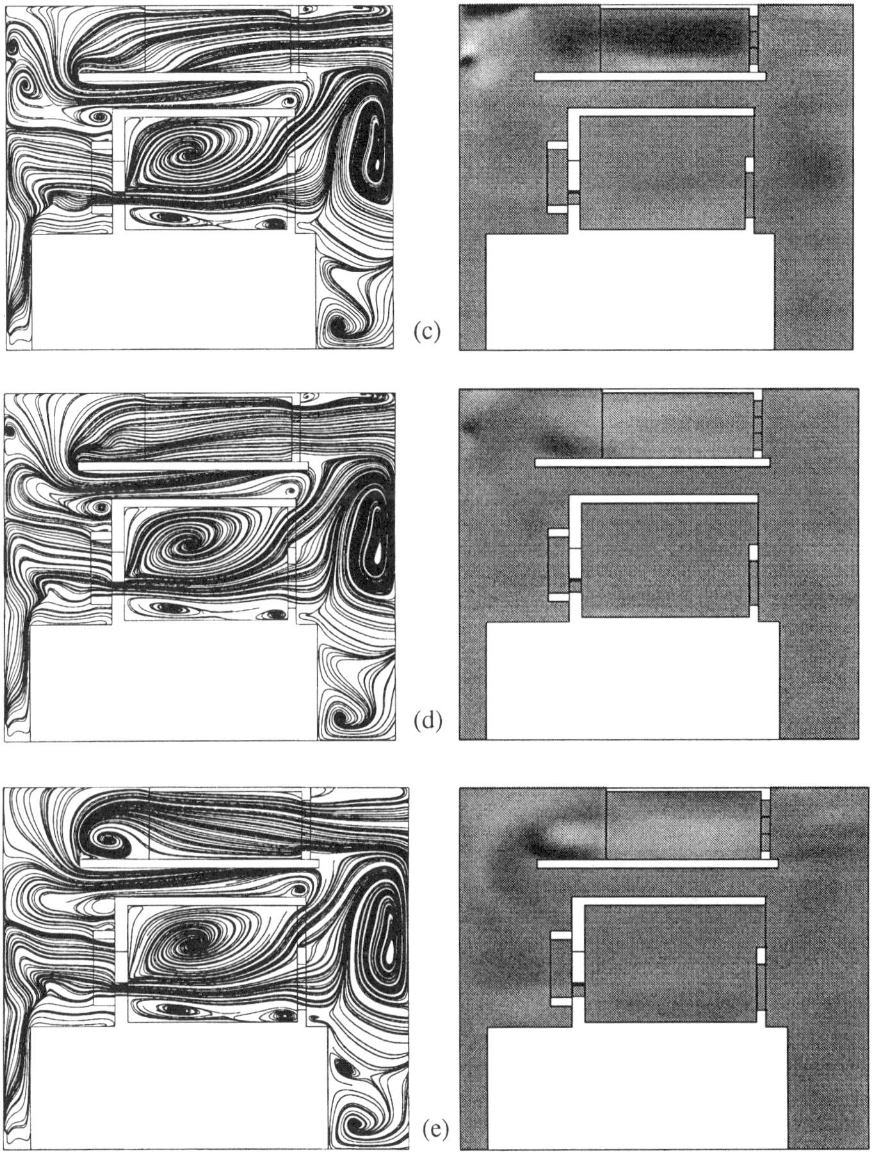


*Figure 10-3.3* Plots of mid *x-y* plane streamlines and horizontal velocity component unsteadiness at: (a) $t^* \cong 0.125$, (b) $t^* \cong 0.25$, (c) $t^* \cong 0.5$, (d) $t^* \cong 0.625$ and (e) $t^* \cong 0.875$

*Temporal Velocity Variations*

Temporal velocity variations are considered for six points on profiles 1-6. Points 1 and 2 are located at $Z = 0.55$, and points 3-6 are at $Y = 0.30$, 0.91, 0.91 and 0.30, respectively. Measurements show there are two dominant unsteady flow component frequencies of about 0.5 and 0.75 Hz. The average amplitude for these is about 0.25 m/s. *Figure 10-3.4*, is a plot of averaged, predicted amplitude against turbulent viscosity for the different turbulence models. Open symbols represent linear turbulence model predictions, closed non-linear models and the full and dotted lines respective best fits to these. The horizontal dashed line gives the measured average amplitude. As can be seen, linear model amplitudes vary significantly, decreasing with increasing $\mu_t$. This has some consistency with the observation of Brackenridge (1960). Zero amplitude linear model points correspond to the $k$-$\varepsilon$ and $k$-$\varepsilon$/$k$-$l_2$ models (However, also $k$-$l_1$ predictions with 1st order upwinding give steady flow and this illustrates the danger of using lower order schemes). As will be seen later, for the former, $k$ and $\mu_t$ are overestimated. The linear model predicted dominant wave frequencies vary significantly (0 - 3.33 Hz), generally decreasing with $\mu_t$. For the non-linear models, some momentum equation, Reynolds stress components are not incorporated through diffusion terms and the trend of decreasing amplitude with increasing turbulent viscosity is no longer observed. The non-linear solutions are steady or quasi-steady with the $Y > 0.75$ flow either fixed in a separated or attached mode, the attached being the preferred.

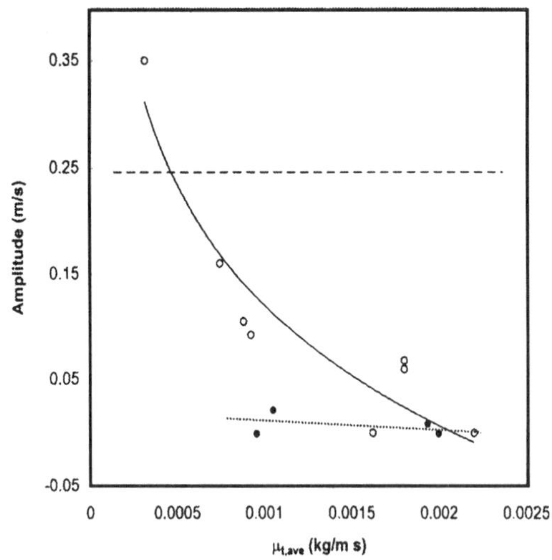


*Figure 10-3.4* Plot of amplitude against turbulent viscosity (o linear models, • non-linear models, —— linear model best fit, ······ non-linear model best fit, - - - - - measured)

*Spatial Turbulence Intensity Variation*

*Figure 10-3.5*, compares total turbulence intensity ($T_i'$) predictions with measurements for profiles 1 to 6. The circle symbols represent the LDA

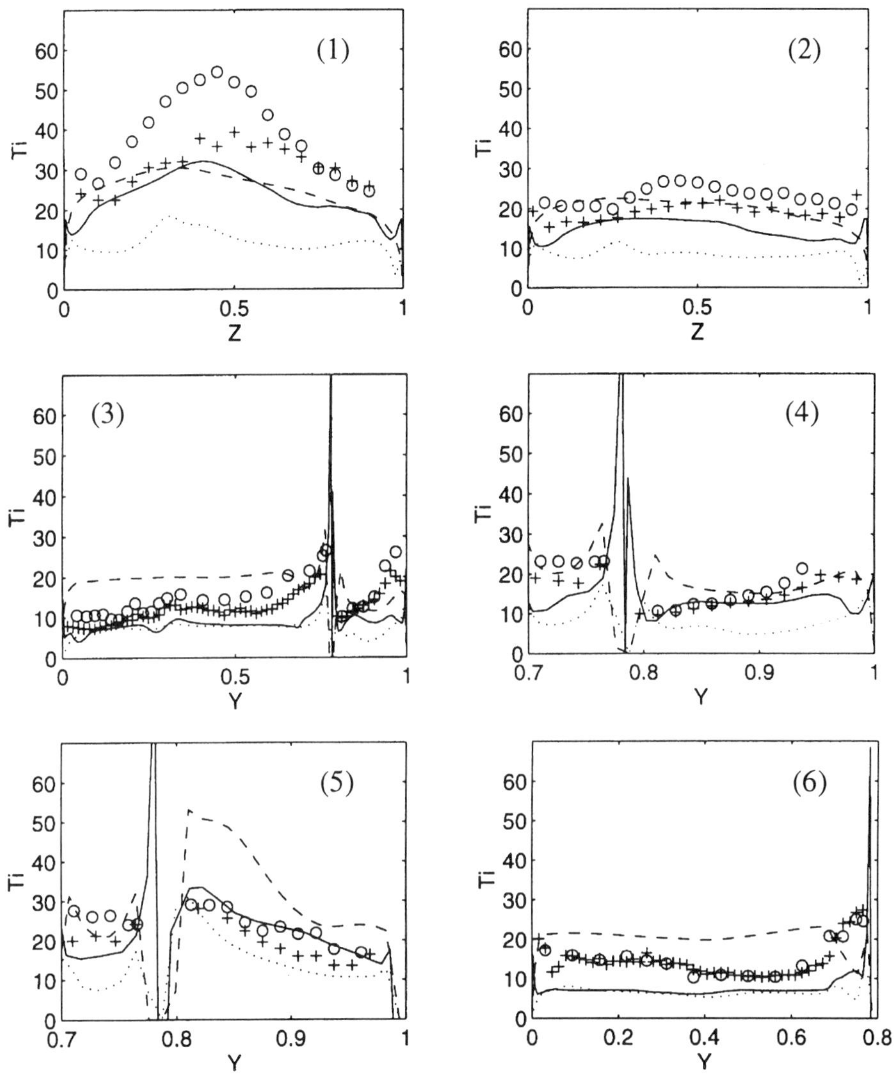


*Figure 10-3.5* Comparison of total intensity predictions with measurement (o LDA Measurements, + Hot wire measurements, —— $k$-$\varepsilon$/$k$-$l_2$, ······· $nl$-$k$-$l_1$, -------- $k$-$\varepsilon$ )

measurements. For interest, (Hot Wire) HW measurements are also presented using the + symbols. The dotted, full and dashed lines represent results for the $nl$-$k$-$l_1$, $k$-$\varepsilon$/$k$-$l_2$ and $k$-$\varepsilon$ models, respectively. The $nl$-$k$-$l_1$ and $k$-$\varepsilon$ computations give the extremes in turbulence intensity predictions for the different models. The significant errors arising when using a HW to measure high intensities are clearly illustrated. It can also be seen that, unlike for velocities, the models give significantly different results. The $nl$-$k$-$l_l$ intensity values are significantly lower than those for the $k$-$\varepsilon$ model. The zonal predictions tend to lay between the above two extremes. Percentage total intensity ($T_i'$) errors for all models are summarized in *Table 10-3.2*.

| Model | 1 | 2 | 3 | 4 | 5 | 6 | Average |
|---|---|---|---|---|---|---|---|
| *(III) $k$-$l_l$* | -66 | -59 | -43 | -43 | -30 | -53 | -49 (-52) |
| *(IV) $k$-$l_2$* | -56 | -50 | -43 | -50 | -24 | -47 | -45 (-48) |
| *(V) $k$-$\varepsilon$* | -33 | -15 | 46 | 27 | 47 | 47 | 35 (35) |
| *(VI) $k$-$\varepsilon$/$k$-$l_2$* | -37 | -36 | -43 | -22 | -12 | -51 | -33 (-33) |
| *(VII) $k$-$\varepsilon$/$ml_l$* | -26 | -24 | - | 35 | -18 | - | -25 (-26) |
| *(VIII) $k$-$\varepsilon$/$ml_2$* | -27 | -25 | - | -34 | -15 | - | -25 (-26) |
| *(IX) $nl$-$k$-$l_l$* | -68 | -62 | -50 | -55 | -37 | -57 | -54 (-54) |
| *(X) $nl$-$k$-$l_2$* | -56 | -52 | -46 | -47 | -29 | -51 | -47(-48) |
| *(XI)$nl$-$k$-$\varepsilon$/$k$-$l_2$* | -33 | -31 | -41 | -22 | -17 | -48 | -32(-32) |
| *(XII)$nl$-$k$-$\varepsilon$/$n$-$lk$-$l_2$* | -33 | -31 | -38 | -21 | -25 | -46 | -32(-32) |

*Table 10-3.2* Summary of total ($T_i'$) percentage error in turbulence intensity

For the $k$-$\varepsilon$/$ml_l$ and $k$-$\varepsilon$/$ml_2$ models blanks in table entries correspond to where only the *ml* models are applied. The far right hand column terms, in parenthesis, represent average $T_i$ values. The table shows the $k$-$\varepsilon$ model tends to over predict turbulence intensities. This is consistent with the observations of Franke and Rodi (1993), where, like here, the excessively dissipative model leads to a steady solution. Tests show that in the present work, wall functions are a significant cause of the $k$-$\varepsilon$ model's intensity over prediction. When first off wall grid node positions become close to walls (which is difficult to avoid everywhere for low Reynolds number flows) intensities increase significantly (the $k$-$l_l$ model shows far less sensitivity in this respect). As in the work of Franki and Rodi, the zonal models, relative to the standard $k$-$\varepsilon$ model, give reduced turbulence intensities. However, the tendency for these zonal results is to give intensities that are too low. Disappointingly, the inclusion of non-linear terms does not really benefit the predictive accuracy for any of the models. Also, all the $k$-$l$ models tend to under predict intensities. Contrary to measurements, the bracketed terms in *Table 10-3.2* indicate $q''$ makes little contribution to $T_i'$.

## 10-4 Adaptive Time-stepping

Performances are compared to $k$-$l_l$ results for the $\Delta t = 0.001$ s as used in the above predictions. For quadratic source terms (Wilcox (1998)) and Neumann boundary condition (Fletcher (1997)), of which there are many for this case (they are used to model both fans and grills), Crank-Nicolson scheme stability is not guaranteed. Therefore, to control any error growth, Crank-Nicolson is just used in the error evaluation. The backwards Euler solution is used for the next *An* step. Also, for stability $\alpha = 2.5$ in Equation (2-4.30). A further measure, motivated by Crank-Nicolson scheme concerns is limiting time-steps so that Courant numbers are below 5. The temporal error estimate (Equation (5-2.16)) is based on the key variable of interest here, which is the fluid axial velocity $u$. Due to computational expense the strategies tested for this case are limited.

As noted before, *Figure 10-3.3* contour plots show most unsteadiness is focused around $Y > 0.75$. Unlike, the cyclic Chapter 7 and 8 cases, there is a significant temporal variation in the volume average of $\partial u/\partial t$. This fact, along with the *Figure 10-3.3* unsteadiness contours suggest, for this case, there is the potential for the adaptive schemes to be effective.

Performances of different schemes, for one unsteadiness cycle, are summarised in *Table 10-4.1*. Average $u$ solution errors, normalised by $U_o$, are evaluated at 6 locations (coinciding with points at the centre of the 6 profiles used in the above profile comparisons) to give $E'_{ave}$ (the average true solution error).

<table>
<tr><th rowspan="3">Scheme</th><th rowspan="3">t<sub>s</sub>(%)</th><th rowspan="3">%Δt fails</th><th colspan="3">% $E'_{ave}$ over one unsteadiness cycle</th></tr>
<tr><th rowspan="2">Full region</th><th colspan="2">$Y > 0.75$ region</th></tr>
<tr><th>Standard Approach</th><th>Pseudo Injection</th></tr>
<tr><td>An (n=0.33)</td><td>86</td><td>0</td><td>0.152</td><td>0.140</td><td>0.141</td></tr>
<tr><td>ASc</td><td>61</td><td>0</td><td>0.065</td><td>0.055</td><td>0.058</td></tr>
<tr><td>AΔt</td><td>87</td><td>0</td><td>0.240</td><td>0.160</td><td>0.163</td></tr>
</table>

*Table 10-4.1* Adaptive scheme performances

The table shows that all schemes give solutions close to the specified 0.1 % error and significant (on average 78 %) computing time savings. Unusually, for this case (as will be seen from considering the other chapters), the *ASc* approach appears least efficient (however, some of this can perhaps be offset against the low $E'_{ave}$ value). Although the *An* method looks most efficient, as noted in the forgoing discussion, rather case specific stability measures are required to gain the tabulated result. On these grounds, for this case, the $A\Delta t$ approach probably should be considered to give the best performance. This is unusual, since, generally, for other cases considered in this text, the $A\Delta t$ approach is worst. The most likely reason for its good performance here, is that for this non-rotating flow, with quadratic momentum sources and sinks, the $A\Delta t$ solution components show greatest stability. The table indicates that focusing on the key flow unsteadiness region increases accuracy whilst reducing storage requirements. Pleasingly, the novel, further storage saving measure, of multilevel related 'pseudo injection' just gives minor (on average 0.0025 %) error increases.

## 10-5 Discussion of Results for Scalar and Particle Transport

*Scalar Transport*

Frequently the transport of scalar quantities, such as smoke, is used to visualise/interpret fluid flows. Often, after a short period, the scalar quantity will completely fill the region of interest. Then the flow structure cannot be seen. Therefore, even for certain steady flows, the use of scalar transport to observe flow structures is carried out in an unsteady sense. The comparison of predictions and measurements for the temporal development of scalar distributions can be of value for solution validation purposes.

*Figure 10-5.1* compares $k$-$\varepsilon$/$k$-$l_2$ model scalar transport predictions with mid $z$ axis, $x$-$y$ plane time dependent smoke flow visualisation evidence. The two fans produce an area of low pressure in the left hand side of the system drawing fluid in through the lower left hand hole. Smoke is released in the vicinity of this hole. The left hand column corresponds to the experiment and the other columns predictions. Computations are made using Equation (4-5.8) with $\phi = C$. Fan 2 swirl is modelled (due to a grill, fan 1 swirl is negligible and so not accounted for). The swirl component is approximately 5.5 m/s. Frames (a), (b), (c), (d) and (e) correspond to $t^* = t\ U_o/x_{max}$ = 1.2, 2,4, 3.6, 4.8 and 6, respectively. For the predictions, the inlet smoke concentration is set at 100%. White areas correspond to where the concentration is greater than 5 %. This scale is chosen so as to achieve agreement with measurements at $t^* = 0.2$. There is some agreement between the experimental smoke flow visualization and measurements.

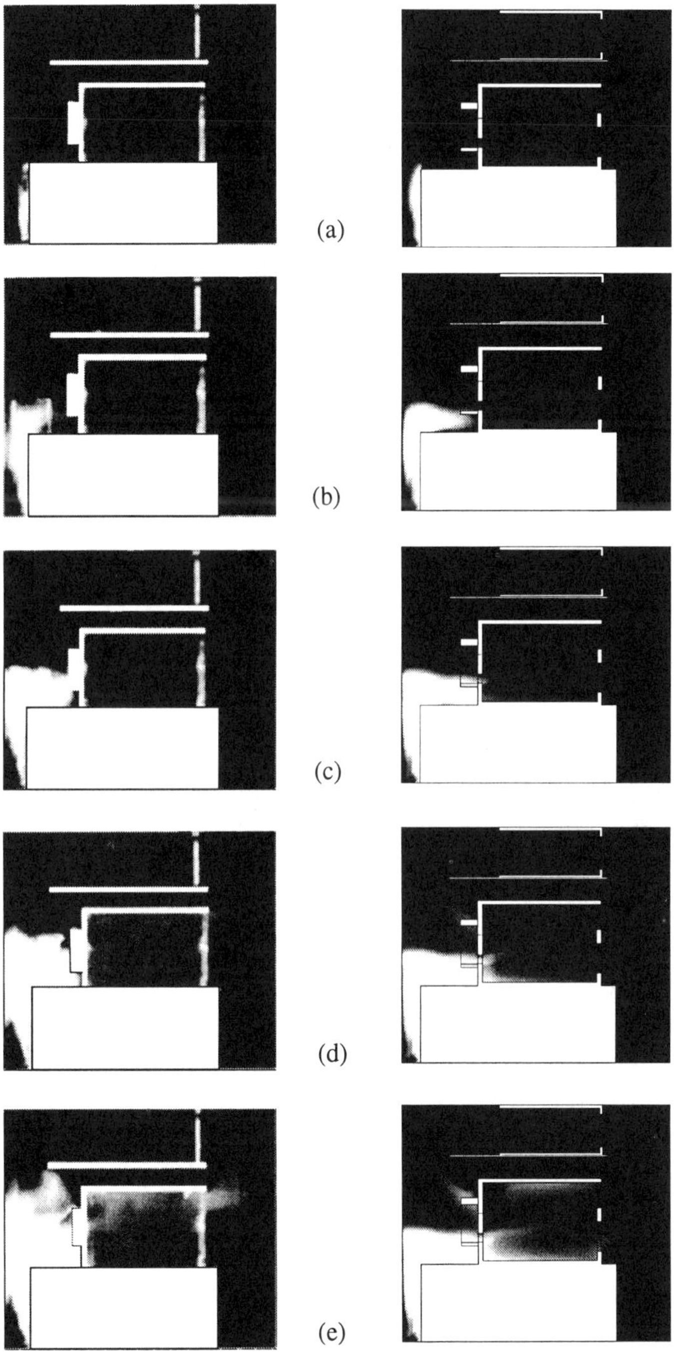

*Figure 10-5.1* Comparison of smoke flow visualisation evidence with linear zonal predictions at the following dimensionless times: (a) $t^* = 1.2$; (b) $t^* = 2.4$; (c) $t^* = 3.6$; (d) $t^* = 4.8$ and (e) $t^* = 6.0$

*Particle Transport*

As noted in Section 2-11, in a Lagrangian frame, the prediction of particle trajectories is an inherently time-dependent process. *Figure 10-5.2* shows the dispersion of particles for the geometry given in *Figure 10-1.1*. Forty particles are seeded at the centre of left lower floor hole and their trajectories integrated through time using a 4$^{th}$ order Runge-Kutta integration procedure (see Section 2-6). The density and diameter of the particles is 1500 kg/m$^3$ and 6 μm, respectively. So that the effects of unsteadiness and other factors on particle trajectories can be isolated, for all predictions, the mean velocity field from the isotropic $k$-$\varepsilon$/$k$-$l_2$ model is used. In *Figure 10-5.2a*, turbulent fluctuations are set to zero. This, in a sense, mimics laminar flow. As expected, there is no dispersion, all 40 particles exactly overlaying. *Figure 10-5.2b* shows the results when the turbulent fluctuations are non-zero. The

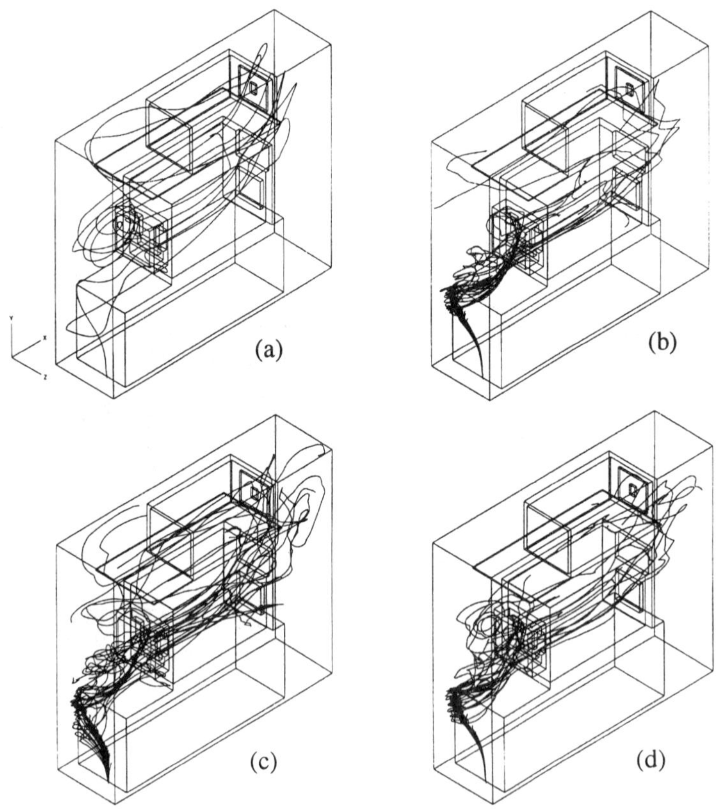


*Figure 10-5.2* Particle dispersion for the complex system: (a) 'laminar' flow; (b) isotropic turbulence ($k$-$\varepsilon$/$k$-$l_2$); (c) anisotropic turbulence and (d) cyclic flow.

significant effect of turbulence on dispersion is clear.
Anisotropy can have a significant effect on particle and passive scalar transport (see Zhou and Leschziner (1997)). In *Figure 10-5.2c*, anisotropy is incorporated into the analysis. The turbulence field is computed using the non-linear Boussinesq approximations (see Equation (6-5.42)). As can be seen, anisotropy has made a significant difference to particle transport. In *Figure 10-5.2d* the effect of the flows periodic component is displayed. Comparison of the root mean square deviations of final particle positions shows that modelling anisotropy alters dispersion by 19 % relative to the linear model and unsteadiness 18 % relative to the steady model. Clearly influence of flow unsteadiness on particle transport can be significant.

**10-6 Conclusions**
URANS predictions have been considered for a high total (turbulence and unsteadiness components) intensity flow in a complex system. Twelve turbulence models were tested. For all models, predicted velocities were similar. The linear zonal models gave the most and $k$-$\varepsilon$ the least accurate values. Inclusion of non-linear terms, through the use of non-linear eddy viscosity models, did not improve predictive accuracy. The predicted temporal velocity variations and turbulence intensities for the different turbulence models varied significantly. All of them, except the $k$-$\varepsilon$, tended to under predict turbulence levels. As with velocities, the linear zonal models gave the best agreement with total intensity measurements. However, even for these predictions the average error was 28 %. Generally, predicted unsteadiness amplitudes were lower than the measured. For the zonal models, increases in amplitudes of the predicted unsteadiness components are likely to yield improvements. However, how this can be achieved in practice is uncertain. Importantly, for the first order diffusive upwind scheme, there was no detectable unsteadiness, illustrating a danger of using it. Results showed that adaptive time-stepping can yield significant savings in computer time and relieves program users of the time-step selection burden. The significant effect of unsteadiness on particle transport was shown.

# CHAPTER 11

# TRANSIENT COMPLEX SYSTEM FLOW

## 11-1 Introduction

The computation of transient turbulent flows has considerable engineering significance. In Chapter 5, a transient wall driven flow is considered. With this, the wall has tangential rotational movement. This produces a relatively simple shear flow with wall aligned streamlines. However, frequently transient flows involve *impingement,* where a suddenly released fluid interacts with downstream mechanical structures. The resulting flow structures can then be complex with significant streamline curvature. Such a flow is considered here for the idealised industrial system shown in *Figure 11-1.1*.

Here, comparison is made with the ±10% accuracy thermistor based velocity measurements of Pan and Tucker (2000). The flow inlet area, shown in *Figure 11-1.1*, has a suddenly opening shutter, giving rise to an inrush of air from an external source. The geometry is quite specific. However, the results produced here have a wider context. They are suggestive of the performance of turbulence models when predicting unsteady flows in systems with similar flow features. These include stagnation resulting from an impinging rectangular jet, streamline curvature and large vortical structures. The turbulence models tested include the standard $k$-$\varepsilon$ model ($k$-$\varepsilon_l$), low and high Reynolds number $k$-$l$ models ($k$-$l_{1,}$), and a standard mixing length ($ml_1$) model. Also, two zonal models are tried using the $k$-$\varepsilon$ model away from walls with the $k$-$l_2$ and $ml_1$ models applied elsewhere. Theses are designated the $k$-$\varepsilon$/$k$-$l_2$ and $k$-$\varepsilon$/$ml_1$ models. Full turbulence model details are given in Section 6-5. The range of models tested is broadly similar to that in the previous chapter. Some of the adaptive time-step approaches described in Chapter 5 are also further tested.

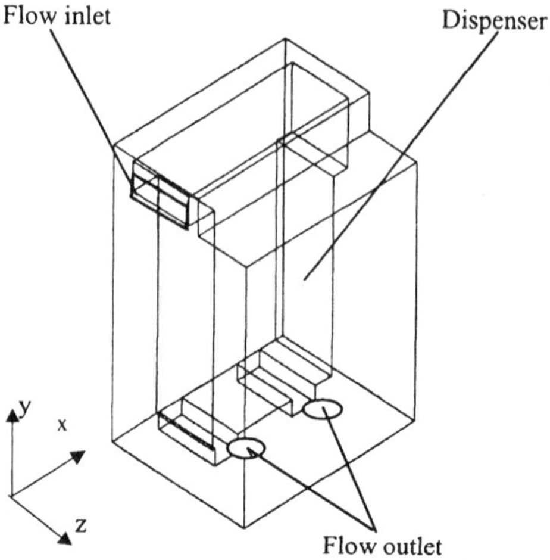


*Figure 11-1.1* Schematic of geometry

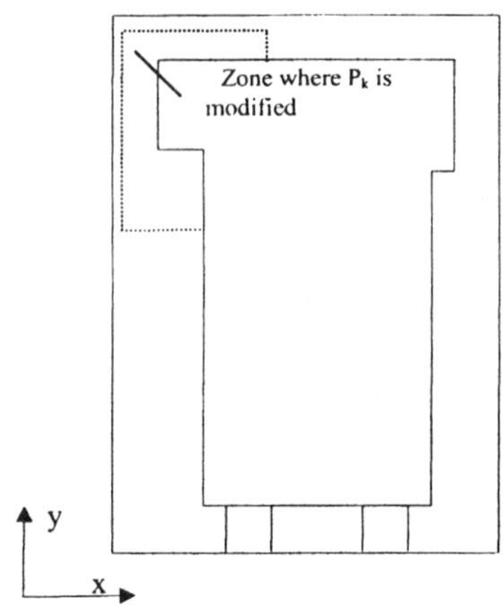


*Figure 11-2.1* Zone where stagnation modification applied

## 11-2 Numerical Details

The general numerical features of the program used are given in Section 4-6. Here, the pressure field (*p*) is computed using the SIMPLE and AVPI schemes (see Section 4-4). For convective terms CONDIF is used (see Section 3-3). Transient terms are discretized using the fully implicit backwards Euler scheme. With this flow, for higher order temporal methods convergence is difficult to secure.

The use of the stagnation modification of Kato and Launder (1993) is explored. With this Equation (6-5.34) (a modified strain and vorticity turbulence energy production, $P_k$, term) is applied in the stagnation region nominally defined in *Figure 11-2.1*.

Time-steps of $\Delta t$ = 0.001 s are used. Halving this value makes no significant difference to the results to be presented. Spatial grid sizes, for sensible grid independence, vary from 88 x 77 x 53 to 99 x 91 x 67 ($x$, $y$, $z$) control volumes. Grids are constructed so that for high and low Reynolds number predictions $y^+_{ave}$ is approximately equal to 100 and 2, respectively.

*Boundary Conditions and Solution Initialization*

Matching experiments, a specified inlet momentum source, $S_u = p_0 - \rho u^2/2$ where $p_0$ =50.0 N/m$^2$ (gauge pressure), is used in the discretized momentum equations (see Equation (2-3.1)). The present three-dimensional flow case is especially computationally expensive. Therefore, some two-dimensional predictions are used to supplement three-dimensional. For the two-dimensional, to ensure greater similarity of case dynamics, $p_0$ is linearly ramped to 50 N/m$^2$ for $t < 0.5$ s. Other boundary conditions are indicated in Section 4-6. For the domain interior, because initially, there is no flow, away from the inlet, at $t = 0$, all turbulence properties ($k$, $\varepsilon$, $\mu_t$) are set to zero along with velocities and $p$.

For the inlet, $k$ and $\varepsilon$ are prescribed based on a 10% intensity. It is expected that for transient flow, results will be more sensitive to this prescribed intensity. This is because, before the mean flow field has evolved at a particular location, there can be no significant turbulence generation. Therefore, turbulence, which has been transported to that location, from the inlet, will play a greater role. However, sensitivity tests show this is not the case.

## 11-3 Discussion of Results

*Steady Flow Comparisons*

In this case, because a representative steady state solution exists, before moving to more expensive transient computations steady predictions are worthwhile. *Figure 11-3.1* shows the predicted $k$-$\varepsilon_l$ model flow in the inlet region for the $x$-$y$ plane. Stagnation and streamline curvature can be clearly seen. Nine horizontal mean velocity profiles are compared with measurements. The results of these comparisons are fully reported by Pan and Tucker (2000). To save space only the key findings are summarised here.

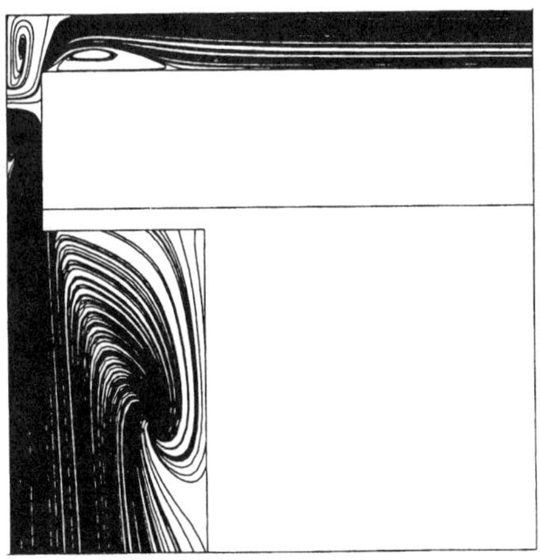

*Figure 11-3.1* Streamline plot downstream of the shutter

These are that, reversing the trend shown in the previous chapter, the $k$-$\varepsilon_l$ model now has the best accuracy and the zonal models the worst. This is probably because, for the zonal models, since the Reynolds numbers are relatively low over a significant area, the $y^+<100$ region, where the zero and one equation models are applied, occupies a large part of the geometry. This point is illustrated in *Figure 11-3.2*. Since, for steady flow the $k$-$\varepsilon_l$ model gives best agreement it would seem sensible to apply it now to the transient case.

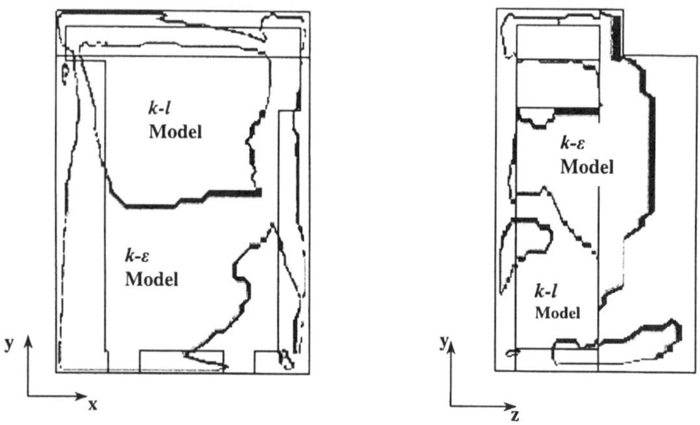


*Figure 11.3.2* Plot showing regions where two-equation model is used in zonal approach

*Transient Flow Comparisons*

*Figure 11-3.3* shows the streamline plots in the $y$-$z$ plane at $X$ = 0.033 ($X = x/x_{max}$, where $x_{max}$ is the maximum $x$ dimension), downstream of the shutter for $t$ = 0.5, 0.75, 1.0, 1.25 s. As can be seen at $t$ = 0.5 s two vortices are established which are convected downwards with time. Again, there is strong streamline curvature.

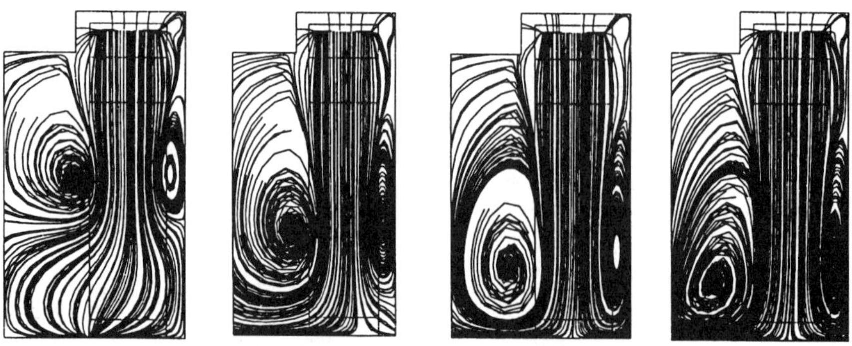

(a) $t$ = 0.5 s (b) $t$ = 0.75 s (c) $t$ =1.0 s (d) $t$ =1.25 s

*Figure 11-3.3* Streamline plot of the transient solution

*Figure 11-3.4* compares predicted normalised velocity time variations with measurements at the following four dimensionless locations: (a) $X$=0.644, $Y$=0.977, $Z$=0.400; (b) $X$=0.039, $Y$=0.527, $Z$=0.287; (c) $X$=0.169, $Y$=0.00, $Z$=0.481 and (d) $X$=0.840, $Y$=0.00, $Z$=0.464. Symbols represent measurements and the full and dashed lines three and two-dimensional predictions, respectively. Some agreement can be observed between the three-dimensional predictions and measurements. The figure also shows two-dimensional predictions (where $Z$ = 0) have some similar temporal characteristics to the three-dimensional.

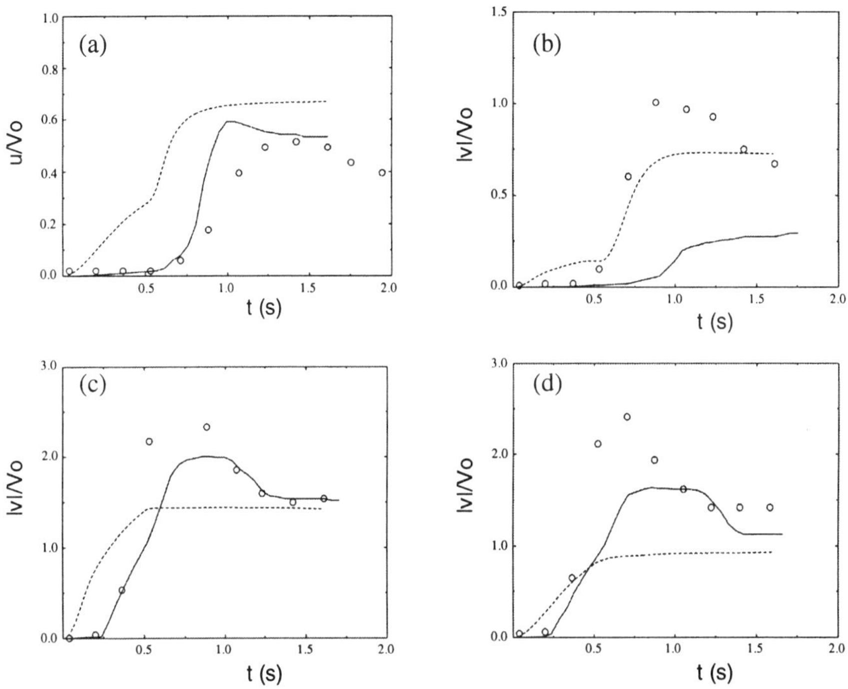


*Figure 11-3.4* Comparison of predicted pointwise temporal velocity variations with measurements (O Measurements —, 3-D Predictions, ---- 2-D Predictions)

By observation, the peaks in the velocity time traces are under predicted. Indeed based on, for computational economy, two-dimensional computations the under prediction is likely to be the case for all models considered in this chapter. For two-dimensional predictions, the other models did not increase the temporal velocity maximum above that predicted by the two-dimensional $k$-$\varepsilon_l$ model. The results suggest more refined turbulence models are required with perhaps the very minimum level of sophistication being non-linear eddy viscosity (see Speziale (1987)) or algebraic stress models (see Gatski and Speziale (1993)). However, due to computational expense, testing these models here, has not been possible. For the $nl$-$k$-$l_l$ model described in Section 6-5, under transient conditions converged solutions could not be gained.

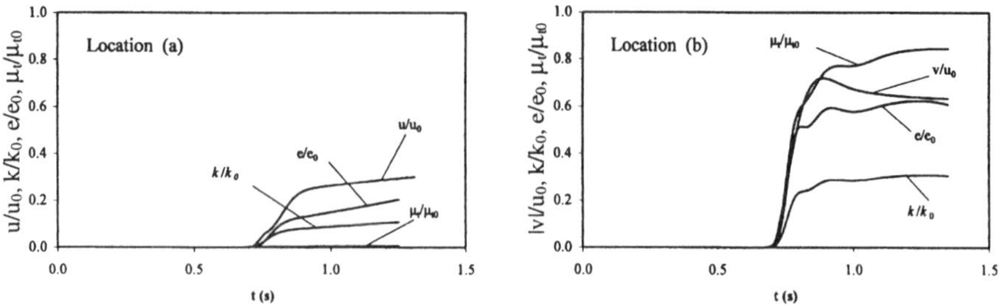


*Figure 11-3.5* Predicted time variations of turbulence properties

*Figure 11-3.5* shows the predicted evolution of $k$, $\varepsilon$, $\mu_t$ at locations (a) and (b). These are normalized by inlet values. In order to see clearly whether or not there is any phase difference between the evolution of these turbulence properties and the velocity field, mean flow velocities are also presented. As can be seen, there is no clear difference in phase. It is interesting to note, for the three-dimensional $k$-$\varepsilon_l$ case, that, although cheaper on storage, the AVPI pressure correction scheme requires 15 % more computing time than SIMPLE.

*Adaptive Time-stepping*

For economy, to assess adaptive time scheme performances, two-dimensional high Reynolds number $k$-$l_l$ predictions are used. Basing temporal error estimates (Equation (5-2.16)) on just $u$ (giving a storage saving) or both $u$ and $v$ makes little difference to results.

Without, the 0.5 s $p_o$ ramp (used in *Figure 11-3.4*) sharper temporal gradients are produced. Then, Equation (5-2.7) variable relaxation parameters, give up to 25 % time savings. *Figure 11-3.6* plots two-dimensional solution streamlines and contours of velocity unsteadiness at 0.2 s intervals with no $p_o$ ramp. The flow is characterised by the formation of two vortices. As the flow accelerates these elongate. A key feature is that the areas of greatest unsteadiness move with the elongating vortex front. This case specific observation precludes the storage saving strategy (described in Chapter 5) of just focusing the temporal error estimate on some stationary restricted region $R'$. Instead $R'$ would have to move. For flows with specific unsteadiness regions or structures that can readily be identified this is feasible. The idea of evolving $R'$ with the flow has some attractions. A possibility would be to solve a scalar transport equation (introducing an additional overhead) and through the temporal variations of the scalar identify a dynamic $R'$ region.

| Scheme | $t_s$ (%) | %$\Delta t$ Fails | % $E'_{ave}$, $t < 1.6$ s |
|---|---|---|---|
| *$\Delta t$=0.01s (Standard k-l)* | NA | NA | 0.18 |
| *$\Delta t$=0.01s (Modified k-l)* | NA | NA | 0.18 |
| *An* | -38 | 0 | 0.12 |
| *ASc* | -1.5 | 0 | 0.15 |
| *A$\Delta t$* | -56 | 0 | 0.12 |
| *An(p=b$_{12}$)* | -28 | - | 0.16 |
| *An(p,k=b$_{12}$)* | -23 | - | 0.15 |
| *A$\Delta t$(p=c$_{12}$)* | -40 | - | 0.21 |
| *A$\Delta t$(p,k=c$_{12}$)* | -39 | - | 0.22 |

*Table 11-3.1* Adaptive scheme performances

*Table 11-3.1* summarises adaptive scheme performances. With the less temporally severe ramp included (despite this, the case still has challenging transients), the destabilising effect of altering relaxation parameters slows convergence. Therefore, the table gives fixed under-relaxation parameter results. Tabulated true average errors ($E'_{ave}$) are based on all 4 velocity traces compared to a datum $\Delta t$=0.001 s solution. For the adaptive approaches $E'_{ave}$ is about 0.15 % (i.e. higher than $\varepsilon$ = 0.1 %). This is slightly better than $E'_{ave}$ for $\Delta t$=0.01s. Therefore, conservatively, $t_s$ (the computing time saving) is calculated relative to the $\Delta t$=0.01 s solution. The table shows the stagnation *k-l* modification has no real influence. As usual (see previous Case Studies and Chapter 5), the *ASc* (Altered Scheme that compares forwards and backwards Euler solutions) method (with the cheap forward Euler component) shows best performance. However, even this is 1.5 % slower than the $\Delta t$=0.01s solution. None-the-less, the *ASc* $E'_{ave}$ is 15 % less than that for the fixed step and so this result should not be considered as negative. Performances for all the other standard approaches (*An* – altered order and *A$\Delta t$* altered step), seem especially disappointing. Therefore, 'subcycling' is tried in the *An*(*p*=b$_{12}$), *An*(*p,k*=b$_{12}$), *A$\Delta t$*(*p*=c$_{12}$) and *A$\Delta t$*(*p,k*=c$_{12}$) frameworks described in Section 5-2. For these, relative to the standard schemes, the table shows clear computing time reductions with just modest $E'_{ave}$ *increases*. The latter increases make these 'subcycling' results slightly inconclusive. However, reminiscent of the Section 5-3 transient square box flow, when comparing the *An*(*p*=b$_{12}$) and *An*(*p,k*=b$_{12}$), results

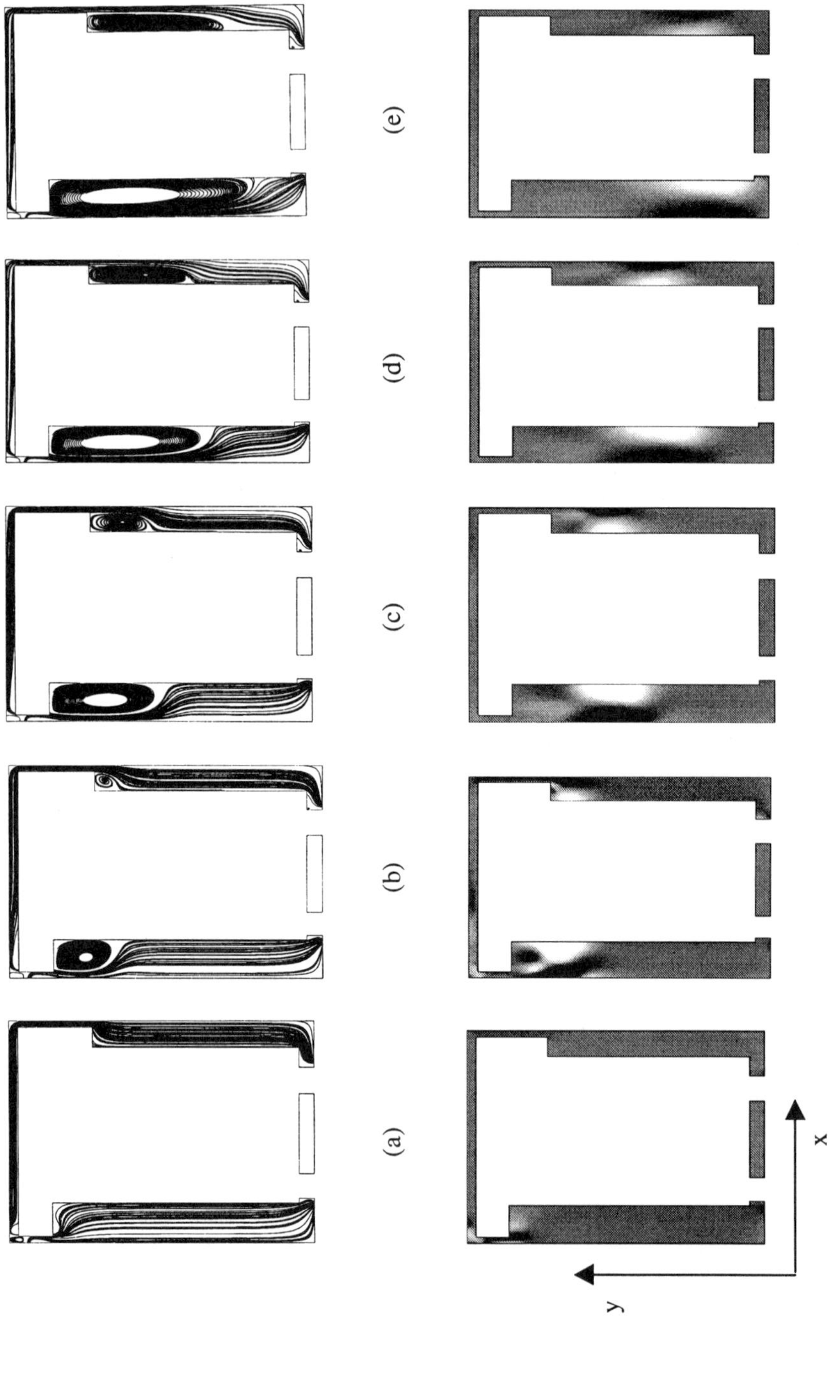


*Figure 11-3.6* Streamlines and contours of velocity unsteadiness for: (a) $t = 0.2$ s, (b) $t = 0.4$ s, (c) $t = 0.6$ s, (d) $t = 0.8$ s and (e) $t = 1$ s.

'subcycling' has given an accuracy improvement. This suggests it has the inherent positive trait of enhancing stability.

One observed difficulty, that explains most of the current poor performance, is initial transient severity causes numerical instabilities. These result in adaptive schemes predicting extremely small time-steps - significant differences between the two comparative solution elements not being due to truncation errors but instability. However, this aspect does have a positive side. The instability gives an in-built conservativeness. Therefore, unlike fixed step solutions, if used in some industrial contexts, the adaptive predictions all give computations that could be used with more reasonable immediate confidence. Hence, in certain circumstances the adaptive solutions could be regarded as efficient.

## 11-4 Conclusions

Broadly, for the present flow, the turbulence model performances observed in the previous chapter are reversed. The $k$-$\varepsilon$ model now gives the best spatial profiles and zonal models the worst. Results suggest a strain rate and vorticity based stagnation region modification does not improve agreement and that all models underpredict the peak transient velocity magnitude. They also show that development of the predicted turbulence parameters is in phase with the transient velocity field.

These steady state and other two-dimensional transient predictions suggest for improved temporal accuracy more refined turbulence models are required. Due to computational expense it was not possible to try such models. However, before starting this further time-dependent work, the steady state predictive accuracy of the turbulence models needs to be substantially improved.

Due to especially severe transients, adaptive scheme performances were disappointing. Instabilities appeared to give rise to especially small time-steps. However, this in built time-step conservativeness could still make the current adaptive solutions worthwhile in certain design contexts.

# CHAPTER 12

# OVERALL CONCLUSIONS WITH FUTURE PERSPECTIVES

Most flows are unsteady. Often, this can be viewed as relating to stochastic components associated with smaller turbulence scales or larger scale more orderly unsteadiness or a combination of the two. In many cases, especially for complex industrial systems, the larger scale, and lower frequency more orderly components are significant. Since many flows are unsteady, there has been much research into the development of fast, accurate algorithms for their modelling. A significant number of these are orientated towards geophysical applications (a Case Study, of a more benchmark nature, used in this area was considered in Chapter 7). These, like industrial CFD cases, have strong spatial resolution demands. Consequently there is much common ground between geophysical and industrial CFD. Hopefully, as in the past, more geophysical flow related algorithm developments will continue to filter through to industrial CFD, especially finding greater use in commercial codes. Unsteady flow modelling is important in many areas of general design. For example, unsteadiness can cause significant heat transfer increases in electronic systems. A Case Study on the modelling of an unsteady electronic system flow was presented in Chapter 10. The modelling of unsteadiness can help explain a hydrodynamic bearing failure mechanism (see Chapter 9). The modelling of unsteady behaviour is also important in aiding the understanding of aero engine cooling flows. Studies on this, which had strong flow structure similarities to the Chapter 7 geophysical case, were given in Chapter 8.

Generally, if an unsteady flow is treated as steady and accurate algorithms used, solution convergence cannot be achieved. Then, frequently, the predicted flow structure and associated parameters may not even be qualitatively correct and be of danger when used as part of a design analysis. The increasing speed and storage of computers along with numerical algorithm improvements is making the accurate simulation of three-dimensional unsteady flows in an industrial design context increasingly feasible. However, despite computing power and algorithm improvements, there is evidence to suggest that, in a significant number of simulations, unsteadiness is ignored through inadequate convergence (modulations in solution variables being ignored) or damped through excessively dissipative discretization schemes or models for turbulence. These points were enforced in Chapter 10 where both the dissipative 1[st] order upwind scheme and $k$-$\varepsilon$ models were shown to erroneously give steady flow solutions. Unsteadiness can be caused by: (a) *Boundary or Component Movement* (as shown in Chapter 9) and *Instability* (see Chapters 7, 8 and 10). Two broad, but key, unsteady flow classifications are: (I) *Cyclic, Periodic or Oscillatory* flows – these being flows that repeat in time and (ii) *Transient flows* – i.e. flows that tend to a steady state. Case studies for the former were presented in chapters 7, 8, 9 and 10. The latter was considered in Chapter 11. From partial differential equation theory it is possible that a flow is *attracted* to the following solutions: (a) steady state attractor; (b) periodic attractor; (c) quasi-periodic attractor and (d) strange attractor. The periodic attractor corresponds to a *limit cycle* where the flow structure modulates between distinct states. A limit cycle flow was presented in Chapter 8. Temporal traces for attractor (c) display an orderly component along with a less orderly element. Therefore, the Chapter 10 periodic turbulent flow, with its superposition of stochastic turbulence and an orderly cycle of separation and reattachment, could be viewed as involving a quasi-periodic attractor. Importantly, for the same boundary conditions, depending on initial values, different solutions (attractors) can be gained. Reminiscent of this, in Chapter 8 iT was noted different steady flow patterns (*pitchfork* bifurcations) could be found depending on initial tangential velocity conditions. The presence of such bifurcations suggests care is required with multipoint temporal scheme starting conditions – several solutions are required and these must present the correct route to the final solution.

As indicated in Chapter 2, there are many temporal schemes available. These mostly originate from the solution of ordinary differential equations. As with many other CFD algorithms, essentially the same non-discrete temporal scheme can be discretized in different ways. Therefore, for the same parameter settings and 'schemes' different codes can have different accuracies and convergence rates. Three-level (multipoint) temporal schemes require more storage, computational effort and attention to the starting solutions. This point was alluded to earlier with respect to solution uniqueness. For industrial computations, as shown in Chapter 4, mostly two-level schemes are used. For increased efficiency these can be cast in the 'Delta' form.

If time-steps are too large, solutions for unconditionally stable schemes may not be physically realistic. For stable, accurate solutions, consideration of Courant and time-step to diffusion scale ratio parameters is important. For the explicit Euler, the latter parameter gives a severe time-step restriction. Therefore, generally, this scheme is only suitable for problems where high temporal resolution is required. For flows where variables are changing slowly, implicit schemes should generally be used. The Chapter 7 Case Study gave an unusual exception to this rule. Although it exhibited relatively slow temporal changes, the first order explicit scheme was most efficient. Generally, implicit schemes are more stable than explicit. For example, in the relatively high Reynolds number Chapter 8 flow it was not possible to gain an explicit solution. For many industrial CFD applications implicit scheme stability is attractive and so such approaches are popular in commercial codes. Generally, the higher the order of an explicit scheme, the more severe the stability restriction. However, explicit methods lend themselves to vector and parallel processing. Therefore, with the increased utilisation of parallel machines featuring individual vector processors, explicit scheme use could further increase. Also, as shown in Chapter 5, their low computational cost makes them effective when used as a comparative solution element in adaptive time-stepping strategies.

The most common multipoint methods are the *Adams* types. A well-known member of the class of *predictor-corrector* methods are the *Runge-Kutta* schemes. These are more stable and accurate than multipoint schemes and have seen use in specialized industrial CFD applications (especially in the aerospace area). For steady flows more stable, non-time accurate Runge-Kutta forms can be used. *Splitting methods* combine implicit and explicit schemes. In this way significant improvements in computational performance can be produced. However, as shown in Chapter 8, the introduction of the explicit element detracts from stability. Except when modelling Lagrangian particle transport, although robust and interesting, the exponential temporal scheme is probably best avoided.

Unless there are obvious reasons, an effective strategy is to model all flows as unsteady. Time marching allows stable (this is especially important when sophisticated turbulence models are to be implemented with higher order spatial discretization schemes) integrations to steady states and also the application of potentially efficient splitting and other temporal schemes. Equations with differing time scales are called *stiff*. Implicit methods can be helpful when modelling such problems. The application of implicit methods to a stiff problem was considered in Chapter 10. A key additional aspect to removing the stiffness was the use of sliding meshes.

Sliding meshes can also be used to reduce complex unsteady moving boundary problems to quite straightforward steady flow cases. For all temporal predictions, to minimise the build up of machine round off error, through time integrations, double precision arithmetic is generally recommended. For economy, preliminary computations to ascertain the optimum time-step (in the time-step independence range) to number of convergence iterations ratio are worthwhile. As with the internal spatial discretization accuracy, it is important that the order of the boundary condition discretization and also the order of the initialisation data is consistent with the temporal discretization. For accurate industrial computations Crank-Nicolson appears a good temporal scheme choice with both reasonable accuracy and robustness.

As discussed in Chapter 3, often unsteadiness is a function of the solution frame of reference. It was noted in this chapter that temporal contributions to the discretized unsteady flow equations can be broadly split into the following three components: the unsteadiness arising from just constant flow movement/drift relative to the grid; that due to fluid dynamic related flow structure vacillations and that arising from geometric distortion. Therefore, the use of inertial coordinate systems, possibly with sliding grids, can give rise to a flow that is steady or quasi-steady relative to a new frame of reference. Such concepts were practically addressed in the chapter 7, 8 and 9 case studies. Through these, it was implied that careful coordinate system selection has the potential to yield significant computational performance gains. Key reasons for these are reductions in degrees/sizes of: temporal derivatives, convective terms, velocities (thus increasing numerical precision) and grid distortion. Also, as noted in Chapter 1, certain problems can be made steady by variable transformations rendering solutions that are similar in time. However, such approaches have found little industrial CFD use.

As part of the convective term discretization, dependent variables are interpolated from grid points to control volume faces. For example, the third order QUICK scheme involves a quadratic fit to three grid point values. When used with ADI the involvement of three grid points detracts from its efficiency. Compact techniques offer the potential of high accuracy without involving extra grid points. This is advantageous for ADI type methods.

Oscillations arising from the use of higher order schemes are associated with *dispersion* errors, which cause phase errors. Generally introducing numerical or false diffusion damps these oscillations but introduces amongst other things amplitude errors. A superior oscillation removal technique is the *Total Variation Diminishing* (TVD) approach. The accuracy of QUICK with first order forward Euler time differencing is poor. Therefore, it is best modified using the *Estimated STreaming* (EST) approach to give QUICKEST. The use of moving grids, in a sense, introduces some EST like behaviour into the spatial discretization and as noted earlier reduces convective term magnitudes. Schemes related to the EST approach can be derived that are accurate for Courant numbers greater than unity. However, these have only been applied to one-dimensional flow with more work required for their extension to two and three-dimensions. *Aliasing* problems can be avoided by using special energy conserving convective term treatments. Although advances have been made in the area of unsteady convective term treatments, as yet, they do not seem to have been incorporated into many commercial codes. A matter of concern is that although certain unstructured grid commercial codes now offer LES, there has been little research into how these grids and their associated convective term treatments impact on temporal accuracy. In the author's opinion more work is required in this area.

For moving boundary problems, the domain movement can either by specified or solved as part of the numerical solution. The latter approach can be far more numerically demanding. Cartesian, trimmed cell methods can be used for problems involving bodies moving through fluid domains. However, despite interpolations between the significantly different grid systems making conservation difficult to enforce, the *Chimera* grid method is more popular. Nonetheless it does not appear to be implemented in the most popular commercial codes (PHOENICS offers a *Chimera* grid like option but this cleverly overcomes much of the interpolation difficulties). With other moving boundary schemes, involving cell distortion, conservation errors can also occur. However, the use of small time-steps or the enforcement of the *Space Conservation Law* can minimize these. Many commercial codes now ensure this law is enforced. If it is not, small time-steps should be used.

A possibility when modelling free surfaces is to have the grid conform to the fluid region and move with it (approaches were also discussed in Chapter 3 – i.e. ICED-ALE and LINC – where the grid moved with the interior flow thus reducing convective term velocity magnitudes and hence increasing accuracy). Another strategy is to use a grid that extends beyond the free surface. The fluid boundary, within the extended domain, is then tracked. This is the basis of the relatively well-known MAC technique. An example showing the use of a MAC like surface tracking technique was given in Chapter 2. There are many free surface-tracking methods. These vary in accuracy, flexibility, computational overhead, ease of implementation and required storage. The choice of optimal approach is therefore problem dependent. Adaptive solutions, where space-steps (or even the order of numerical algorithms used to resolve these) are automatically adjusted as the flow evolves are attractive. With curvilinear grids, time moving grid transforms can be used to cluster nodes in areas of high gradients. Grid point insertion and repositioning approaches have been available for some years but have found relatively little commercial code industrial application. Generally the technique is highly compatible with multilevel (multigrid) convergence acceleration (a simple illustration was given in Chapter 6). Future increases in the use of adaptive mesh refinement in commercial code CFD application should be expected.

As discussed in Chapter 4 and indicated above, explicit schemes do not require a simultaneous equation solver and can be easily parallelized. More stable, implicit techniques require such a solver. This is generally a major solution overhead. Helpfully, approximate factorization or splitting methods, like ADI, only require one-dimensional simultaneous equation solvers. For most basic solvers, the number of iterations to convergence increases with about the square of the grid point number. However, with geometric multilevel methods, in theory, the iterations stay constant (see Appendix C). Explicit Euler solvers, when used to time march to steady states, can greatly benefit from multilevel convergence acceleration. This is because the Courant-Friedrich-Lewy condition is circumvented. Multilevel algorithms also have the potential to yield computational enhancements for evolution problems and also solution adaptive time-step approaches. For unsteady flows, Newton methods can be viable. Also, when parallel processing, temporal domain decomposition techniques have found use. However, the success of this appears, amongst other factors, problem dependent.

With unsteady flows, it is desirable to avoid under-relaxation. Instead, generally, to gain stability, time-steps can be reduced giving the bonus of increased temporal accuracy. In this way, even extremely difficult to converge equation systems can become stable. Relative to those for momentum, the convergence rate of Poisson type pressure equations can be disappointing. With iterative pressure prediction methods, to avoid temporal error growth, an adequate number of continuity enforcing pressure sweeps is desirable. For unsteady solutions, convergence level care is required. For each time-step, errors will grow. Hence, over a long integration period, relatively small local convergence errors can become highly significant.

For industrial CFD, SIMPLE (including PISO) related methods are still the most popular pressure prediction approaches. An increasing number of commercial codes are offering LES options. Hopefully, with the greater use of these more efficient pressure prediction approaches will begin to see commercial code use. SIMPLE based schemes excessively restrict the stable solution time-step range. The dominance of the time derivative term, in the momentum equation diagonal elements, allows use of the storage saving AVPI (a simplification to SIMPLE) scheme. It can be combined with other pressure correction schemes to form hybrid solution approaches. For example, in Chapter 8, a hybrid AVPI/SIMPLEC pressure correction scheme was implemented. Fractional step methods are an alternative, unsteady flow, pressure field prediction approach. They have found little use in industrial CFD. Without modification, the use of artificial compressibility methods, to gain pressures, for time accurate solutions is dubious. The penalty function method has similarities to the artificial compressibility method. It is mostly used in Finite Element (e.g. FIDAP) programs and gives rise to stiff equations that can produce initial time-step non-physical transients. These can be damped using the dissipative backward Euler scheme. Once this damping has been carried out more accurate schemes can be invoked. Fully coupled solvers naturally overcome pressure prediction problems but need extra computer storage. Since the cost of latter is decreasing, it is likely that coupled solvers (with multilevel convergence) will see more widespread future commercial code use. At the moment the commercial CFX-5 program has a coupled solver. For turbulent flows, leaving the turbulence model equations as a sequential coupled solver element can be the most computationally efficient approach. This has similarities to multilevel convergence acceleration where generally the turbulence equations are best treated on a single level/grid. Although staggered grids have many desirable properties, for general geometry codes they introduce excessive programming complexities. Therefore, collocated variable methods are popular. However, convergence rates for these can be sensitive to both time-step and under-relaxation parameter values.

As discussed in Chapter 5, solution adaptive time-step approaches are relatively easy to implement and can yield dramatic savings in computing time. Overall, methods that use a relatively cheap to compute (but unstable) explicit comparative solution element are best. Usefully, the instability of the explicit component gives rise to conservative time-step values. As might be expected, step doubling approaches are generally least efficient. Adaptive time-stepping approaches can be readily combined with 'subcycling' strategies. This can yield further performance improvements. When using multilevel convergence acceleration, storage reductions can be made by saving solution elements on coarse grids. Then, comparative solution elements can be advanced through time on coarser mesh levels, the forcing function being used to include fine grid information. In this way further computing time savings can be made. Sometimes, with adaptive time-stepping, the use of step clipping is necessary – especially to prevent excessively small steps. These can instead be limited by the choice of appropriate reference values. Whether clipping or reference values are used partly depends on data availability. Adaptive step cases reveal that solver robustness is a key issue. Time-steps and hence diagonal dominance can vary dramatically over a solution. With severe transients, adapting relaxation parameters to maintain diagonal dominance was found, for one case (see Chapter 11), to be effective. With increasing solver robustness and affordable computer storage, the use of adaptive time-stepping is likely to increase. Conjugate gradient type solvers, that are less sensitive to diagonal dominance, might yield further efficiency gains. Also, with parallel architecture machines, the two comparative solution elements could be completed concurrently. This is likely to yield dramatic performance gains and so is worthy of future study. Generally, the first solution component should be a good guess to the second comparative element. Then a Newton type solver would be highly efficient. Therefore, testing the use of a Newton type solver for the second solution element also appears worthwhile future work.

URANS/VLES computations offer the potential for greater accuracy over RANS. However, as shown in chapters 6 and 10, computed temporal behaviour is highly sensitive to turbulence model choice. Also, comparison of results for chapters 10 and 11 show that turbulence model selection is far from straightforward. In Chapter 10 zonal models are best and the $k$-$\varepsilon$ worst. In Chapter 11 the reverse is true. Increases in turbulence model sophistication seldom appear to directly equate to increased temporal accuracy. For example, although tending to improve turbulence intensities, all the Chapter 10 non-linear models reduced temporal accuracy. However, as shown in Chapter 6, relative to eddy viscosity models, for simple geometries/flows, second moment closures mostly show some temporal accuracy improvements. Unfortunately, for routine complex geometry industrial computations, with presently available computing resources Reynolds stress models are still, in general, excessively expensive. LES, which places less emphasis on the turbulence model, shows promise, but at present, for realistic engineering flows is also too computationally expensive. DES represents an interesting zonal compromise, but is insufficiently mature and untested to be a URANS alternative. Consequently, for the next few years, URANS modelling appears to be the most viable option with the occasional use of LES/DES for specialised more research orientated applications.

Generally, with steady flows, turbulence models and convective term treatments are regarded as the two weakest CFD links. For unsteady flows some reasonable progress has been made in the area of convective term treatments. However, as indicated in Section 6-2, when modelling such flows, relative to steady problems, turbulence-modelling deficiencies become much clearer. This is to a large extent because, unlike steady mean flow velocity variations, temporal behaviour is far more sensitive to eddy viscosity levels. Another difficulty is that many unsteady flow mechanisms involve separation. This is a strong function of the near wall turbulence modelling and grid structure. For complex geometries control of the latter, to ensure grid points are in their ideal positions, is not always straightforward. Also, when using wall functions, eddy viscosity levels can be sensitive to near wall node positions and this further impacts on temporal accuracy. During strong non-equilibrium wall functions are inappropriate. A more specific deficiency is that since the production term in the $k$-$\varepsilon$ model is the square of the mean strain rate, it cannot predict turbulence decay during the expansion phase in positive displacement systems. Therefore, it can be concluded that although with respect to general solver performance clear progress is being made, the lack of widely applicable turbulence models is a matter of concern. Although turbulence modelling research is vibrant, much work involves producing relatively minor steady flow model extensions. These often do not greatly benefit the industrial practitioner, giving rise to ever greater uncertainty regarding what model to use. Also, many extensions (which can be quite case specific) yield little worthwhile gains in predictive accuracy. It is possible that, improvements in computing power and general solver performance will be such that the more promising turbulence modelling techniques like second moment closures, DES and LES will have become common tools prior to any significantly improved turbulence modelling advent. For geophysical applications work has started using traditional non-linear turbulence models to account for subgrid LES components. Such an approach acknowledges and to an extent overcomes grid density demands. Therefore, techniques such as this would seem a sensible area of exploration for industrial CFD applications. A key LES attraction is that solution sensitivity with respect to turbulence model is weakened. This should give rise to significant increases in temporal accuracy. At the moment (as clearly suggested by the Chapter 10 results) generally it is the URANS turbulence model that totally controls temporal accuracy making the choice of time differencing scheme of relatively little importance. However, as suggested by Goshal (1999), before LES/DES can be confidently used in industrial applications, more work is required considering the errors arising from strong grid stretching (with this derivative and filter operators do not commute) and typically used complex geometry discretization schemes. Although commercial codes now offer LES options, when using them it would appear some caution is required. It could be that, unlike with URANS modelling, the spatial and to an extent temporal discretization now exerts too strong a solution accuracy influence.

# APPENDIX

# A

# DISCRETIZATION OF THE GOVERNING EQUATIONS AND MULTILEVEL CONVERGENCE OPERATORS

## A-1 General Equation Discretization

All the equations in this appendix relate to a cylindrical polar coordinate system. Transformation to a Cartesian system involves setting radial coordinates to unity (*r=1*). Also, the following substitutions should be made, $\Delta x = \Delta z$, $\Delta y = \Delta r$, $\Delta z = \Delta\theta$, $u = w$, $v = u$ and $w = v$. The energy equation discetization neglects both compressive and frictional heating. Therefore, no source terms arise with consequent savings in computing time. Also, the formal stagnation enthalpy equation is substituted by one in which the temperature is the dependent variable. The equivalent equations for a general curvilinear system are given by Rayner (1993).

The general transport equation can be written in the form of Equation (A-1.1) where the time derivative $\partial(\rho\phi)/\partial t$ is incorporated in the source term $S_\phi$.

$$\frac{1}{r}\frac{\partial(J_r)}{\partial r}+\frac{\partial(J_z)}{\partial z}+\frac{1}{r}\frac{\partial(J_\theta)}{\partial\theta}=S_\phi \qquad \text{(A-1.1)}$$

The terms $J_r, J_z$ and $J_\theta$ can be considered to be total fluxes (convection + diffusion) given by Equations (A-1.2) to (A- 1.4) below.

$$J_r = \rho r u \phi - \Gamma_r r \frac{\partial \phi}{\partial r} \tag{A-1.2}$$

$$J_z = \rho w \phi - \Gamma_z \frac{\partial \phi}{\partial z} \tag{A-1.3}$$

$$J_\theta = \rho v \phi - \Gamma_\theta \frac{1}{r}\frac{\partial \phi}{\partial \theta} \tag{A-1.4}$$

The positions of the discretized variables are given in *Figure 2-3.1.* Equation (A-1.1) can be rewritten as Equation (A-1.5).

$$[J_n - J_s] + [J_e - J_w] + [J_f - J_b] = S_\phi \tag{A-1.5}$$

where,

$$J_n = \frac{1}{r_P \Delta r}\left[\rho_n r_n u_n \phi_n - \Gamma_{rn} r_n \left.\frac{\partial \phi}{\partial r}\right|_n\right] \tag{A-1.6}$$

$$J_s = \frac{1}{r_P \Delta r}\left[\rho_s r_s u_s \phi_s - \Gamma_{rs} r_s \left.\frac{\partial \phi}{\partial r}\right|_s\right] \tag{A-1.7}$$

$$J_e = \frac{1}{\Delta z}\left[\rho_e w_e \phi_e - \Gamma_{ze} \left.\frac{\partial \phi}{\partial z}\right|_e\right] \tag{A-1.8}$$

$$J_w = \frac{1}{\Delta z}\left[\rho_w w_w \phi_w - \Gamma_{zw} \left.\frac{\partial \phi}{\partial z}\right|_w\right] \tag{A-1.9}$$

$$J_f = \frac{1}{r_P \Delta \theta}\left[\rho_f v_f \phi_f - \Gamma_{\theta f} \frac{1}{r_P}\left.\frac{\partial \phi}{\partial \theta}\right|_f\right] \tag{A-1.10}$$

$$J_b = \frac{1}{r_P \Delta \theta}\left[\rho_b v_b \phi_b - \Gamma_{\theta b} \frac{1}{r_P}\left.\frac{\partial \phi}{\partial \theta}\right|_b\right] \tag{A-1.11}$$

The following convection coefficients can be defined,

$$C_n = \frac{\rho_n r_n u_n}{r_p \Delta r} \tag{A-1.12}$$

$$C_s = \frac{\rho_s r_s u_s}{r_p \Delta r} \tag{A-1.13}$$

$$C_e = \frac{\rho_e w_e}{\Delta z} \tag{A-1.14}$$

$$C_w = \frac{\rho_w w_w}{\Delta z} \tag{A-1.15}$$

$$C_f = \frac{\rho_f v_f}{r_p \Delta \theta} \tag{A-1.16}$$

$$C_b = \frac{\rho_b v_b}{r_p \Delta \theta} \tag{A-1.17}$$

The following diffusion coefficients are also defined,

$$D_n = \frac{\Gamma_{rn} r_{rn}}{\delta r_n r_p \Delta r} \tag{A-1.18}$$

$$D_s = \frac{\Gamma_{rs} r_s}{\delta r_s r_p \Delta r} \tag{A-1.19}$$

$$D_e = \frac{\Gamma_{ze}}{\delta z_e \Delta z} \tag{A-1.20}$$

$$D_w = \frac{\Gamma_{zw}}{\delta z_w \Delta z} \tag{A-1.21}$$

$$D_f = \frac{\Gamma_{\theta f}}{r_P^2 \delta \theta_f \Delta \theta} \tag{A-1.22}$$

$$D_b = \frac{\Gamma_{\theta b}}{r_P^2 \delta \theta_b \Delta \theta} \tag{A-1.23}$$

The convection coefficients $C$ multiply $\phi$ at the control volume face itself. Therefore a value for $\phi$ must be interpolated from the main grid points onto the respective control volume faces. Numerous schemes for doing this are available, as discussed in Section 3-3. For the simpler schemes, an interpolation involving the two neighbouring grid points either side of the control volume face is used. So, for the general variable $\phi$ at the 'w' control volume face, the following interpolation is used:

$$\phi_w = Wf_W\phi_W + Wf_P\phi_P \tag{A-1.24}$$

Using the HYBRID scheme results in the following main grid point coefficients:

$$A_N = \left[\!\left[\; D_n, \; \left|\frac{C_n}{2}\right| \;\right]\!\right] - \frac{C_n}{2} \tag{A-1.25}$$

$$A_S = \left[\!\left[\; D_s, \; \left|\frac{C_s}{2}\right| \;\right]\!\right] + \frac{C_s}{2} \tag{A-1.26}$$

$$A_E = \left[\!\left[\; D_e, \; \left|\frac{C_e}{2}\right| \;\right]\!\right] - \frac{C_e}{2} \tag{A-1.27}$$

$$A_W = \left[\!\left[\; D_w, \; \left|\frac{C_w}{2}\right| \;\right]\!\right] + \frac{C_w}{2} \tag{A-1.28}$$

$$A_F = \left[\!\left[\; D_f, \; \left|\frac{C_f}{2}\right| \;\right]\!\right] - \frac{C_f}{2} \tag{A-1.29}$$

$$A_B = \left[\!\left[\; D_b, \; \left|\frac{C_b}{2}\right| \;\right]\!\right] + \frac{C_b}{2} \tag{A-1.30}$$

where the construct $\left[\!\left[\; A, B \;\right]\!\right]$ means that the maximum value of A and B is taken. The following discretization is then obtained,

$$A_{P,\phi}\phi_P = A_{N,\phi}\phi_N + A_{S,\phi}\phi_S + A_{W,\phi}\phi_W + A_{E,\phi}\phi_E + A_{F,\phi}\phi_F + A_{B,\phi}\phi_B + S_{1,\phi}\phi_P + S_{2,\phi}\phi_P \tag{A-1.31}$$

The central grid point coefficients for axial velocity, $w$, radial velocity, $u$, tangential velocity,$v$, and temperature, $T$, are as follows,

$$A_{P,w} = \sum_{NB} A_{NB} + \frac{\rho_P^{\ old}}{\Delta t} \tag{A-1.32}$$

$$A_{P,u} = \sum_{NB} A_{NB} + \frac{2\mu_P}{r_P^{\ 2}} + \alpha_G \left| \frac{v_P}{r_P \rho_P} \right| + \frac{\rho_P^{\ old}}{\Delta t} \tag{A-1.33}$$

$$A_{P,v} = \sum_{NB} A_{NB} + \frac{\mu_P}{r_P^{\ 2}} + \frac{\rho_P^{\ old}}{\Delta t} \tag{A-1.34}$$

$$A_{P,T} = \sum_{NB} A_{NB} + \frac{\rho_P^{\ old}}{\Delta t} \tag{A-1.35}$$

The source terms are given by the following equations:

$$\begin{aligned} S_{1,w} = {} & \frac{p_w - p_e}{\Delta z} + \frac{\lambda_e \left(r_n u_{ne} - r_s u_{se}\right)}{r_P \Delta z \Delta r} - \frac{\lambda_w \left(r_n u_{nw} - r_s u_{sw}\right)}{r_P \Delta z \Delta r} \\ & + \frac{\lambda_e \left(v_{fe} - v_{be}\right)}{r_P \Delta z \Delta \theta} - \frac{\lambda_w \left(v_{fw} - v_{bw}\right)}{r_P \Delta z \Delta \theta} \\ & + \frac{\mu_n r_n \left(u_{ne} - u_{nw}\right)}{r_P \Delta z \Delta r} - \frac{\mu_s r_s \left(u_{se} - u_{sw}\right)}{r_P \Delta z \Delta r} \\ & + \frac{\mu_f \left(v_{fe} - v_{fw}\right)}{r_P \Delta z \Delta \theta} - \frac{\mu_{bw} \left(v_{be} - v_{bw}\right)}{r_P \Delta z \Delta \theta} \\ & + \frac{\rho_P^{\ old} w_P^{\ old}}{\Delta t} + \rho_P g \sin\theta_g \end{aligned} \tag{A-1.36}$$

$$S_{2,w} = 0 \tag{A-1.37}$$

$$
\begin{aligned}
S_{1,u} = & \frac{p_s - p_n}{\Delta r} + \frac{\mu_e\left(w_{ne} - w_{se}\right)}{\Delta z \Delta r} - \frac{\mu_w\left(w_{nw} - w_{sw}\right)}{\Delta z \Delta r} \\
& + \frac{\lambda_n\left(w_{ne} - w_{nw}\right)}{\Delta z \Delta r} - \frac{\lambda_s\left(w_{se} - w_{sw}\right)}{\Delta z \Delta r} \\
& + \frac{\lambda_n\left(v_{nf} - v_{nb}\right)}{r_n \Delta\theta \Delta r} - \frac{\lambda_s\left(v_{sf} - v_{sb}\right)}{r_s \Delta\theta \Delta r} \\
& + \frac{\mu_f}{\Delta\theta \Delta r}\left(\frac{v_{nf}}{r_n} - \frac{v_{sf}}{r_s}\right) \\
& - \frac{\mu_b}{\Delta\theta \Delta r}\left(\frac{v_{nb}}{r_n} - \frac{v_{sb}}{r_s}\right) \\
& - \frac{2\mu_p}{r_p^{\,2}}\frac{v_f - v_b}{\Delta\theta} + \frac{\rho_p v_p}{r_p} + \frac{\rho_p^{\;old} v_p^{\;old}}{\Delta t} + 2\rho_p v \Omega \\
& + \alpha_G \left|\frac{v_p}{\rho_p r_p}\right| u_p + \rho_p g \cos\left(\theta_g\right)\cos\left(\theta_p - \Omega t\right)
\end{aligned} \tag{A-1.38}
$$

$$
S_{2,u} = \frac{1}{\Delta r}\left(\frac{\lambda_n}{r_n} - \frac{\lambda_s}{r_s}\right) \tag{A-1.39}
$$

$$
\begin{aligned}
S_{1,v} = & \frac{p_b - p_f}{r_P \Delta\theta} + \frac{\mu_e\left(w_{fe} - w_{be}\right)}{\Delta z \Delta\theta r_P} - \frac{\mu_w\left(w_{fw} - w_{bw}\right)}{\Delta z \Delta\theta r_P} \\
& + \frac{\mu_n r_n\left(u_{nf} - u_{nb}\right)}{\Delta\theta \Delta r\, r_P^{\,2}} - \frac{\mu_s r_s\left(u_{sf} - u_{sb}\right)}{\Delta\theta \Delta r\, r_P^{\,2}} \\
& + \frac{\lambda_f\left(w_{ef} - w_{wf}\right)}{\Delta\theta \Delta z\, r_P} - \frac{\lambda_b\left(w_{eb} - w_{wb}\right)}{\Delta\theta \Delta z\, r_P} \\
& + \frac{\lambda_f\left(r_n u_{nf} - r_s u_{sf}\right)}{\Delta\theta \Delta r\, r_P^{\,2}} - \frac{\lambda_b\left(r_n u_{nb} - r_s u_{sb}\right)}{\Delta\theta \Delta r\, r_P^{\,2}} \\
& + \frac{2\left(\mu_f u_f - \mu_b u_b\right)}{\Delta\theta\, r_P^{\,2}} - 2\rho_P u_P \Omega + \frac{\rho_P^{old} v_P^{old}}{\Delta t} \\
& + \frac{\rho_P g \cos\theta_g\left(\cos\left(\theta_f - \Omega t\right) - \cos\left(\theta_b - \Omega t\right)\right)}{\Delta\theta}
\end{aligned} \tag{A-1.40}
$$

$$S_{2,v} = -\frac{\rho_P u_P}{r_P} - \frac{\mu_n - \mu_s}{r_P \Delta r} \tag{A-1.41}$$

$$S_{1,T} = \frac{\rho_P^{old} T_P^{old}}{\Delta t} \tag{A-1.42}$$

$$S_{2,T} = 0 \tag{A-1.43}$$

For steady state flows, under certain conditions $\sum_{NB} A_{NB} = 0$, which will result in a division by zero. Therefore, the following term is added to the left and right hand side of equation (A-1.31),

$$\left| C_{n,\phi} - C_{s,\phi} + C_{f,\phi} - C_{b,\phi} + C_{e,\phi} - C_{w,\phi} \right| \phi_P \tag{A-1.44}$$

This term will also increase diagonal dominance (the quotient $A_P / \sum A_{NB}$) of Equation (A-1.31), which improves convergence.

*Coefficients and Source Terms for Pressure Correction Equation*

The pressure field is computed iteratively by solving an equation to predict pressure corrections, $p'$, to an initial guessed pressure field. The form of the pressure correction equation is the same as Equation (A-1.31), with the coefficients and source terms defined as follows:

$$A_{N,p'} = \frac{\rho_n d_{n.u} r_n}{r_P \Delta r \delta r_n} \tag{A-1.45}$$

$$A_{S,p'} = \frac{\rho_s d_{s.u} r_s}{r_P \Delta r \delta r_s} \tag{A-1.46}$$

$$A_{E,p'} = \frac{\rho_e d_{e,w}}{\Delta z \delta z_e} \tag{A-1.47}$$

$$A_{W,p'} = \frac{\rho_w d_{w,w}}{\Delta z \delta z_w} \tag{A-1.48}$$

$$A_{F,p'} = \frac{\rho_f d_{f,v}}{r_P \Delta\theta r_P \delta\theta_f} \tag{A-1.49}$$

$$A_{B,p'} = \frac{\rho_b d_{b,v}}{r_P \Delta\theta r_P \delta\theta_b} \tag{A-1.50}$$

$$A_{P,p'} = \sum_{NB} A_{NB,p'} \tag{A-1.51}$$

$$S_{1,p'} = \frac{\rho_s u_s r_s - \rho_n u_n r_n}{r_P \Delta r} + \frac{\rho_w w_w - \rho_e w_e}{\Delta z} + \frac{\rho_b v_b - \rho_f v_f}{r_P \Delta\theta} + \frac{\rho_P^{old} - \rho_P^{new}}{\Delta t} \tag{A-1.52}$$

$$S_{2,p'} = 0. \tag{A-1.53}$$

## A.2 Treatment of Variables at the Axis of Rotation

At the axis of rotation the values of pressure, $p$, temperature, $T$, axial velocity, $w$, and tangential velocity, $v$, have no influence on the solution. This is because the coefficients at the axis $(A_{S,\phi})$ for these variables are multiplied by the radius (as can be seen from the previous discretization) and are therefore zero. Values of the variables, $p$, $T$, $w$ and $v$ at the axis are set as the average of the values at the radial node just away from the axis.

As a consequence of the grid staggering, the "South" face of the control volume for the radial velocity is not coincident with the axis of rotation. Therefore, the value set at the axis for this variable does influence the solution. The value, for each axial grid position, is set by resolving the tangential and radial velocity at node points adjacent to the axis into one velocity vector, passing through the axis. This vector is then resolved into a radial velocity component at each radial velocity control volume face coincident with the axis. Algebraic expressions for the treatment of variables at the axis are given below.

The variable $\phi_{i,j,k}$ is defined as the value of pressure, $p$, $T$, or $w$ at the cavity axis of rotation. The subscript $i$ represents the axial location, $j$ the radial, $k$ the tangential and $nk$ the number of grid nodes in the tangential direction. The value of $\phi_{i,j,k}$ at the axis is then defined by Equation (A-2.1),

$$\phi_{i,j,k} = \frac{1}{nk} \sum_{i=1,nk} \phi_{i,j+1,k} \tag{A-2.1}$$

The tangential velocity is assumed to be zero at the axis. The radial velocity is evaluated by defining the average horizontal velocity component,

$$E_{1i}=\frac{1}{nk}\sum_{i=1,nk}\left[v_{i,j+1,k}\cos(\theta_k)-w_{i,j+1,k}\sin\left(\frac{\theta_k+\theta_{k+1}}{2}\right)\right] \tag{A-2.2}$$

and the average vertical velocity component,

$$E_{2i}=\frac{1}{nk}\sum_{i=1,nk}\left[v_{i,j+1,k}\sin(\theta_k)+w_{i,j+1,k}\cos\left(\frac{\theta_k+\theta_{k+1}}{2}\right)\right] \tag{A-2.3}$$

A resultant vector,

$$E_{3i}=\left(E_{1i}^2+E_{2i}^2\right)^{1/2} \tag{A-2.4}$$

which passes through the axis of rotation is defined and this is resolved as a radial velocity vector at the axis using Equation (A-2.5),

$$v_{i,j,k}=E_{3i}\,cos\left[\theta_k-tan^{-1}\left(\frac{E_{2i}}{E_{1i}}\right)\right] \tag{A-2.5}$$

**A.3 Discretization of Earth's Acceleration due to Gravity**

For the axial momentum equation, $S_{1,w}$ is augmented by $\rho_P g\,\sin\theta_g$, where $\theta_g$ is the angle of the gravity vector relative to the $r-\theta$ plane of the cavity. For the radial momentum equation, $S_{1,u}$ is augmented by $\rho_P g\,\cos(\theta_g)\cos(\theta_P)$.

For the tangential momentum equation, the term $S_{1,u}$ can be augmented by $-\rho_P g\,\cos(\theta_g)\sin(\theta_P)$, but the curvature of the tangential control volume makes discretization of the body force non-conservative in this form. This can be seen if the hydrostatic pressure equation $p=h\rho g$ is applied across a control volume having a circumferential extent of $\pi/2$, one face being in the horizontal and the other in the vertical plane with $\theta_g=0$. The difference in hydrostatic pressure across the volume control faces, given by this equation, is $\rho\ g\ r$. The hydrostatic pressure difference given by using the non-conservative term is $\pi\rho gr/2$ which is 11% in error. However, since $d(\cos(\theta_P))/d\theta = -\sin(\theta_P)$, the term

$$\rho_P g \cos\theta_g \frac{\lfloor \cos(\theta_f) - \cos(\theta_b) \rfloor}{\Delta\theta} \tag{A-3.1}$$

which is conservative can be added to the tangential momentum equation. As can be seen, a discretization of this form has no error for the example discussed.

## A.4 The multilevel restriction and prolongation operators

*Restriction Operators*

To restrict variables from a mesh, m, onto a coarser mesh, *m + 1*, weighting functions are defined as:

$$Wf_w = \frac{\delta z_w}{2\Delta z},\quad Wf_e = \frac{\delta z_e}{2\Delta z},\quad Wf_b = \frac{\delta\theta_b}{2\Delta\theta},\quad Wf_f = \frac{\delta\theta_f}{2\Delta\theta},\quad Wf_p = \frac{\Delta z}{2\Delta z},$$

$$Wf_s = \frac{\delta r_s}{2\Delta r},\quad Wf_n = \frac{\delta r_n}{2\Delta r},\quad Wf_{pp} = \frac{r_p}{r_p},\quad Wf_{NP} = \frac{r_N}{r_p},\quad Wf_{SP} = \frac{r_S}{r_p}, \tag{A-4.1}$$

where $\Delta z = \delta z_e + \delta z_w, \Delta r = \delta r_n + \delta r_s$ *and* $\Delta\theta = \delta\theta_f + \delta\theta_b$. Therefore, for a one-dimensional flow, on a uniform grid, the restriction from the grid level, *m*, to the grid level, *m + 1*, is, $\phi^{m+1}{}_i = \phi^m{}_i/2 + (\phi^m{}_{i-1}/2 + \phi^m{}_{i+1}/2)/2$.

The following subscript functions are also defined: *ip1 = i + 1, im1 = i - 1, jp1 = j + 1, jm1 = j - 1, kp1 = k + 1, km1 = k - 1.* If the grid is not coarsened in the axial direction *ip1 = im1 = i.* If the grid is not coarsened in the radial direction *jp1= jm1 = j.* If the grid is not coarsened in the tangential direction, *kp1 = km1 = k.* For the main (as opposed to staggered) grid node variables, where $\phi = p$ or *T*, the 27 point weighted restriction given by Equation (A-4.2) is used,

$$\phi^{m+1}{}_P = Wf_{PP}Wf_P\phi^{m+1}{}_j + Wf_{NP}Wf_s\phi^{m+1}{}_{jp1} + Wf_{SP}Wf_n\phi^{m+1}{}_{jm1}, \tag{A-4.2}$$

where

$$\begin{aligned}\phi^{m+1}{}_j = [&Wf_P(Wf_P\phi^m{}_{i,j,k} + Wf_w\phi^m{}_{ip1,j,k} + Wf_e\phi^m{}_{im1,j,k}) + \\ &Wf_b(Wf_P\phi^m{}_{i,j,kp1} + Wf_w\phi^m{}_{ip1,j,kp1} + Wf_e\phi^m{}_{im1,j,kp1}) + \\ &Wf_f(Wf_P\phi^m{}_{i,j,km1} + Wf_w\phi^m{}_{ip1,j,km1} + Wf_e\phi^m{}_{im1,km1})]\end{aligned}$$

and $j = jm1, j$ or $jp1$.

The variables for velocity ($u$, $v$ and w) are staggered relative to the main grid and so an eighteen point weighted restriction is used. For the radial velocity, $u$, the following expression is defined,

$$\begin{aligned}u^{m+1}{}_k = [&Wf_p(Wf_s u^m{}_{i,j,k} + Wf_n\, u^m{}_{i,j,m1,k}) + \\ &Wf_w(Wf_s u^m{}_{ip1,j,k} + Wf_n u^m{}_{ip1,jm1,k}) + \\ &\quad Wf_e(Wf_s u^m{}_{im1,j,k} + Wf_n u^m{}_{im1,jm1,k})]\end{aligned}$$

and $k = km1$, $k$ or $kp1$.

The radial velocity on the coarse grid $m + 1$ is then given by Equation (A-4.3).

$$u^{m+1}{}_p = Wf_p u^{m+1}{}_k + Wf_b\, u^{m+1}{}_{kp1} + Wf_f\, u^{m+1}{}_{km1}. \tag{A-4.3}$$

For the tangential velocity, the following expression is defined,

$$\begin{aligned}v^{m+1}{}_i = [&Wf_p(Wf_b\, v^m{}_{i,j,k} + Wf_f\, v^m{}_{i,j,km1}) + \\ &Wf_s(Wf_b\, v^m{}_{i,jp1,k} + Wf_f\, v^m{}_{i,jp1,km1}) + \\ &\quad Wf_n\,(Wf_b\, v^m{}_{i,jm1,k} + Wf_f\, v^m{}_{i,jm1,km1})]\end{aligned}$$

and $i = im1$, $i$ or $ip1$.

The tangential velocity on the coarse grid node $m + 1$ is then given by Equation (A-4.4)

$$v^{m+1}{}_p = Wf_p\, v^{m+1}{}_i + Wf_w v^{m+1}{}_{ip1} + Wf_e v^{m+1}{}_{im1}. \tag{A-4.4}$$

For the axial velocity, the following expression is a defined,

$$\begin{aligned}w^{m+1}{}_k = [&Wf_P(Wf_w w^m{}_{i,j,k} + Wf_e w^m{}_{im1,j,k}) + \\ &Wf_s(Wf_w\, w^m{}_{i,jp1,k} + Wf_e w^m{}_{im1,jp1,k}) + \\ &\quad Wf_n(Wf_w w^m{}_{i,jm1,k} + Wf_e w^m{}_{im1,jm1,k})]\end{aligned}$$

and $k = km1$, $k$ or $kp1$.

The axial velocity on the coarse grid node $m + 1$ is then given by Equation (A-4.5).

$$w^{m+1}{}_P = Wf_P w^{m+1}{}_k + Wf_b w^{m+1}{}_{kp1} + Wf_f w^{m+1}{}_{km1}. \tag{A-4.5}$$

*Prolongation Operators*

This is the process used to interpolate corrections from the coarse onto the fine grid. The general expression (A-4.6) is used everywhere except where a coarse grid node point is coincident with a fine grid point.

$$\begin{aligned}\phi'^{m}{}_P = (Wf_f [Wf_s (Wf_w \phi'^{m+1}{}_{i,j,k} + Wf_e \phi'^{m+1}{}_{im1,j,k}) +\\ Wf_n (Wf_w \phi'^{m+1}{}_{i,jm1,k} + Wf_e \phi'^{m+1}{}_{im1,jm1,k})] +\\ Wf_b [Wf_s (Wf_w \phi'^{m+1}{}_{i,j,kp1} + Wf_e \phi'^{m+1}{}_{im1,j,kp1}) +\\ Wf_n (Wf_w \phi'^{m+1}{}_{i,jm1,kp1} + Wf_e \phi'^{m+1}{}_{im1,jm1,kp1})]).\end{aligned} \tag{A-4.6}$$

When the grid is not coarsened in the axial direction, all terms having the subscript $im1$ are set to zero and $Wf_w = Wf_e = 1$. When not coarsened in the radial direction, terms having the subscript $jm1$ are set to zero and $Wf_n = Wf_s = 1$. When not coarsened in the tangential direction, terms having the subscript $kp1$ are set to zero and $Wf_f = Wf_b = 1$.

# APPENDIX B

# DERIVATION OF THE CONDIF SCHEME FOR A VARIABLE DENSITY FLUID

The discretization is demonstrated using the one-dimensional Euler type Equation (B.1). It is made about the control volume shown in *Figure 2-2.2*. The scheme is easily extended to three-dimensional flows.

$$\frac{\partial(\rho u\phi)}{\partial x} = S_\phi \tag{B.1}$$

Equation (B.1) can be discretized using central differences to give Equation (B.2)

$$A_{P\phi}\phi_P = A_{E\phi}\phi_E + A_{W\phi}\phi_W + S_\phi, \tag{B.2}$$

where the coefficients are given by Equations (B.3) and (B.4), and $C = \rho u / \Delta x$.

$$A_E = \frac{-C_e}{2} \tag{B.3}$$

$$A_W = \frac{C_w}{2} \tag{B.4}$$

For a convection/diffusion equation terms denoted by say $D_e$ and $D_w$ ($D = \Gamma/\Delta x$, where $\Gamma$ is a diffusion coefficient) should be added to the right hand side of equations (B.3) and (B.4), respectively. The coefficients $A_E$ and $A_W$ are redefined in equations (B.5) and (B.6) as

$$A_E = \mathbf{A}_E - a_e \tag{B.5}$$

$$A_W = \mathbf{A}_W - a_w \tag{B.6}$$

where:

$$\mathbf{A}_E = \frac{\left(|C_e| - C_e\right)}{4} \tag{B.7}$$

$$\mathbf{A}_W = \frac{\left(|C_w| + C_w\right)}{4} \tag{B.8}$$

$$a_e = \frac{\left(|C_e| + C_e\right)}{4} \tag{B.9}$$

$$a_W = \frac{\left(|C_w| - C_w\right)}{4} \tag{B.10}$$

Equations (B.3) and (B.4) are exactly equivalent to (B.5) and (B.6). Equation (B.2) can now be stated as (B.11),

$$\left(\mathbf{A}_E + \mathbf{A}_W\right)\phi_P = \mathbf{A}_{E,\phi}\phi_E + \mathbf{A}_{W,\phi}\phi_W + a_e\left(\phi_P - \phi_E\right) + a_W\left(\phi_P - \phi_W\right) + S_\phi \tag{B.11}$$

If the terms involving $a_e$ and $a_w$ are amalgamated with $S_\phi$, the equation will always have positive coefficients giving a useful central difference based discretized equation form. However, if *Pe*, the Peclet number, becomes too large, the equation will become source-term dominated. To overcome this, the term involving $a_e$ in Equation (B.11) can be written as Equation (B.12),

$$a_e\left(\phi_P - \phi_E\right) = a_e\mathbf{R}\left(\phi_W - \phi_P\right), \tag{B.12}$$

where $\boldsymbol{R}$ is given by Equation (B.13),

$$\mathbf{R} = \frac{(\phi_E - \phi_P)}{(\phi_P - \phi_W)} \tag{B.13}$$

Similarly, the term involving $a_w$ can be written as Equation (B.14)

$$a_W(\phi_P - \phi_W) = \frac{a_W(\phi_E - \phi_P)}{\mathbf{R}} \tag{B.14}$$

Equation (B.11) can now be rewritten as (B.2), with the coefficients redefined as (B.15) and (B.16):

$$A_E = \mathbf{A}_E + \frac{a_W}{\mathbf{R}} \tag{B.15}$$

$$A_W = \mathbf{A}_W + a_e\mathbf{R} \tag{B.16}$$

Since the whole procedure has involved manipulation of the second-order central difference scheme, second order accuracy is maintained. In regions where, $\phi$ (the variable being solved for) is spatially invariant, the parameter $\mathbf{R}$ becomes equal to zero, and $A_E$ will become infinite. Where there are steep gradients, $\mathbf{R}$ can become large. Therefore, limits must be set on the parameter $\mathbf{R}$, and Runchal suggests that $1/\mathbf{R}_{\max} \le \mathbf{R} \le \mathbf{R}_{\max}$. For a one-dimensional flow of compressible fluid moving from left to right, the coefficients given in Equations (B.15) and (B.16) reduce to those given in Equations (B.17) and (B.18):

$$A_E = 0, \tag{B.17}$$

$$A_w = \frac{C_W(1+\mathbf{R})}{2} \tag{B.18}$$

The coefficients for the first order upwind scheme correspond to

$$A_E = 0 \tag{B.19}$$

$$A_W = C_W \tag{B.20}$$

and so for $\boldsymbol{R}$ = *1*, the CONDIF can be seen to be equivalent to this upwind scheme.

The CONDIF scheme for a three-dimensional convection-diffusion equation can be stated using equations (B.21) and (B.22) below.

$$A_{Q1,\phi} = D_{q1} + W_{f\,Q1} \tag{B.21}$$

$$A_{Q2,\phi} = D_{q2} + W_{f\,Q2} \tag{B.22}$$

For a cylindrical polar coordinate system, grid point locations are shown in *Figure 2-3.1*. For this, the axial momentum equation coefficients correspond to: *Q1=W; q1=w; Q2=E;* and *q2=e*. Similarly for the radial momentum equation*: Q1=S; q1=s; Q2=N* and *q2=n*. For the tangential momentum equation: *Q1=B*; *q1=b; Q2=F* and *q2=f.* Suitable rearrangement of these subscripts will give the form of the coefficients for a Cartesian system.

| **R** | Pe | $W_f$ |
|---|---|---|
| < 0 | - | $w_{fQ1} = \left[\!\left[ C_{q1}, 0 \right]\!\right]$ |
| .>0 | $\lvert Pe \rvert \le 2$ | $W_{fQ1} = C_{q1}/2$ |
| >0 | $\lvert Pe \rvert > 2$ | $W_{fQ1} = \left(\lvert C_{q1} \rvert + C_{q1}\right)/4 + \mathbf{R}\left(\lvert C_{q2} \rvert + C_{q2}\right)/4$ |
| <0 | - | $W_{fQ2} = \left[\!\left[ -C_{q2}, 0 \right]\!\right]$ |
| >0 | $\lvert Pe \rvert \le 2$ | $W_{fQ2} = -C_{q2}/2$ |
| >0 | $\lvert Pe \rvert > 2$ | $W_{fQ2} = \left(\lvert C_{q2} \rvert - C_{q2}\right)/4 + \mathbf{R}^{-1}\left(\lvert C_{q1} \rvert - C_{q1}\right)/4$ |

*Table B.1* Convection coefficients for the CONDIF scheme

The values for $W_{f\,Q1}$ and $W_{f\,Q2}$, for the CONDIF scheme, are given in *Table (B.1)*. The construct $\left[\!\left[ A, B \right]\!\right]$ means the maximum value of *A* and *B* is taken. From the table, it is notable that for $\mathbf{R} > 0$ (which will occur at very few grid points) first order upwinding is used. Also, for $Pe \le 2\,(\mathbf{R} > 0)$ the normal unmodified central difference discretization is used. For all other conditions the modified more elaborate CONDIF coefficients are applied.

# APPENDIX C

# MULTILEVEL CONVERGENCE PERFORMANCE EXEMPLARS

Aiding clarity, multilevel convergence's performance and wide potential is illustrated here using mostly steady flows.

## C.1 Driven Box Flow

The first case considered is for a three-dimensional square wall driven cavity flow with $Re$ = 200. The *FLO_ELEC* program is used for this prediction. *Table C.1* summarizes the single and multilevel convergence performances. As can be seen, the multilevel algorithm exhibits near perfect behaviour, the number of iterations to convergence remaining virtually constant with increasing number of grid nodes. Any variations are attributed to the number of grid nodes chosen, which, in some instances, prevented maximum grid coarsening. For the $11^3$, $17^3$, $21^3$, $33^3$ and $65^3$ grids, the number of permissible levels are 2, 4, 3, 6 and 6 respectively. Central processor unit (cpu) times are for a 120 MHz Indigo 2 Impact 1000 computer (unless otherwise stated, this is true for all cases in this Appendix).

| Grid | Iterations (Single level) | Cpu (s) (Single level) | Iterations (Multilevel) | Cpu (s) (Multilevel) |
|---|---|---|---|---|
| $11^3$ | 81 | 6.5 | 11 | 3.2 |
| $17^3$ | 157 | 61.7 | 10 | 15 |
| $21^3$ | 214 | 200.4 | 11 | 39.2 |
| $33^3$ | 405 | 2062.2 | 10 | 203.7 |
| $65^3$ | 946 | 51273.2 | 10 | 2216 |

*Table C.1* Convergence details for $Re$ = 200 driven cavity

For the $65^3$ grid the computing time saving factor is around 24. *Table C.2* shows a repeat of the above prediction, but this time, when the cavity wall is stationary and heated to give a Grashof (*Gr*) number of 3 x $10^5$. Again, for the multilevel method, the number of convergence iterations stays virtually constant with increasing numbers of grid points.

| Grid | Iterations (Single level) | Cpu (s) (Single level) | Iterations (Multilevel) | Cpu (s) (Multilevel) |
|---|---|---|---|---|
| $11^3$ | 65 | 4.82 | 9 | 3.31 |
| $17^3$ | 116 | 42.79 | 9 | 13.08 |
| $21^3$ | 145 | 147.7 | 9 | 36.77 |
| $33^3$ | 269 | 1379 | 9 | 184.97 |
| $65^3$ | 656 | 34741.76 | 10 | 2234.83 |

*Table C.2* Convergence details for $Gr = 3 \times 10^5$ heated cavity

The influence of the number of grid levels on multilevel convergence is illustrated in *Figure C.2,* for an increased $Gr = 1.44 \times 10^6$ with the $65^3$ grid. The single grid solution corresponds to $m = 1$. As can be seen, even for $m = 2$ the number of iterations to convergence drops significantly. For $m = 3$ the computing time saving factor is over 4.5 (however, for $m = 1$ the convergence trace is rather monotonic suggesting higher under-relaxation factors could have been used).

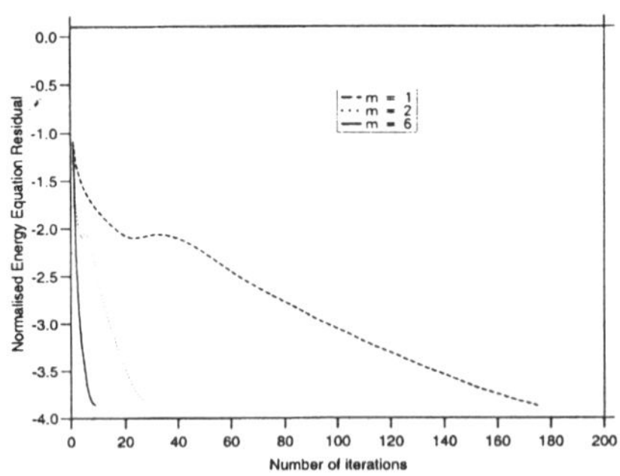


*Figure C.2* Plot of number of iterations against energy residual for $Gr = 1.44 \times 10^6$ and a $65^3$ grid

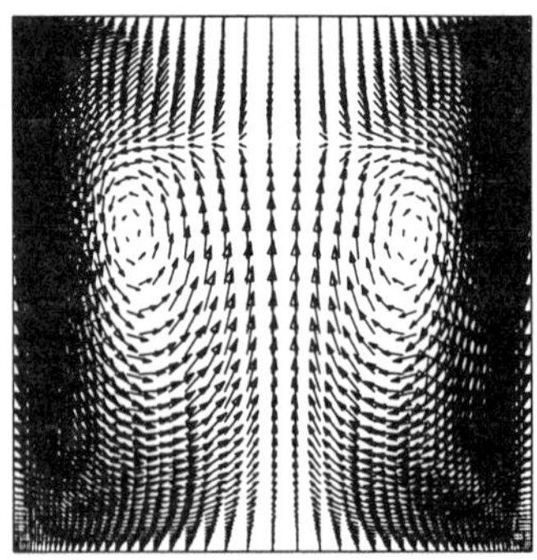

*Figure C.3* Mid-plane flow structure for $Gr = 3 \times 10^5$ heated cavity

*Figure C.3* shows the mid-plane flow structure for a plane parallel to the heated wall. As can be seen there are two recirculations. Again, from the table, impressive multilevel performance can be seen. This time, for the $65^3$ grid, the saving factor is around 15.

## C.2 Grid Embedding

Since the multilevel method uses a series of varying resolution grids it lends itself to grid embedding/adaptation. The availability of computations on the different grids enables local truncation error estimates to be easily made. In fact (see Brandt (1980) and Chapter 4 for nomenclature), the truncation error is proportional to $S'_{m+1} - S_{m+1}$.

Grid embedding can be achieved locally by limiting the extent of the fine grid to specific regions. Thus, unlike with other techniques, fine and hence generally slower to converge, grid regions will naturally be subjected to multilevel convergence acceleration. Fuchs (1986) and Thompson and Ferziger (1989) demonstrate the use of multilevel convergence based grid embedding using Cartesian grid systems. *Figure C.4* (reproduced from Chapter 5) shows a simple embedded grid used for a two-dimensional $Re$ = 400 wall driven cavity flow. The embedded region is nominally set in the half of the cavity adjacent to the moving wall. This is where it is expected that the velocity gradients are steepest. *Figure C.5* (again reproduced from Chapter 5), shows a comparison with the grid independent predictions of Rayner (1990) and the current HYBRID scheme predictions with and without refinement from grid embedding adjacent to the moving wall. The solution is made by the straight forward adaptation of the *FLO_ELEC* program. The improvement in accuracy from grid embedding is clear.

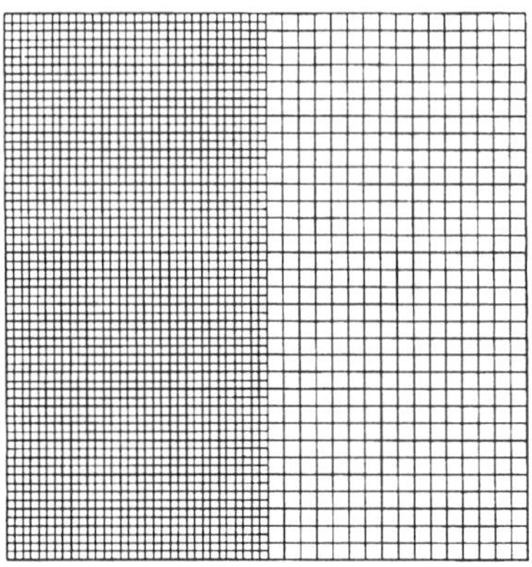

*Figure C.4* Simple embedded grid used for two-dimensional wall driven cavity flow

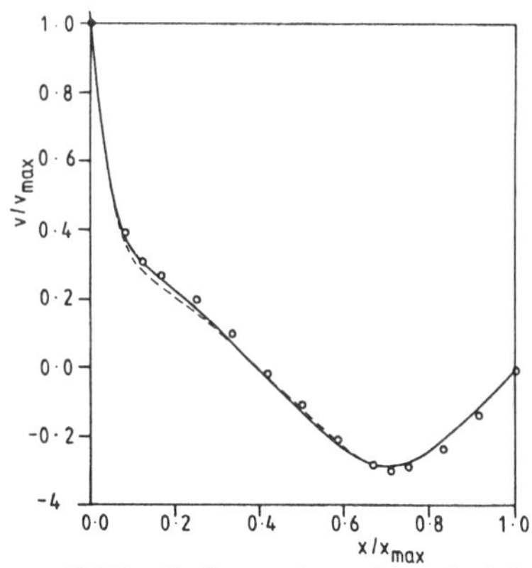


*Figure C.5* Predictions using grid embedding for a wall driven cavity with $Re$ = 400

## C.3 Cluttered Geometries

For complex and cluttered geometries linear algebraic multilevel methods are preferred (see Raw (1995)). With such methods the 'coarser' grid (*m + 1*) equations are obtained by summing surrounding coefficients at the lower, *m*, level. Hence, the method is no longer grid based and is only applicable to linear equations. A key problem with complex cluttered geometries is that if excessively coarse, higher level grids cannot support the geometry (see Ruge and Stuben (1986)). However, grid coarsening is required for the multilevel method's success. The point to be made can be more clearly appreciated when considering the modelling of the two-dimensional geometry shown in *Figure C.6.*

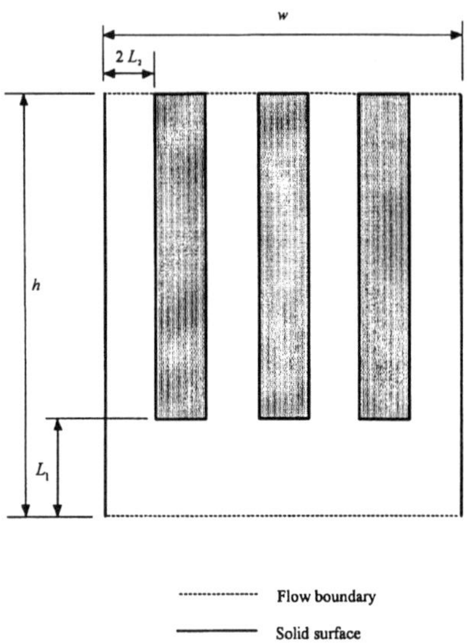


*Figure C.6* Two-dimensional geometry with three interior solid components

This geometry consists of two vertical walls with two horizontal flow inlets and outlets. The flow inlet being at the top. Inside the geometry are three solid components. For a strictly Cartesian, structured grid solver these can be modelled by rendering control volumes inside the solid components inactive. For turbulent flow, according to Shy *et al.* (1993) grid nodes nearest solid walls should be fixed at all levels. Also, to prevent solid boundaries moving, grid nodes either side of them must remain fixed. Therefore, to maintain the geometry shown in *Figure C.6*, (whilst keeping one node at the centre of each flow channel to resolve the flow) in the $x$ direction around 20 nodes would be required. Clearly, by industrial standards this geometry is not highly complex (electronic systems can have around 100 surfaces in each coordinate direction), but does suggest that, for most complex systems, the coarsest grids may require many nodes. Consequently with an optimal two to one refinement ratio (see Brandt (1977)) between grids, the finest grid is likely to be excessively fine and too expensive on both storage and computing time.

The algebraic multilevel technique, which is not concerned with these geometry issues, overcomes the problem, but is less efficient. The question arises regarding how much of the geometry on the coarse grid must be resolved for geometric multilevel convergence to be of use. This is considered by modelling flow for the *Figure C.6* geometry. For this geometry $L_2/w = 0.071$, $L_1/h = 0.23$ and $w/h = 0.38$ (note: for presentation purposes $w$ has been scaled). A suitable solution grid set, refined at solid surfaces, using shifted Chebyshev polynomial functions, is shown in *Figure C.7*. The grid has 65 interior nodes in the $x$ and $y$ directions and the maximum number of six levels ($m_{max} = 6$) are shown. To avoid distortion of the geometry for $m > 1$, grid lines adjacent to solid boundaries should be fixed. This would be quite straightforward for this system. However, to study the effect of geometric distortion, here, lines adjacent to the interior geometry are left free. Therefore, as can be seen for $m > 1$ the solid component positions and shapes change. In frames (e) and (f) the components have merged, forming one component and the only solution nodes are at the boundary.

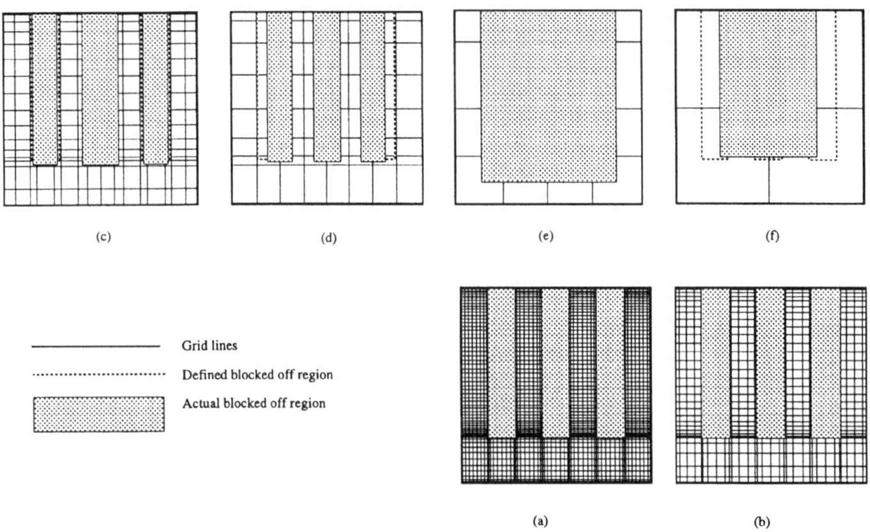


*Figure C.7* Solution grid for the two-dimensional geometry with three interior solid components

To validate the solution, comparison is made with an analytical laminar flow solution to Equation (C.1) presented by Schlichting (1979). This gives the horizontal variation of dimensionless fully developed vertical velocity $v/V$ as

$$\frac{v}{V}=\frac{3}{2}\left(1-\frac{x^2}{L_2^2}\right) \tag{C.1}$$

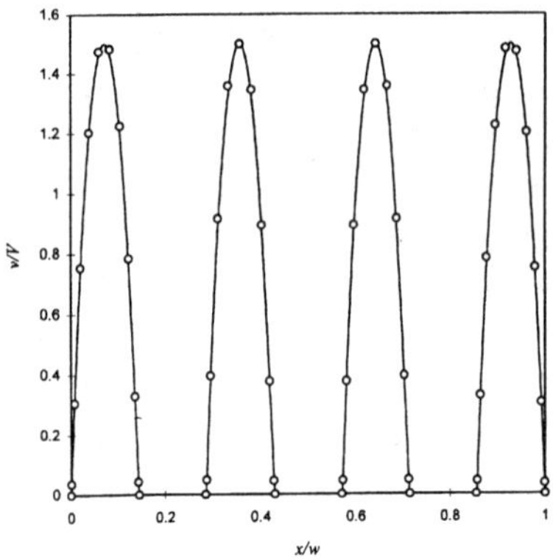


*Figure C.8* Plot of $v/V$ against $x/w$ at $y/h \cong 0.75$ for $\text{Re}=2\rho VL_2/\mu=200$.
(o Numerical prediction, — Analytical solution)

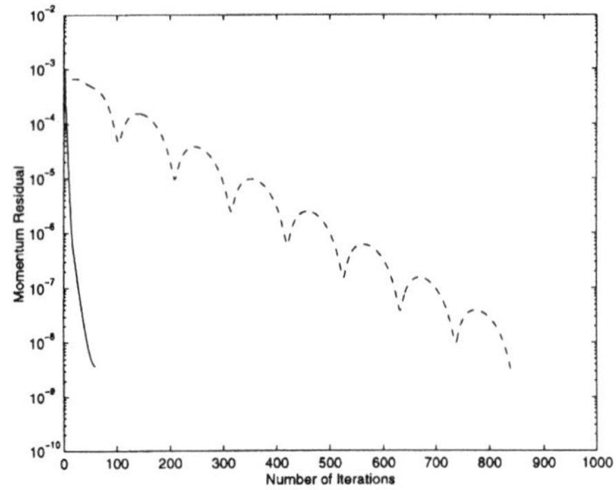


*Figure C.9* Plot of averaged horizontal and vertical momentum residual against number of iterations

Note, $x$ is measured from each of the channel centre lines. *Figure C.8* shows a plot of $v/V$ against $x/w$ at $y/h \cong 0.75$ for $\text{Re}=2\rho VL_2/\mu=200$. The symbols represent the predictions and the lines the analytical solution. As can be seen there is agreement. The multilevel algorithm performance will now be studied for this low Reynolds number case. *Figure C.9* shows a plot of averaged horizontal and vertical momentum residual against number of iterations. The dashed and full lines correspond to single and multilevel (with $m_{\max} = 6$) solutions, respectively. As can be seen, despite the geometrical distortion, the multilevel convergence is effective, successfully reducing residuals by over five orders of magnitude. Due to geometrical distortion the residual for the multilevel convergence solution cannot be driven much below $1 \times 10^{-8}$, but this corresponds to a negligible error.

| Levels ($m_{max}$) | Iterations | Computing time (s) |
|---|---|---|
| 1 | 373 | 38.8 |
| 2 | 52 | 22.4 |
| 3 | 23 | 10.2 |
| 4 | 19 | 8.6 |
| 5 | 18 | 8.3 |
| 6 | 18 | 8.3 |

*Table C.3* Performance for different grids and values of $m_{max}$.

*Table C.3* summarizes multilevel algorithm performance for different numbers of levels $m_{max}$. For $m_{max}$ = 5, the best multilevel algorithm performance is gained. For $m > 5$ (where only equations at boundaries are being solved), despite the significant geometrical distortions, the algorithm performance is not greatly deteriorated. Despite the significant distortions, the higher levels improve, or at worst do no harm to convergence.

| Grid size | Multilevel Predictions | | Single Level Predictions | |
|---|---|---|---|---|
| | Iterations | Computing time (s) | Iterations | Computing time (s) |
| $17^2$ | 15 | 0.4 | 82 | 0.5 |
| $33^2$ | 18 | 1.9 | 208 | 4.8 |
| $65^2$ | 18 | 8.3 | 373 | 38.8 |
| $129^2$ | 18 | 41.9 | 867 | 494 |

*Table C.4* Performances for different numbers of grid nodes

Therefore, it is seen that the multilevel method can be surprisingly robust. *Table C.4* shows the multi and single level performances for different numbers of nodes. Predictions are made using similar grid structures to that for the $65^2$ grid. For the $17^2$, $33^2$, $65^2$ and $129^2$ grids $m_{max}$ is taken as 4, 5, 6 and 7 respectively. Pleasingly (see *Table C.4*), despite geometrical distortion, the number of iterations to convergence is reasonably grid independent and so significant computing time savings are made.

## C.4 Grid Aspect Ratio

For high aspect ratio grids, the coefficients of the discretized equations will become strongly anisotropic (e.g. for a two-dimensional flow $A_{W\phi}$ and $A_{E\phi}$ may be much larger than $A_{N\phi}$ and $A_{S\phi}$). This anisotropy can affect the smoothing properties of certain simultaneous equation solvers and significantly reduce multilevel convergence (see Raw (1995), Dimitriadis and Leschziner (1989) and Ferziger and Peric (1997, p. 124)). The effect of coarse grid aspect ratio will be briefly considered here. In this book several time dependent rotating flows are modelled.

Therefore, it is worth considering the effect of coarse grid aspect ratio on such flows. This is addressed for the rotor stator geometry shown in *Figure C.10*. The inner shaded disc rotates to give a rotational Reynolds number $\mathrm{Re}_\phi = \rho\Omega h^2/\mu \cong 4.5 \times 10^6$. The geometry is such that $L_1/h = 0.1$, $L_2/h = 0.0064$ and $(w - L_3)/h = 0.051$.

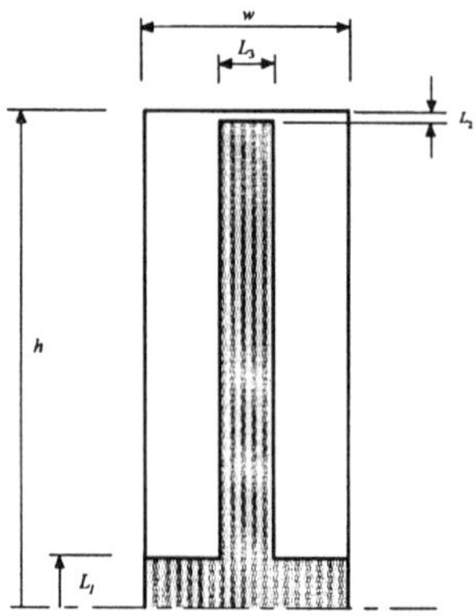

*Figure C.10* Rotor-stator geometry

To control coarse grid aspect ratio, the *FLO_CAV* program is modified so that grid coarsening in a particular co-ordinate direction can be stopped after a specified level. This procedure also enables grids to be coarsened to the maximum extent when there are different numbers of grid points in each of the co-ordinate directions. For use of this technique in three-dimensional computations see Long and Tucker (1994). Once grid coarsening, in a particular co-ordinate direction, has stopped, the multilevel methods performance at reducing low frequency errors is impeded. However, a reasonable coarse grid aspect ratio and hence smoothing properties can be maintained. Because of symmetry, only one half of the system is studied. Also, the grid is such that there is no geometrical distortion for $m > 1$. Computations are made on a 97 × 97 grid with $m_{max} = 4$. The influence of not coarsening in the $r$ or $x$ co-ordinate directions, after specified levels $m_{r,max}$ and $m_{x,max}$, on computing time is shown in *Table C.5*.

| $m_{x,max}$ | $m_{r,max}$ | $m_{r,max} / m_{x,max}$ | Computing Time |
|---|---|---|---|
| 4 | 2 | ½ | 109 |
| 4 | 3 | ¾ | 47 |
| 4 | 4 | 1 | 35 |
| 3 | 4 | 4/3 | 32 |
| 2 | 4 | 2 | 40 |

*Table C.5* Influence of $m_{z,max}$ and $m_{r,max}$ on computing time

Computing times are for a Solbourne SE/905-128 mainframe. Surprisingly, the table shows the worst computing time is for the best coarse grid aspect ratio, corresponding to $m_{r,max}/m_{x,max} = 1/2$. Also, best convergence occurs at a poor coarse grid aspect ratio, $m_{r,max}/m_{x,max} = 4/3$ - corresponding to a grid aspect ratio of approximately 100:1, on the highest grid level. For this grid structure, the sharp disc boundary layer regions are relatively well resolved on the coarser grid levels. This may suggest that for this flow type, gradient resolution is more important than aspect ratio. However, for an aspect ratio of greater than approximately 100:1 convergence finally begins to deteriorate. Clearly, from the computing time range shown in the table, attention should be paid to restricted as well as fine grid structures.

## C.5 Heated Rotating Cavity Flow

*Figure C.11* shows, for the *Figure 8-3.2b* rotating heated cavity flow, plots of average momentum residuals against number of iterations through one $\Delta t$ = 0.05 s, fully implicit scheme, time-step. The full line represents the multilevel solution. The broken and dash/dot lines correspond to single level solutions with ($\alpha_0 = 2$) and without the Gosman term (Equation (4-2.8)), respectively. As can be seen, like multilevel convergence, the Gosman term significantly improves convergence.

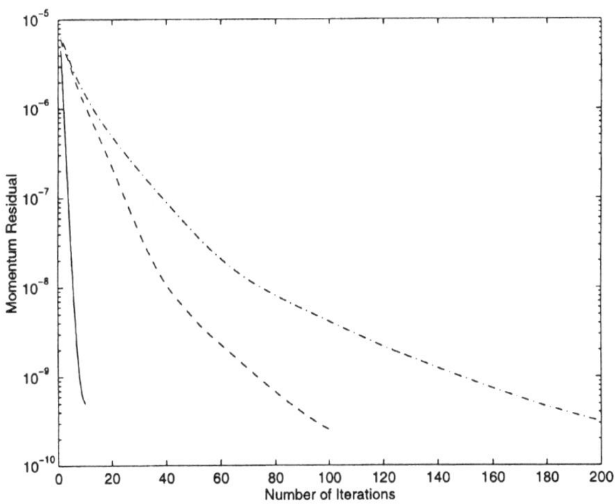


*Figure C.11* Plot of average momentum residual against number of iterations ( $m = 3$ ——, $m = 1$, $\alpha_0 = 0$ -·-·- , $m = 1$, $\alpha_0 = 2$ - - - )

## C.6 Conclusions

As can be seen from the above discussion, the multilevel method has many advantages. However, key disadvantages are that it is difficult to implement and does not generally efficiently accelerate convergence of the $k$-$\varepsilon$ equations (see Tucker (1990)). When this is the case, the $k$-$\varepsilon$ equations are best solved on a single grid, the remaining equations being subjected to multilevel convergence acceleration.

# APPENDIX D

# FURTHER CHAPTER 10 CASE DETAILS

*Boundary Conditions and Solution Initialization*

Boundary condition designations given on *Figure 10-2.1* are summarised below where the velocity components $u$, $v$ and $w$ correspond to the $r$, $\theta$, $z$ directions, respectively. The subscripts $s$, $b$ and $a$ are used to refer to shaft, bearing and ambient conditions respectively, and $z = 0$ taken to correspond with the bearing mid-axial position.

*Fluid-Journal Boundary (bc 1).*

Velocity, centripetal acceleration, and thermal boundary conditions on $r = a$ imply.

$$u = \frac{\partial h}{\partial t}(\theta, t), \quad v = r\Omega + \frac{\partial h}{\partial t}\left(\theta + \frac{1}{2}\pi, t\right), \quad w = 0,$$

$$\frac{\partial p}{\partial r} = \frac{\rho v^2}{r}, \quad T = T_s, \quad k\frac{\partial T}{\partial r} = k_s \frac{\partial T_S}{\partial r} \qquad \text{(D.1)}$$

*Fluid-Bearing Boundary, Excluding Inlet (bc 2).*

These are similar to those for the lubricant-journal boundary, with $r = a + h$:

$$u = v = w = 0, \frac{\partial p}{\partial r} = \frac{\rho v^2}{r},$$

$$T = T_b, k\frac{\partial T}{\partial r} = k_b \frac{\partial T_b}{\partial r} \qquad \text{(D.2)}$$

*Fluid-Groove Interface (bc 3).*
The lubricant-groove interface corresponds to the boundary $r = a + h$, over the area of groove such that:

$$u \text{ by continuity}, \quad v = w = 0, \quad p = p_i, \quad T = T_i \tag{D.3}$$

*Fluid at Bearing Sides (bc 4).*
At the bearing sides $(|z| = L/2)$:

$$\frac{\partial u}{\partial z} = \frac{\partial v}{\partial z} = \frac{\partial T}{\partial z} = 0, \quad w \text{ by continuity}, \quad p = p_a \tag{D.4}$$

*Shaft External to Bearing Width (bc 5).*
On the surface of the shaft exterior to the bearing width $(r = a, |z| > L/2)$ and at the nominal ends $(|z| = L_s/2)$, heat transfer coefficient boundary conditions are imposed:

$$k_s \frac{\partial T_s}{\partial r} = -h_s (T_s - T_a) \tag{D.5}$$

*Bearing Outer Surface and sides (bc6).*
On the outer bearing surface $r = b_o$ and at the bearing sides $(|z| = L_s/2)$

$$k_b \frac{\partial T_b}{\partial r} = -h_b (T_b - T_a) \tag{D.6}$$

Typically at $t = 0$ $T = T_s = T_b = T_a$, $p = p_a$ and $u = v = w = 0$, everywhere.

*Dynamic Cavitation Modelling*
The cavitation region consists of an oil, oil vapour and/or air mixture. The overall large-scale properties of the cavitation region are fairly homogenous and it is reasonable to use averaged fluid properties, based on the fraction of lubricant and gas/vapour. The fraction of fluid $\Phi(\theta, z, t)$ can be obtained from conservation of mass considerations using the following equation (see Tucker and Keogh (1996)).

$$\frac{1}{a}\frac{\partial}{\partial \theta}\left(\int_a^{a+h} v \, dr \, \Phi\right) + \frac{\partial}{\partial z}\left(\int_a^{a+h} w dr \, \Phi\right) + \frac{\partial}{\partial t}(h\Phi) = 0 \tag{D.7}$$

This equation can be used either with Couette velocity components (velocity components arising when pressure gradients are assumed zero) or the naturally predicted velocity components (Couette and Poiseuille). In journal bearing cavitation modelling the Couette component is generally preferred and so this is used in Case Studies. The mean properties of a lubricant-vapor mixture in a cavitated region need also to be specified. For density, a mass conserving mean is used

$$\rho = (1-\Phi)\rho_{\upsilon} + \Phi\rho_{l} \tag{D.8}$$

where the subscripts v and $l$ refer to vapour and lubricant properties, respectively. However, a harmonic mean (Knight and Niewiarowski. (1990))

$$\beta^{-1} = (1-\Phi)\beta_{v}^{-1} + \Phi\beta_{l}^{-1} \tag{D.9}$$

is used to define viscosity, specific heat and thermal conductivity $(\beta = \mu, c_{v}, k)$. Equations (D.7) – (D.9) provide an approximate representation of the thermal and fluid dynamic effects of cavitation. Equation (D.7) is solved here for $p < p_{c}$ (the fluid's cavitation pressure) using the explicit forward difference Euler scheme.

Data for the Lund and Tonnesen (1984) geometry is given in *Table D.1.*

| | | |
|---|---|---|
| a = 0.05$m$ | b = 0.1 $m$ | L = 0.055$m$ |
| c = 0.0686$mm$ | $\Omega = 367\, rad/s$ | $e_0 = 0.0398mm$ |
| $\rho_l = 850 kg/m^3$ | $\rho_v = 1.2\ kg/m^3$ | $\rho_b = \rho_s = 7850\ kg/m^3$ |
| $k_l = 0.15 W/m\,K$ | $k_v = 0.025 W/m\,K$ | $k_b = k_s = 50 W/m\,K$ |
| $c_{vl} = 2000 J/kg\,K$ | $c_{vv} = 1000 J/kg\,K$ | $c_{vb} = c_{vs} = 460\ J/kg\,K$ |
| $h_b = h_s = 50 W/m^2 K$ | $T_a = 35^\circ C$ | $T_i = 50^\circ C$ |
| $p_c = 0\ bar$ | $p_i = 1.2,\ 1.5\ bar$ (inlets (1,2)) | $L_s = 0.45m$ |
| $\theta_i = 9^\circ$ | $L_i = 0.0275$ m | |

The lubricant viscosity relation is such that $\mu_l = 0.0182$ kg/m s at 50°C, $\mu_l = 0.0056$ kg/m s at 90 °C. The vapour viscosity is set at $\mu_v$ = 2 x 10-5 kg/m s.

*Table D.1* Data for non-isothermal Lund and Tonnesen comparison

APPENDIX

# E

# FLO_CAV PROGRAM USER INFORMATION

## E.1 *FLO_CAV* Program

*FLO_CAV* is a CFD program for the prediction of three-dimensional flows in a cylindrical polar coordinate system. This section provides a description of the program structure, subroutines, data input format and boundary conditions. The programs general use is also described.

## E.2 Description of Program Structure and Subroutines

*Figure E.1*, shows a simplified diagram of the program structure. The term RELAX is used to describe the process of setting up the coefficients of the discretised equations and applying the TDMA solver a specified number a times. The multigrid (or multilevel) algorithm and the other algorithms used in this program are described in detail in Chapters 2-7. Sections of the program that are redundant when not using the multigrid algorithm (i.e. making a single grid solution) are indicated using the superscript m.

### INITIALISE

1. Read in problem definition.
2. Set up constants.
3. Call **GRID3** - (Calculates the fine grid mesh) or read in mesh from external file.
4. Call **INIT3** - initialises variables or reads in old data and sets up previous values for time dependent solution. Also repositions reference point on the nearest grid node.
5. Checks storage sufficient.

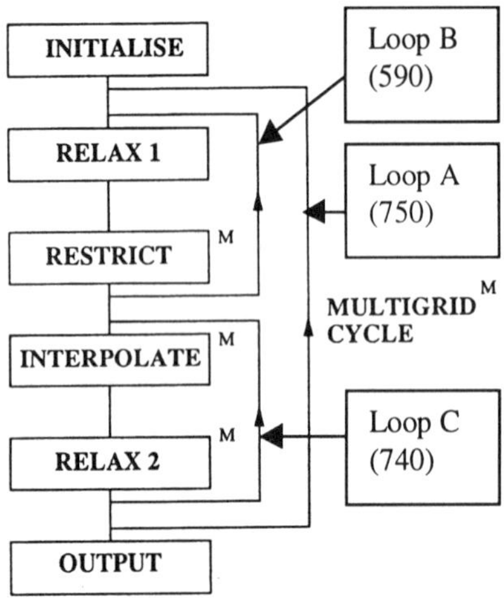


*Figure E.1* Simplified diagramof program structure

6. Call **GRIDC3**$^{m}$ - calculates all coarse meshes (every other main grid node omitted).
7. If incompressible, constant laminar viscosity, sets density and viscosity on all meshes.
8. Moves sector positions.
9. Sets indicators for derivative boundary conditions.
10. Checks whether reference point is on a coarse grid node.
11. Call **GRIDB3** - works out coordinates of first node within each sector.
12. Call **BLOCK3** - works out coordinates of first node beyond block. Also assigns indicator to each node to define whether solid, fluid or derivative boundary.
13. Sets up indicator for boundary nodes to say whether boundary should be treated as a wall for purposes of mixing length model.
14. Write to screen problem definition.
15. Call **BOUND3** - sets up all fixed boundary conditions.
16. Call **BLKBND3** - sets up values of variables on blocks.
17. Call **CVISL3** - calculates laminar viscosity.
18. Call **CDEN3** - calculates density from gas law for air, or correlation for water. If using gas law, sets up all boundary pressures.
19. Call **BNDCOM3** - resets boundary velocities for fixed profiles but variable density so as to maintain specified flow rate.
20. Restricts previous values of time dependent variables onto coarse grids (routines described later).
21. Adjusts angular velocity when using spin-up option.
22. Starts single grid solution cycle or multigrid cycle.

**RELAX 1** - (start of solution cycle) SIMPLEC algorithm, starting on finest grid.

1. Call **TRANSW3** - switches tangential velocity into rotating frame, if required.
2. Call **COEFFW3** - sets up coefficients for tangential momentum equation.
3. Call **SOLVE3** - TDMA solver (repeating boundary in tangential direction).
4. Call **TRANSW3** - switch back to stationary frame.
5. Call **DBNDW3** - reset boundary tangential velocities for derivative conditions.
6. Call **PSWB3** - resets 'fixed' pressure condition so that dp/dr = (rho*w**2)/r if w (tangential velocity) varying on boundary.
7. Call **AXIS3** - calculates radial velocity on axis by resolving radial velocity and tangential velocity and set values of other axis values when making 3-d solution.
8. Call **COEFFV3** - sets up coefficients for radial momentum equation.
9. Call **SOLVE3** - for derivative boundary nodes, call **SOLVE3** for main domain.
10. Call **AXIS3**
11. Call **COEFFU3** - sets up coefficients for axial momentum equation.
12. Call **SOLVE3** - for derivative boundary nodes - call SOLVE3 for main domain.
13. Call **DBNDUV3** - calculates boundary values of axial and radial velocities for derivative conditions not in their own directions.
14. Call **COEFFP3** - sets up coefficients for pressure correction equation.
15. Call **SOLVE3** - for specified number of sweeps.
16. Call **PRC3** - corrects velocities and pressure.
17. Call **COEFFT3** - sets up coefficients for temperature equation.
18. Call **SOLVE3** - for a specified number of sweeps.
19. Call **DBNDT3** - calculates boundary temperatures for derivative conditions.
20. Call **CVISL3**
21. Call **CDEN3**
22. Call **BNDCOM3**
23. Call **CVIST3** - calculates turbulent viscosity from mixing length model.
24. Repeat for specified number of sweeps, then go to RESTRICT unless on coarsest grid, then go to INTERPOLATE.

**RESTRICT**

1. Call **RESTP3** - restricts variables on main nodes to next coarse grid, call for p, T and turbulent viscosity$^m$.

2. Call **RESTU3** - restricts variables on axial velocity nodes, call for axial velocity$^{m}$.
3. Call **RESTV3** - restricts variables on radial velocity nodes, call for radial velocity$^{m}$.
4. Call **RESTW3** - restricts variables on tangential velocity nodes, call for tangential velocity$^{m}$.
5. Call **BLKBND3**$^{m}$
6. Call **DBNDUV3**$^{m}$
7. Call **AXIS3**$^{m}$
8. Call **DBNDT3**$^{m}$
9. Call **CDEN3**$^{m}$
10. Call **CVISL3**$^{m}$
11. Call **DBNDW3**$^{m}$
12. Call **PSWB3**$^{m}$
13. Work out defects/residuals for each variable.
14. Calculate right hand side of defect equation$^{m}$.
    i.e Call **COEFF** - fine grid.
        Call **RES3** - works out defects/residuals - fine grid.
        Call **REST** - restrict to coarse grid$^{m}$.
        Call **RHS** - calculates right hand side on coarse grid.
    do for axial, radial and tangential velocity and pressure and temperature$^{m}$.
15. Call **OLD3** - saves restricted values of all principal variables$^{m}$.
16. Store residuals and rms values for finest grid.
17. Go to **RELAX**$^{m}$

**INTERPOLATE**

1. Call **INTERP3** - calculates multigrid corrections for each variable, under-relaxes them, interpolates onto next finest grid and adds them onto variable$^{m}$.
2. Call **BLKBND3**$^{m}$
3. Call **DBNDT3**$^{m}$
4. Call **CVISL3**$^{m}$
5. Call **CDEN3**$^{m}$
6. Call **BNDCOM3**$^{m}$ - if of finest grid.
7. Call **CVIST3**$^{m}$ - if on finest grid.

**RELAX 2**$^{m}$

1. As before, for a specified number of sweeps then go to **INTERPOLATE** unless on finest grid (then one multigrid cycle complete)$^{m}$.
2. Print out residuals and r.m.s and values at monitoring point.

**OUTPUT** stage at end of run or time-step.

1. Call **OUTPUT1** - writes data out (this first call is used when running unsteady cases to write out the data at the end of each time-step).
2. Call **CHANGE3** - makes all variables at the current time-step ($t^{new}$) old ($t^{old}$) values ready to compute the next time-step.
3. Call **OUTPUT2** - calculates stream function, shear stresses, moment coefficients, Nusselt numbers, momentum and energy balances.

## E.3 Program Input

In this section the program input is formally introduced. Generally only one input file is required to run the *FLO_CAV* program. The program can be run on a UNIX machine by typing the following,

flow_cav.e < flow_cav.com,

where flow_cav.com (sample command files are included on disc with file names corresponding to case figure numbers used in the main text) is the input file and flow_cav.e is the executable file. Generally, the solution grid is generated in the main program (neat-b.e). However, *FLO_CAV* has the facility to add block like components emanating from solid boundary surfaces. When this is the case, grids should be refined at the block and other boundary surfaces. To do this, the grid is best generated outside the *FLO_CAV* program (if done internally the grid is not refined at block surfaces). The data file (generated using grid.f) containing grid coordinates can be read into the program. To do this, the grid data file name should be specified in the flow_cav.com file. The input file parameter layout, followed by a description of the parameters is given below. Program input should be in M.K.S units.

**ADAPTIVE TIME-STEP PARAMETERS**
nadapt, dtmin, dtmax, tmax, nadaptv

**FALSE DIFFUSION PARAMETER, CONVECTIVE SCHEME INDICATOR**
rupper, ncond

**IBM**
ivc

**TIME DEPENDENCE PARAMETERS**
itd, nts, dt, tf, nstep,dummy

**'ALTERNATING DIRECTION IMPLICIT' CONTROLS (Z,R,THETA)**
fqwe, fqns, fqfb

**SOLUTION TYPE**
lt, dummy, incc, isnis, icp, iflu

**GEOMETRY**
ndr(1), rb(m,1) for m = 1 to ndr(1)+1
ndr(2), rb(m,2) for m = 1 to ndr(2)+1
ndz(1), zb(m,1) for m = 1 to ndz(1)+1
ndz(2), zb(m,2) for m = 1 to ndz(2)+1
thmax1

**BLOCK DEPTHS**
zbl(m,1) for m = 1 to ndr(1)
zbl(m,2) for m = 1 to ndr(2)
rbl(m,1) for m = 1 to ndz(1)
rbl(m,2) for m = 1 to ndz(2)

**MOMENTUM BOUNDARY CONDITIONS**
cruv(m,1) for m = 1 to ndr(1)
cruv(m,2) for m = 1 to ndr(2)
czuv(m,1) for m = 1 to ndz(1)
czuv(m,2) for m = 1 to ndz(2)

**SWIRL BOUNDARY CONDITIONS**
crw(m,1) for m = 1 to ndr(1)
crw(m,2) for m = 1 to ndr(2)
czw(m,1) for m = 1 to ndz(1)
czw(m,2) for m = 1 to ndz(2)

**TEMPERATURE BOUNDARY CONDITIONS**
crt(m,1) for m = 1 to ndr(1)
crt(m,2) for m = 1 to ndr(2)
czt(m,1) for m = 1 to ndz(1)
czt(m,2) for m = 1 to ndz(2)

**MASS FLOW RATES**
cwr(m,1), scwr1(m,1), scwr2(m,1) for m = 1 to ndr(1)
cwr(m,2), scwr1(m,2), scwr2(m,2) for m = 1 to ndr(2)
cwz(m,1), scwz1(m,1), scwr2(m,1) for m = 1 to ndz(1)
cwz(m,2), scwz1(m,2), scwr2(m,2) for m = 1 to ndz(2)

**ANGULAR VELOCITIES AT CAVITY WALLS**
omr(m,1) for m = 1 to ndr(1)
omr(m,2) for m = 1 to ndr(2)
omz(m,1) for m = 1 to ndz(1)
omz(m,2) for m = 1 to ndz(2)

**PRESSURES AT CAVITY WALLS**
mpr(1), ( pcr(m,i,1) for m = 0 to mpr(1) ) for i = 1 to ndr(1)
mpr(2), ( pcr(m,i,2) for m = 0 to mpr(2) ) for i = 1 to ndr(2)
mpz(1), ( pcz(m,i,1) for m = 0 to mpz(1) ) for i = 1 to ndz(1)
mpz(2), ( pcz(m,i,2) for m = 0 to mpz(2) ) for i = 1 to ndz(2)

**TEMPERATURES**
mtr(1), ( tcr(m,i,1) for m = 0 to mpr(1) ) for i = 1 to ndr(1)
mtr(2), ( tcr(m,i,2) for m = 0 to mpr(2) ) for i = 1 to ndr(2)
mtz(1), ( tcz(m,i,1) for m = 0 to mpz(1) ) for i = 1 to ndz(1)
mtz(2), ( tcz(m,i,2) for m = 0 to mpz(2) ) for i = 1 to ndz(2)

**FLOW PARAMETERS**
rc, cp, pr, prt, grv
omref, pref, tref, dtnu, dtnorm
rref, thref1, zref
gvt, gva

**MULTIGRID PARAMETERS**
ng, ngt, ni(1), nj(1), nk(1)
na, nb, nc, np, nntmp
ic, inr, dummy

**UNDER-RELAXATION PARAMETER**
urfu, urfv, urfw, urfp, urft
urfvt, ex, dummy, dummy, dummy, dummy, dummy, c5
alpha, beta

**CONVERGENCE CRITERION**
maxit1, maxit2, nwrite

**ROOT MEAN SQUARE TOLERANCE AND RESIDUAL TOLERANCE**
tol, tolresid
nuspec, nvspec, nwspec, npspec, ntspec
vnormu, vnormv, vnormw, vnormp, vnormt
imon, jmon, kmon

**GRID PARAMETERS**
igty
efz,efr

**DATA FILE NAMES**
ndfn1
dfn
gfn

**WALL TEMPERATURE FILE SWITCH**
itempe

**WALL TEMPERATURE FILE**
tfn

**INITIALIZATION**
omi, pini
dummy, dummy
ireset
inpu1

A description of the variables shown in the input file above will now be given.

**ADAPTIVE TIME-STEP PARAMETERS**
**nadapt** = 0 No adaptation
**nadapt** = 1 Crude gradient based
**nadapt** = 2 Altered order
**nadapt** = 3 Altered scheme
**nadapt** = 4 Altered step

**dtmin, dtmax** = Time-step clipping parameters
**tmax** = Maximum simulated time period
**nadaptv** = 1 Adapt time-step on u
**nadaptv** = 2 Adapt time-step on v
**nadaptv** = 3 Adapt time-step on w
**nadaptv** = 4 Adapt time-step on T
The program can be readily modified to allow time-steps to be set on combinations of the above.

**FALSE DIFFUSION PARAMETER, CONVECTIVE SCHEME INDICATOR**
The variable **ncond** is used to set the type of convective term treatment used.
**ncond** = 0 – First order upwind
**ncond** = 1 - HYBRID
**ncond** = 2 - CONDIF
**ncond** = 3 - Standard CD
**ncond** = 4 - Positive coefficient CD

**rupper** = parameter to control false diffusion for CONDIF scheme. This parameter should be greater than or equal to one. The larger rupper, the less the false diffusion (i.e the more accurate the solution) but the penalty is greater instability of the solution.

**IBM**
Parameter used to flag if running on an IBM computer.
**ivc** = 0 non-IBM computer.
**ivc** > 0 IBM computer + modifications.

**TIME DEPENDENCE PARAMETERS**
**itd** = 0 - steady solution.
**itd** = 1 - unsteady solution.
**nts** = 0 - not a new time-step (i.e converge current time-step further).
**nts** = 1 - new time-step.
**dt** = time-step.
**tf** = spin up time. The program will spin rotating surfaces up from stand still to the angular velocity set in the command file over the time period tf.
**nstep** = number of time-steps.

**'ALTERNATING DIRECTION IMPLICIT' CONTROLS (Z,R,THETA)**
**(fqwe, fqns, fqfb) –** Caution is required for 3d cases (see Chapter 2)

**fq..** = 1.0 - implicit time scheme.
= 0.6 - Galerkin time scheme.
= 0.5 - Crank-Nicolson.
= 0.0 - explicit.

**fqwe** - time scheme used in axial direction.
**fqns** - time scheme used in radial direction.
**fqfb** - time scheme used in tangential direction.

**SOLUTION TYPE**
**lt** = 0 - laminar solution.
**lt** = 1 - turbulent solution using mixing length model.
**dummy** - not used (set to zero).
**incc** = 0 - incompressible fluid.
**incc** = 1 - compressible fluid.
**isnis** = 0 - isothermal flow.
**isnis** = 1 - nonisothermal flow.
**icp** = 0 - constant viscosity.
**icp** = 1 - variable viscosity.
**iflu** = 0 - air.
**iflu** = 1 - water.
**isr** = 0 - solve in a stationary coordinate system.
**isr** = 1 - solve in a rotating coordinate system.

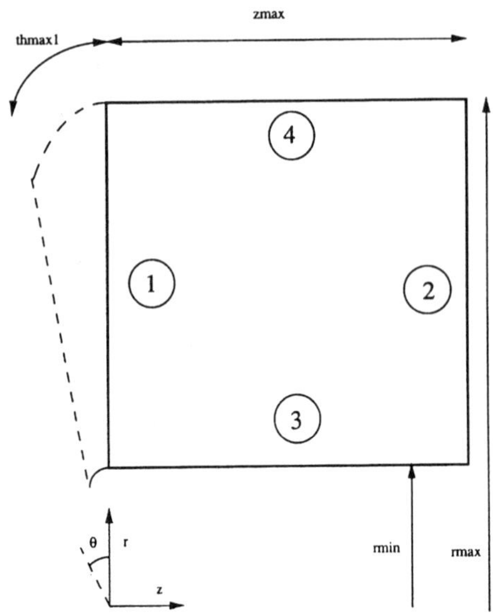


*Figure E.2* Program input structure

**GEOMETRY**

The program input is based on the idea that the fluid region is a box with sides numbered as shown in *Figure E.2*. Each side (1, 2, 3, 4) can be broken down into several pieces or sectors with different boundary conditions. The variable ndr(1) is the number of sectors on side 1, ndr(2) the number of sectors on side 2, ndz(1) the number on side 3 and ndz(2) the number on side 4.

On side 1 the radial coordinate of the start point or end point of each sector is stored in the array rb(m,1) where m = 1 .. to ndr(1)+1.

On side 2 the radial coordinate of the start point or end point of each sector is stored in the array rb(m,2) where m = 1 .. to ndr(2)+1

On side 3 the axial coordinate of the start point or end point of each sector is stored in the array zb(m,1) where m = 1 .. to ndz(1)+1

On side 4 the axial coordinate of the start point or end point of each sector is stored in the array zb(m,2) where m = 1 .. to ndz(2)+1

**thmax1** = tangential extent of region (generally be 360°). If thmax1 < 360° a repeating boundary condition will be used in the z-r planes. The value of thmax1 in the input file must be in degrees.

**BLOCK DEPTHS**

Obstructions, or blocks, which stick out from either of the sides 1, 2, 3 or 4 can be modelled. If a block is to be added, a separate sector must be defined for the block. Hence the width of the block is defined by the sector width. The height, or extent to which the block goes into the region, is defined using the variables below.

**zbl(m,1)** = height of block from surface 1
**zbl(m,2)** = height of block form surface 2
**rbl(m,1)** = height of block form surface 3
**rbl(m,2)** = height of block from surface 4

When not using this option set all block heights to zero.

**MOMENTUM BOUNDARY CONDITIONS**

For each sector, boundary conditions must be defined. In this section the boundary condition type for axial and radial velocity is set. Either fixed boundary conditions (Dirichlet) or differential boundary conditions can be set. The letter f is used to denote flat or fixed boundary conditions and d for differential. There are variations on the basic d type boundary condition and these will be discussed in a separate section.

**cruv(m,1)** = boundary conditions on wall 1
**cruv(m,2)** = boundary conditions on wall 2
**czuv(m,1)** = boundary conditions on wall 3
**czuv(m,2)** = boundary conditions on wall 4

**SWIRL BOUNDARY CONDITIONS**

In this section the tangential velocity boundary conditions are specified. Again, f or d type boundary conditions can be used. Variations on the basic d type boundary condition will, again, be discussed later.

**crw(m,1)** = tangential boundary condition on wall 1 etc.
**crw(m,2)** =
**czw(m,1)** =
**czw(m,2)** =

**TEMPERATURE BOUNDARY CONDITIONS**
Differential and fixed boundary conditions can also be specified for temperature, again, using either the letter d or f. Other temperature boundary conditions will discussed later.

**crt(m,1)** = temperature boundary condition on wall 2 etc.
**crt(m,2)** =
**czt(m,1)** =
**czt(m,2)** =

**MASS FLOW RATES**
The mass flow rate, through each sector, is specified in this section. If the mass flow rate is set as zero and the f type boundary condition used, the sector will be modelled as a solid wall. Generally, if a differential momentum boundary condition is specified, for a sector, the mass flow rate through the sector should be set to zero. Three input parameters are used for each sector. The first parameter 'cwr(m,*)' is the mass flow rate. The remaining two parameters can be used to specify an input velocity profile suitable for turbulent flow. These two parameters will be discussed later in the section on boundary conditions, but generally should be set to zero.

**cwr(m,1), scwr1(m,1), scwr2(m,1)=** Mass flow rates for wall 1 etc.
**cwr(m,2), scwr1(m,2), scwr2(m,2)=**
**cwz(m,1), scwz1(m,1), scwr2(m,1)=**
**cwz(m,2), scwz1(m,2), scwr2(m,2)=**

**ANGULAR VELOCITIES AT CAVITY WALLS**
The angular velocity for each sector is specified in this section.

**omr(m,1)** = Angular velocities for wall 1 etc.
**omr(m,2)** =
**omz(m,1)** =
**omz(m,2)** =

**PRESSURES AT CAVITY WALLS**
The pressure in each sector can be specified as a polynomial of the form,

$$p = \Sigma\ pc^*(m,i,q)\ x^m\ (m = 0, \ldots\ mp^*(q)),$$

where x is either the axial or radial coordinate, q is used to define the wall and i the sector number of the wall.

**mpr(1), pcr(m,i,1)** for m = 0 to mpr(1) = pressure distribution on wall 1 etc.
**mpr(2), pcr(m,i,2)** for m = 0 to mpr(2)
**mpz(1), pcz(m,i,1)** for m = 0 to mpz(1)
**mpz(2), pcz(m,i,2)** for m = 0 to mpz(2)

For a wall with 1 sector and a constant pressure distribution (the most commonly used option) of say 10 $N/m^2$ (the most commonly used option) the input for the wall would be - 0 10.

**TEMPERATURES**
The temperature in each sector can be specified as a polynomial of the form,

$$T = \Sigma \, tc^*(m,i,q)\, x^m \,(m = 0,\ldots.\, mt^*(q)),$$

where, again, x is either the axial or radial coordinate, q is used to define the wall and i the sector number of the wall.

For a wall with 1 sector and a constant temperature distribution (the most commonly used option) of say 293 K the input for the wall would be - 0 293. When a differential temperature boundary condition is used at a surface with fluid flow the temperature is best set to zero. Where a solid wall is defined and a differential temperature boundary condition set, the temperature input, in this section, will be treated as a heat flux. Therefore, an adiabatic boundary condition could be set by defining a differential temperature boundary condition with the temperature specified as zero. When using fine grids differential temperature conditions can be slow to converge. The slow convergence can result in a low rms change in temperature which can wrongly suggest convergence. Therefore, care must be taken when judging convergence.

**mtr(1), tcr(m,i,1)** for m = 0 to mpr(1) = temperature distribution on wall 1 etc.
**mtr(2), tcr(m,i,2)** for m = 0 to mpr(2)
**mtz(1), tcz(m,i,1)** for m = 0 to mpz(1)
**mtz(2), tcz(m,i,2)** for m = 0 to mpz(2)

**FLOW PARAMETERS**
General flow parameters are specified in this section.
**rc** = gas constant (i.e $p = \rho \, rc \, T$ ).
**cp** = specific heat capacity.
**pr** = Prandtl number.
**prt** = turbulent Prandtl number.

**grv** = gravity. This is generally ignored and so grv = 0. To model the earth's acceleration due to gravity grv = 9.81 m/s$^2$. This gravity parameter can also be used to model the effect of an accelerating fluid system, i.e. that found in aircraft components.
**omref** = reference angular velocity. Used to evaluate the rotational Reynolds number written to screen. When solving in a rotating coordinate system omref is taken as the angular velocity of the coordinate system.
**pref** = reference point pressure.
**tref** = reference temperature for evaluating reference properties and initialising fluid temperature.
**dtnu** = temperature difference used when evaluating Nusselt numbers.
**dtnorm** = temperature difference used when normalising energy equation.
**rref** = reference point radius.
**threfl** = reference point tangential coordinate.
**zref** = reference point axial coordinate.
**gvt** = tangential angle of gravity vector.
**gva** = axial angle of gravity vector.

**MULTIGRID PARAMETERS**
**ng** = number of multigrid levels in the axial and radial directions.
**ngt** = number of tangential multigrid levels.
**ni(1)** = number of axial grid nodes.
**nj(1)** = number of radial grid nodes .
**nk(1)** = number of tangential grid nodes.
**na** = number of sweeps on finest grid.
**nb** = number of sweeps on intermediate grid.
**nc** = number of sweeps on coarsest grid.
**np** = number of times TDMA solver used for pressure correction equation.
**nntmp** = number times TDMA solver used for energy equation.
**ic** = compatibility for multigrid algorithm.
**inr** = Newton-Raphson for turbulent viscosity.

**UNDER-RELAXATION PARAMETER**
**urfu** = under-relaxation parameter for axial momentum equation.
**urfv** = under-relaxation parameter for radial momentum equation.
**urfw** = under-relaxation parameter for tangential momentum equation.
**urfp** = under-relaxation parameter for pressure.
**urft** = under-relaxation parameter for energy equation.
**urfd** = under-relaxation parameter for density.
**urfvis** = under-relaxation parameter for viscosity.
**urfvt** = under-relaxation parameter for turbulent viscosity.
**ex** = exponent for turbulent viscosity.

**c5** = Richardson number.
**alpha** = Gosman factor.
**beta** = multigrid under-relaxation parameter.

**CONVERGENCE CRITERION**
**maxit1** = maximum number of iterations for a time-step.
**maxit2** = maximum number iterations specified for whole run. For a steady state solution set maxit1 = maxit2.
**nwrite** = number of time-steps computed before data is written out, i.e. if nwrite = 10, data is written out to a file every 10th time-step. Several different data file names, to be written to, can be supplied in the input file.

**ROOT MEAN SQUARE TOLERANCE AND RESIDUAL TOLERANCE**
**tol** = rms change at which solution considered converged
**tolresid** = value of normalised residual at which solution considered converged (note, residuals printed to screen are expressed as percentages).
**nuspec** = 1 - normalise axial momentum equation residuals by normalising value (vnormu) specified in command file,
= 2 - normalise axial momentum residuals by computed values (please study program when using this option – it is case specific)
**nvspec** = As for nuspec but applied to radial momentum equation.
**nwspec** = As for nuspec but applied to tangential momentum equation.
**npspec** = As for nuspec but applied to pressure correction equation.
**ntspec** = As for nuspec but applied to energy equation.
**vnormu** = Normalising value for axial momentum equation (only used when nuspec = 1).
**vnormv** = Normalising value for radial momentum equation (only used when nvspec = 1).
**vnormw** = Normalising value for tangential momentum equation (only used when nwspec = 1)
**vnormp** = Normalising value for pressure correction equation (only used when npspec = 1).
**vnormt** = Normalising value for energy equation (only used when ntspec = 1).

A monitoring point should be defined. The values of the variables being solved for, at this point, are printed on the computer screen. The location of the monitoring point is defined by inputting grid point numbers.

**imon** = axial location of monitoring point.
**jmon** = radial location of monitoring point.
**kmon** = tangential location of monitoring point.

If you are not concerned about the monitoring point position then set imon = jmon = kmon = 1.

**GRID PARAMETERS**
**igty** = 0 - internally generated grid with grid refinement at walls based on cosine functions.
**igty** = 1 - internally generated grid with grid refinement at walls based on geometric expansions.
**igty** = 2 - grid read in externally from a data file.
**efz** = axial grid expansion factor (used when igty = 1).
**efr** = radial grid expansion factor (used when igty = 1).

**DATA FILE NAMES**
**ndfn1** = number of data file names. Several data file names can be specified in the command file. This is useful when running time dependent problems, since the solution at different points in time can be written to a different data file.
**dfn(i)** = data file names (for i = 1 ...to ndfn1).
**gfn** = external grid data file name.

**WALL TEMPERATURE FILE SWITCH**
An external data file containing the temperatures at the grid points on walls 1 and 2 can be read in. This is useful if the temperature distribution on walls 1 and 2 is complex and not easily described by a polynomial.

**itempe** = 0 - no external temperature data file.
**itempe** = 1 - external temperature data file to be read in.

**WALL TEMPERATURE FILE**
**tfn** = external data file name containing the grid point temperatures for walls 1 and 2.

**INITIALIZATION**
**omi** = initial angular velocity for a new solution.
**pini** = initial pressure.
**ireset** = 0 - iteration counter not reset.
**ireset** = 1 - iteration counter reset.
**inpul** = new - there is no existing solution data file.
**inpul** = old - read in existing solution data file and continue solution. This option will be used if a solution has not converged, a new time-step of an unsteady solution is to be computed or a data file for another solution is to be used as an initial guess for a new solution.

## E.4 Boundary Conditions

The main boundary conditions for the *FLO_CAV* program and discussed in this section.

### Axial Momentum Boundary Conditions

**f** = flat velocity profile. When mass flow rate set to zero a solid wall is modelled.

**p** = parabolic velocity profile for laminar flow. For turbulent flow a flat core profile with 1/7th power law boundary layers can be set. In the mass flow rate input section of the program, 3 parameters must be specified. When using the turbulent velocity profile option the first parameter is the mass flow rate, the second the inner boundary layer thickness and the third the outer.

**d0** (or **d**) - v = 0, u - set to conserve mass, p = $p_{spec}$ (specified pressure value) or

$$\frac{\partial p}{\partial r} = \frac{\rho w^2}{r}$$

**d1** - same as d0 except,

$$\frac{\partial v}{\partial z} = 0$$

**d2** - same as d0 except,

$$\frac{\partial^2 v}{\partial z^2} = 0$$

**s** - symmetry boundary condition, u = 0,

$$\frac{\partial v}{\partial z} = \frac{\partial p}{\partial z} = 0$$

### Radial Momentum Boundary Conditions

The f and p boundary conditions can be used in the same way as described for axial momentum. Also, the differential boundary conditions listed below can be specified.

**d** -

$$\frac{\partial u}{\partial r} = 0$$

$p = p_{spec}.$

**ds** -

$$\frac{\partial u}{\partial r} = \frac{\partial p}{\partial r} = 0$$

**Tangential Momentum Boundary Conditions**
**f** - specified angular velocity

**d** -

$\frac{\partial w}{\partial z} = 0$ for radial boundaries and
$\frac{\partial wr}{\partial r} = 0$ for axial boundaries.

**E.5 Post Processing**
The program fv.f converts the *FLO_ELEC* output into a form that can be imported into the commercial *FIELDVIEW* postprocessor. The program read.f will read in the output file. Coding can be added at the end of this program to extract variables of interest. These can be incorporated into a basic graph plotting or graphics package. *Figure E.3* indicates the way in which variables are stored in the program. Frame (a) corresponds to a finest grid level m = 1. Data is just stored for this level. Frame (b) indicates the grid structure at m = 2.

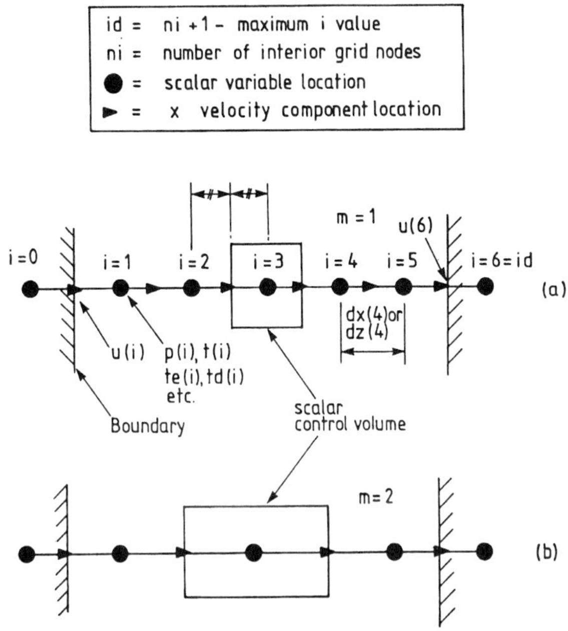


*Figure E.3* Variable storage locations and general grid structure: (a) finest grid level (m=1) and (b) next level m=2

# APPENDIX
# F
# FLO_BEARING PROGRAM USER INFORMATION

### F.1 *FLO_BEARING* Program

The *FLO_BEARING* program is specifically intended to solve the full three-dimensional Navier-Stokes and energy equations for a bearing with an orbiting shaft. It has the capability of producing coupled solutions where the temperature field in both the fluid and solid bearing components is modelled.

The methods used in the program are outlined in Chapters 2, 3 and 4. This section provides a description of the program structure, subroutines, data input format and boundary conditions.

### F.2 Description of Program Structure and Subroutines

The program structure is similar to that for *FLOW_CAV* discussed in Appendix E. However, the present program has more subroutines and a coordinate system transformation to allow bearing modelling. *Figure E.1* shows a simplified diagram of the program structure. The term RELAX is used to describe the process of setting up the coefficients of the discretised equations and applying the TDMA solver a specified number a times. For Petrov bearings (i.e. concentric journal and bearing (bush)) the program's multigrid (or multilevel) algorithm can be used. Sections of the program that are redundant when not using the multigrid algorithm (i.e. making a single grid solution) are indicated using the superscript m.

**INITIALISE**

1. Read in problem definition.

2. Set up constants.
3. Call **GRID3** - (Calculates the fine grid mesh) or read in mesh from external file.
4. Call **INIT3** - initialises variables or reads in old data and sets up previous values for time dependent solution. Also repositions reference point on the nearest grid node.
5. Checks storage sufficient.
6. Call **GRIDC3**[m] - calculates all coarse meshes (every other main grid node omitted).
7. Call **GRID3** - extra call to update grid when attitude angle altered in the command file.
8. If incompressible, constant laminar viscosity, sets density and viscosity on all meshes.
9. Call **GRIDS3** - calculates shaft mesh.
10. Call **GRIDBE3** - calculates bearing (bush) mesh.
11. Moves sector positions .
12. Sets indicators for derivative boundary conditions.
13. Checks whether reference point on a coarse grid node.
14. Call **GRIDB3** - works out coordinates of first node within each sector (in fluid).
15. Call **BLOCK3** - works out coordinates of first node beyond block, also assigns indicator to each node to define whether solid, fluid or derivative boundary.
16. Call **BLOCKS3** - assigns indicator to each shaft node to define whether it is part of solid region or boundary (this subroutine would be useful for modelling grooves/key ways etc. in the shaft).
17. Call **BLOCKS3** - called again to do same for bearing (bush).
18. Sets up indicator for boundary nodes to say whether boundary should be treated as a wall for purposes of mixing length model.
19. Write to screen problem definition.
20. Call **BOUND3** - sets up all fixed boundary conditions, except for slots.
21. Call **BLKBND3** - sets up values of variables within blocks.
22. Call **BSLOT** - sets up slot boundary conditions.
23. Call **VSLOT3** - sets radial velocity in slots.
24. Call **CVISL3** - calculates laminar viscosity.
25. Call **CDEN3** - calculates density from gas law for air, or correlation for water. If using gas law, sets up all boundary pressures. Subroutine not used when solving for oil.
26. Call **BNDCOM3** - resets boundary velocities for fixed profiles but variable density so as to maintain specified flow rate.
27. Restricts previous values of time dependent variables onto coarse grids, routines described later.

28. If solving just shaft in a rotating coordinate system calculates number of grid nodes the shaft grid has moved relative to the fluid grid.
29. Call **GRIDUPD3** - update film heights and velocity boundary conditions for new time-step (only necessary if shaft orbiting).
30. Call **FRACGT** - Calculates fraction of gas in cavitation zone when solving time-dependently.
31. Adjust shaft angular velocity when using spin-up option.
32. Starts single grid solution cycle or multigrid cycle.

**RELAX 1** - (start of solution cycle) SIMPLEC algorithm, starting on finest grid

1. Call **TRANSW3** - switches tangential velocity into rotating frame, if required. Only suitable if bearing and shaft concentric.
2. Call **COEFFW3** - sets up coefficients for tangential momentum equation.
3. Call **SOLVE3** - TDMA solver (repeating boundary in tangential direction).
4. Call **TRANSW3** - switch back to stationary frame.
5. Call **DBNDW3** - reset boundary tangential velocities for derivative conditions.
6. Call **PSWB3** - resets 'fixed' pressure condition so that dp/dr = (rho*w**2)/r if w (tangential velocity) varying on boundary.
7. Call **AXIS3** - calculates radial velocity on axis by resolving radial velocity and tangential velocity and set values of other axis values when making 3-d solution (not generally required).
8. Call **COEFFV3** - sets up coefficients for radial momentum equation.
9. Call **SOLVE3** - for derivative boundary nodes and next line in, call **SOLVE3** for main domain.
10. Call **AXIS3** (not generally required).
11. Call **COEFFU3** - sets up coefficients for axial momentum equation.
12. Call **SOLVE3** - for derivative boundary nodes - call SOLVE3 for main domain.
13. Call **DBNDUV3** - calculates boundary values of axial and radial velocities for derivative conditions not in their own directions.
14. Call **COEFFP3** - sets up coefficients for pressure correction equation.
15. Call **SOLVE3** - for specified number of sweeps.
16. Call **PRC3** - corrects velocities and pressure, resets derivative boundary conditions, sets pressures when using Reynolds condition. This is not recommended since algorithm will not conserve mass when the Reynolds condition is used.
17. Call **VSLOT3** - sets radial velocity in slots.
18. Call **COEFFT3** - sets up coefficients for fluid temperature equation.

19. Call **SOLVE3** - for a specified number of sweeps.
20. Call **DBNDT3** - calculates boundary temperatures for derivative conditions.
21. Call **CVISL3**
22. Call **FRACG** - calculates fraction gas in cavitation zone, when solution is steady.
23. Call **FRACDEN** - calculates mass weighted mean density in cavitation zone based on fraction gas and fraction lubricant.
24. Call **FLUXS3** - calculates the heat flux entering the shaft (calculation made in fluid).
25. Call **FLUXSHIFT3** - re-orientates heat fluxes computed on stationary fluid grid to be compatible with rotating shaft grid (only required when shaft solved in a rotating coordinate system).
26. Call **COEFFTS3** - sets up coefficients for shaft temperature equation.
27. Call **DBNDTS3** - uses heat flux from fluid as a boundary condition to solve the shaft temperature and calculates boundary temperatures for heat transfer coefficient boundary conditions.
28. Call **SOLVE3**
29. Call **SWAPS3** - enforces compatibility of temperature at the journal/fluid interface. This is done by making the computed shaft temperature the boundary temperature for the fluid.
30. Call **TSHIFT3** - re-orientates temperatures computed on the rotating shaft grid to be consistent with those computed on the stationary fluid grid (only required if shaft temperature solved for in a rotating coordinate system).
31. Call **FLUXB3** - calculates the heat flux entering the bearing (calculation made in fluid).
32. Call **COEFFTB3** - sets up coefficients for bearing temperature equation.
33. Call **DBNDTB3** - uses heat flux from fluid as a boundary condition to solve the bearing temperature and calculates boundary temperatures for heat transfer coefficient boundary conditions.
34. Call **SOLVE3**
35. Call **SWAPB3** - enforces compatibility of temperature at the bearing/fluid interface. This is done by making the computed bearing temperature the boundary temperature for the fluid.
36. Call **CDEN3**
37. Call **BNDCOM3**
38. Call **CVIST3** - calculates turbulent viscosity from mixing length model (only coded for Petrov bearing).
39. Repeat for specified number of sweeps, then go to RESTRICT unless on coarsest grid, then go to INTERPOLATE.

**RESTRICT**

1. Call **RESTP3** - restricts variables on main nodes to next coarse grid, call for p, T and turbulent viscosity[m].
2. Call **RESTU3** - restricts variables on axial velocity nodes, call for axial velocity[m].
3. Call **RESTV3** - restricts variables on radial velocity nodes, call for radial velocity[m].
4. Call **RESTW3** - restricts variables on tangential velocity nodes, call for tangential velocity[m].
5. Call **BLKBND3**[m]
6. Call **DBNDUV3**[m]
7. Call **AXIS3**[m]
8. Call **DBNDT3**[m]
9. Call **CDEN3**[m]
10. Call **CVISL3**[m]
11. Call **DBNDW3**[m]
12. Call **PSWB3**[m]
13. Work out defects/residuals for each variable.
14. Calculate right hand side of defect equation[m].
    i.e Call **COEFF** - fine grid.
        Call **RES3** - works out defects/residuals - fine grid.
        Call **REST** - restrict to coarse grid[m].
        Call **RHS** - calculates right hand side on coarse grid.
    do for axial, radial and tangential velocity and pressure and temperature[m.]
15. Call **OLD3** - saves restricted values of all principal variables[m].
16. Store residuals and rms values for finest grid.
17. Go to **RELAX**[m].

**INTERPOLATE**

1. Call **INTERP3** - calculates multigrid corrections for each variable, under-relaxes them, interpolates onto next finest grid and adds them onto variable[m].
2. Call **BLKBND3**[m]
3. Call **DBNDT3**[m]
4. Call **CVISL3**[m]
5. Call **CDEN3**[m]
6. Call **BNDCOM3**[m] - if on finest grid.
7. Call **CVIST3**[m] - if on finest grid

**RELAX 2$^m$**

1. As before, for a specified number of sweeps then go to **INTERPOLATE** unless on finest grid (then one multigrid cycle complete)$^m$.
2. Call **MASSR** - calculates maximum tangential velocity and tangential mass flow rate for normalisation of residuals and bearing side flow for mass conservation check.
3. Print out residuals and r.m.s and values at monitoring point.

**OUTPUT** stage at end of run or time-step

1. Call **OUTPUT1** - writes data out (this first call is used when running unsteady cases to write out the data at the end of each time-step).
2. Call **CHANGE3** - makes all variables at the current time-step ( $t^{new}$ ) old ($t^{old}$) values ready to compute the next time-step.
3. Call **OUTPUT2** - calculates bearing load, stream function, shear stresses, moment coefficients, Nusselt numbers, momentum and energy balances (some aspects just coded for Petrov bearing).

## F.3 Program Input

In this section the program input is formally introduced. Generally only one input file is required to run the *FLO_BEARING* program. The program can be run on a UNIX machine by typing the following, flo_bearing.e < flo_bearing.com, where flo_bearing.com is the input file and flo_bearing.e is the executable file. Generally the solution grid is generated in the main program (flo_bearing.e). For more complex grids a data file containing grid coordinates can be read into the program. This data file name should be specified in the flo_bearing.com file. The input file parameter layout, followed by a description of the parameters is given below. Program input should be in M.K.S units with, unless otherwise stated, angles represented in radians. However, generally, any consistent set of units can be used.

**FALSE DIFFUSION PARAMETER, CONVECTIVE SCHEME INDICATOR**
rupper, ncond

**IBM**
ivc

**TIME DEPENDENCE PARAMETERS**
itd, nts, dt, tf, nstep, dummy

**'ALTERNATING DIRECTION IMPLICIT' CONTROLS (Z,R,THETA)**
fqwe, fqns, fqfb

**SOLUTION TYPE**
lt, dummy, incc, isnis, icp, iflu, isr, isrs

**SWITCH FOR FRICTIONAL HEATING**
ifr

**GENERAL BEARING OPERATING PARAMETERS**
ecc, ecc1, omorbit, attang

**CAVITATION MODEL CONTROLS 1**
nswifts, pcav, ngbm, urffg

**CAVITATION MODEL CONTROLS 2**
nrbf, pctol, nvel

**NUMBER OF TANGENTIAL NODES IN SLOTS**
nsnode

**SLOT BOUNDARY CONDITION PARAMETERS**
ksl, nbslot(ksl), pslot(ksl), vslot(ksl), tslot(ksl)

**GEOMETRY**
ndr(1), rb(m,1) for m = 1 to ndr(1)+1
ndr(2), rb(m,2) for m = 1 to ndr(2)+1
ndz(1), zb(m,1) for m = 1 to ndz(1)+1
ndz(2), zb(m,2) for m = 1 to ndz(2)+1
thmax1

**BLOCK DEPTHS**
zbl(m,1) for m = 1 to ndr(1)
zbl(m,2) for m = 1 to ndr(2)
rbl(m,1) for m = 1 to ndz(1)
rbl(m,2) for m = 1 to ndz(2)

**MOMENTUM BOUNDARY CONDITIONS**
cruv(m,1) for m = 1 to ndr(1)
cruv(m,2) for m = 1 to ndr(2)
czuv(m,1) for m = 1 to ndz(1)
czuv(m,2) for m = 1 to ndz(2)

**SWIRL BOUNDARY CONDITIONS**
crw(m,1) for m = 1 to ndr(1)
crw(m,2) for m = 1 to ndr(2)
czw(m,1) for m = 1 to ndz(1)
czw(m,2) for m = 1 to ndz(2)

**TEMPERATURE BOUNDARY CONDITIONS**
crt(m,1) for m = 1 to ndr(1)
crt(m,2) for m = 1 to ndr(2)

czt(m,1) for m = 1 to ndz(1)
czt(m,2) for m = 1 to ndz(2)

**MASS FLOW RATES**
cwr(m,1), scwr1(m,1), scwr2(m,1) for m = 1 to ndr(1)
cwr(m,2), scwr1(m,2), scwr2(m,2) for m = 1 to ndr(2)
cwz(m,1), scwz1(m,1), scwr2(m,1) for m = 1 to ndz(1)
cwz(m,2), scwz1(m,2), scwr2(m,2) for m = 1 to ndz(2)

**ANGULAR VELOCITIES AT CAVITY WALLS**
omr(m,1) for m = 1 to ndr(1)
omr(m,2) for m = 1 to ndr(2)
omz(m,1) for m = 1 to ndz(1)
omz(m,2) for m = 1 to ndz(2)

**PRESSURES AT CAVITY WALLS**
mpr(1), ( pcr(m,i,1) for m = 0 to mpr(1) ) for i = 1 to ndr(1)
mpr(2), ( pcr(m,i,2) for m = 0 to mpr(2) ) for i = 1 to ndr(2)
mpz(1), ( pcz(m,i,1) for m = 0 to mpz(1) ) for i = 1 to ndz(1)
mpz(2), ( pcz(m,i,2) for m = 0 to mpz(2) ) for i = 1 to ndz(2)

**TEMPERATURES**
mtr(1), ( tcr(m,i,1) for m = 0 to mpr(1) ) for i = 1 to ndr(1)
mtr(2), ( tcr(m,i,2) for m = 0 to mpr(2) ) for i = 1 to ndr(2)
mtz(1), ( tcz(m,i,1) for m = 0 to mpz(1) ) for i = 1 to ndz(1)
mtz(2), ( tcz(m,i,2) for m = 0 to mpz(2) ) for i = 1 to ndz(2)

**THERMAL CONDUCTIVITY AND SPECIFIC HEAT CAPACITY COEFFICIENTS FOR OIL**
ck1, ck2, cc1, cc2

**FLOW PARAMETERS**
rc, cp, pr, prt, grv
omjunk, pref, tref, dtnu, dtnorm
rref, thref1, zref
gvt, gva

**MULTIGRID PARAMETERS**
ng, ngt, ni(1), nj(1), nk(1)
na, nb, nc, np, nntmp
ic, inr, dummy

**UNDER-RELAXATION PARAMETER**
urfu, urfv, urfw, urfp, urft, urfd, urfvis
urfvt, ex, dummy, dummy, dummy, dummy, dummy, c5
alpha, beta

**CONVERGENCE CRITERION**
maxit1, maxit2, nwrite

**ROOT MEAN SQUARE TOLERANCE AND RESIDUAL TOLERANCE**
tol, tolresid
nuspec, nvspec, nwspec, npspec, ntspec
vnormu, vnormv, vnormw, vnormp, vnormt
imon, jmon, kmon

**GRID PARAMETERS**
igty
efz,efr

**DATA FILE NAMES**
ndfn1
dfn
gfn

**WALL TEMPERATURE FILE SWITCH**
itempe

**WALL TEMPERATURE FILE**
tfn

**INITIALIZATION**
omi, pini
dummy, dummy
ireset
inpu1

****SHAFT CONDUCTION INPUT PARAMETERS****
isols

**MARKERS FOR SHAFT BOUNDARY CONDITION TYPES**
nhtls, nhtrs, nhtts

**SHAFT HEAT TRANSFER COEFFICIENTS AND THERMAL CONDUCTIVITY**
htls, htrs, htts, tkshaft

**SHAFT LENGTH FOLLOWED BY OTHER PARAMETERS**
shaftl
nis, njs, efzs, efrs
imons, jmons, kmons
urfswaps, urfqs, nntmps, dts

****CONDUCTION IN BEARING PARAMETERS****
isolb

**MARKERS FOR BEARING BOUNDARY CONDITION TYPES**
nhtlb, nhtrb, nhttb

**BEARING HEAT TRANSFER COEFFICIENTS AND THERMAL CONDUCTIVITY**
htlb, htrb, httb, tkbearing
rmaxb
njb, erfb
imonb, jmonb, kmonb
urfswapb, urfqb, nntmpb, dtb

A description of the variables shown in the input file above will now be given.

**FALSE DIFFUSION PARAMETER, CONVECTIVE SCHEME INDICATOR**
The variable **ncond** is used to set the type of convective term treatment used.
**ncond** = 1 - HYBRID.
**ncond** = 2 - CONDIFB (bounded CONDIF scheme).
**ncond** = 3 - CONDIF.
**ncond** = 4 - central difference.
**ncond** = 14 - central differencing in the radial direction but HYBRID scheme in axial and tangential directions.
**ncond** = 41 - central differencing in the radial and tangential directions but HYBRID scheme in tangential direction.

**rupper** = parameter to control false diffusion for CONDIF and CONDIFB schemes. This parameter should be greater than or equal to one. The larger rupper the less the false diffusion (i.e the more accurate the solution) but the penalty is greater instability of the solution.

**IBM**
Parameter used to flag if running on an IBM computer.
**ivc** = 0 non-IBM computer.
**ivc** > 0 IBM computer with other changes.

**TIME DEPENDENCE PARAMETERS**
**itd** = 0 - steady solution.
**itd** = 1 - unsteady solution.
**nts** = 0 - not a new time-step (i.e converge current time-step further).
**nts** = 1 - new time-step.
**dt** = time-step for lubricant solution.

**tf** = spin up time. The program will spin rotating surfaces up from stand still to the angular velocity set in the command file over the time period tf.
**nstep** = number of time-steps.

**ALTERNATING DIRECTION IMPLICIT CONTROLS (Z,R,THETA) (fqwe, fqns, fqfb)**

**fq..** = 1.0 - implicit time scheme.
= 0.6 - Galerkin time scheme.
= 0.5 - Crank-Nicolson.
= 0.0 - explicit.

**fqwe** - time scheme used in axial direction.
**fqns** - time scheme used in radial direction.
**fqfb** - time scheme used in tangential direction.

**SOLUTION TYPE**
**lt** = 0 - laminar solution.
**lt** = 1 - turbulent solution using mixing length model.
**dummy** - not used (set to zero).
**incc** = 0 - incompressible fluid.
**incc** = 1 - compressible fluid.
**isnis** = 0 - isothermal flow.
**isnis** = 1 - nonisothermal flow.
**icp** = 0 - constant viscosity.
**icp** = 1 - variable viscosity.
**iflu** = 0 - air.
**iflu** = 1 - water.
**iflu** = 2 - oil.
**isr** = 0 - solve fluid flow in a stationary coordinate system.
**isr** = 1 - solve fluid flow in a rotating coordinate system.
**isrs** = 0 - solve shaft temperature in a stationary coordinate system.
**isrs** = 1 - solve shaft temperature in a rotating coordinate system.

**SWITCH FOR FRICTIONAL HEATING**
**ifr** = 0 - no viscous dissipation in energy equation.
**ifr** = 1 - viscous dissipation in energy equation.

**GENERAL BEARING OPERATING PARAMETERS**
**ecc** = static eccentricity or, when solving orbiting case, centre of journal orbit.
**ecc1** = orbit radius.

**omorbit** = orbit angular velocity.
**attang** = attitude angle (in radians).

**CAVITATION MODEL CONTROLS 1**
**nswifts** = 0 - do not alter pressure in cavitation zone.
**nswifts** = 1 - set pressures less than fluid cavitation pressure to cavitation pressure.
**pcav** = fluid cavitation pressure.
**ngbm** = 0 switch mean film properties cavitation model off.
**ngbm** = 1 switch model on.
**urffg** = under-relaxation parameter for the fraction of gas predicted in the cavitation region (only used for steady state cavitation model).

**CAVITATION MODEL CONTROLS 2**
In this section, parameters that control the mean film properties cavitation model are described. The cavitation region is defined as the area where the pressure is less than the fluid cavitation pressure, specified in the input file. In this region the fluid properties are taken as the average of those for oil and gas. Mass weighted averages are taken, where the weightings are selected to conserve mass.

**nrbf** = 0 - during the iterative solution predicted pressures may fluctuate around the cavitation pressure, resulting in sudden fluid property changes which can make solutions diverge. If nrbf = 0 the program will not let the predicted cavitated region return to uncavitated if the local pressure becomes greater than the cavitation pressure. It is useful to set nrbf = 0, for the initial iterations of a solution, if difficulties with convergence are found.
**nrbf** = 1 - allows fluid to change to a fluid/gas mixture and back again depending on the cavitation pressure.
**pctol** = phase change pressure tolerance. If p ≥ (pcav + ptol) (and nrbf =1) then fluid properties are set, and if p < pcav cavitated properties are set. Setting ptol > 0, will damp fluctuations in fluid properties and so can be a useful parameter to stabilise solutions which initially diverge.

The mass balance, used to find the gas fraction, can be based on either, both the Couette and Poiseuille velocity components, or just the Couette component. The switch nvel is used to select the desired velocity components.

**nvel** = 0 - Couette velocity in cavitation region.
**nvel** = 1 - velocity obtained from integration of the predicted velocity field (i.e. Couette + Poiseuille). The Couette option is iteratively more stable.

**NUMBER OF TANGENTIAL NODES IN SLOTS**
**nsnode** = number of tangential nodes which are coincident with the slot. To calculate this the total number of tangential grid nodes, slot angle and number of slots must be known.

**SLOT BOUNDARY CONDITION PARAMETERS**
The following parameters are read in for ksl = 1 to nsnode.

**nbslot(ksl)** = 1 - velocity boundary condition (i.e. velocity specified).
**nbslot(ksl)** = 2 - pressure boundary condition (i.e pressure specified).
**pslot(ksl)** = inlet pressure at node ksl (only used if nbslot(ksl) = 2).
**vslot(ksl)** = inlet velocity at node ksl (only used if nbslot(ksl) = 1).
**tslot(ksl)** = inlet temperature at node ksl.

**GEOMETRY**
In this section the main geometry input for the lubricant region is discussed. The program input is based on the idea that the lubricant region is a box with sides numbered as shown in *Figure E.2*. Each side (1, 2, 3, 4) can be broken down into several pieces or sectors with different boundary conditions. The variable ndr(1) is the number of sectors on side 1, ndr(2) the number of sectors on side 2, ndz(1) the number on side 3 and ndz(2) the number on side 4.

On side 1 the radial coordinate of the start point or end point of each sector is stored in the array rb(m,1) where m = 1 .. to ndr(1)+1.

On side 2 the radial coordinate of the start point or end point of each sector is stored in the array rb(m,2) where m = 1 .. to ndr(2)+1

On side 3 the axial coordinate of the start point or end point of each sector is stored in the array zb(m,1) where m = 1 .. to ndz(1)+1

On side 4 the axial coordinate of the start point or end point of each sector is stored in the array zb(m,2) where m = 1 .. to ndz(2)+1

**thmax1** = tangential extent of bearing (generally be 360°). If thmax1 < 360° a repeating boundary condition will be used in the z-r planes. <u>The value of thmax1 in the input file must be in degrees.</u>

**BLOCK DEPTHS**
Obstructions, or blocks, which stick out from either of the sides 1, 2, 3 or 4 can be modelled. If a block is to be added, a separate sector must be defined

for the block. Hence the width of the block is defined by the sector width. The height, or extent to which the block goes into the region, is defined using the variables below.

**zbl(m,1)** = height of block from surface 1
**zbl(m,2)** = height of block form surface 2
**rbl(m,1)** = height of block form surface 3
**rbl(m,2)** = height of block from surface 4

When not using this option set all block heights to zero.

**MOMENTUM BOUNDARY CONDITIONS**
For each sector, boundary conditions must be defined. In this section the boundary condition type for axial and radial velocity is set. Either fixed boundary conditions (Dirichlet) or differential boundary conditions can be set.

The letter f is used to denote flat or fixed boundary conditions and d for differential. There are variations on the basic d type boundary condition and these are discussed in Section F.4.

**cruv(m,1)** = boundary conditions on wall 1
**cruv(m,2)** = boundary conditions on wall 2
**czuv(m,1)** = boundary conditions on wall 3
**czuv(m,2)** = boundary conditions on wall 4

**SWIRL BOUNDARY CONDITIONS**
In this section the tangential velocity boundary conditions are specified. Again, f or d type boundary conditions can be used. Variations on the basic d type boundary condition are, again, discussed in Section F.4.

**crw(m,1)** = tangential boundary condition on wall 1 etc.
**crw(m,2)** =
**czw(m,1)** =
**czw(m,2)** =

**TEMPERATURE BOUNDARY CONDITIONS**
Differential and fixed boundary conditions can also be specified for temperature, again, using either the letter d or f. Other temperature boundary conditions will discussed later.

**crt(m,1)** = temperature boundary condition on wall 2 etc.
**crt(m,2)** =

**czt(m,1)** =
**czt(m,2)** =

**MASS FLOW RATES**
The mass flow rate, through each sector, is specified in this section. If the mass flow rate is set as zero and the f type boundary condition used, the sector will be modelled as a solid wall. Generally, if a differential momentum boundary condition is specified, for a sector, the mass flow rate through the sector should be set to zero. Three input parameters are used for each sector. The first parameter 'cwr(m,*)' is the mass flow rate. The remaining two parameters can be used to specify an input velocity profile suitable for turbulent flow. These two parameters generally should be set to zero.

**cwr(m,1), scwr1(m,1), scwr2(m,1)**= Mass flow rates for wall 1 etc.
**cwr(m,2), scwr1(m,2), scwr2(m,2)**=
**cwz(m,1), scwz1(m,1), scwr2(m,1)**=
**cwz(m,2), scwz1(m,2), scwr2(m,2)**=

**ANGULAR VELOCITIES AT CAVITY WALLS**
The angular velocity for each sector is specified in this section.

**omr(m,1)** = Angular velocities for wall 1 etc.
**omr(m,2)** =
**omz(m,1)** =
**omz(m,2)** =

**PRESSURES AT CAVITY WALLS**
The pressure in each sector can be specified as a polynomial of the form,

$$p = \Sigma \ pc^*(m,i,q) \ x^m \ (m = 0, \ldots \ mp^*(q)),$$

where x is either the axial or radial coordinate, q is used to define the wall and i the sector number of the wall.

**mpr(1), pcr(m,i,1)** for m = 0 to mpr(1) = pressure distribution on wall 1 etc.
**mpr(2), pcr(m,i,2)** for m = 0 to mpr(2)
**mpz(1), pcz(m,i,1)** for m = 0 to mpz(1)
**mpz(2), pcz(m,i,2)** for m = 0 to mpz(2)

For a wall with 1 sector and a constant pressure distribution (the most commonly used option) of say 10 N/m$^2$ (the most commonly used option) the input for the wall would be - 0 10.

**TEMPERATURES**

The temperature in each sector can be specified as a polynomial of the form,

$$T = \Sigma\ tc^*(m,i,q)\ x^m\ (m = 0,\ldots.\ mt^*(q)),$$

where, again, x is either the axial or radial coordinate, q is used to define the wall and i the sector number of the wall.

For a wall with 1 sector and a constant temperature distribution (the most commonly used option) of say 293 K the input for the wall would be - 0 293. When a differential temperature boundary condition is used at a surface with lubricant flow the temperature is best set to zero. Where a solid wall is defined and a differential temperature boundary condition set, the temperature input, in this section, will be treated as a heat flux. Therefore, an adiabatic boundary condition could be set by defining a differential temperature boundary condition with the temperature specified as zero. When using fine grids, differential temperature conditions can be slow to converge. The slow convergence can result in a low rms change in temperature which can wrongly suggest convergence. Therefore, care must be taken when judging convergence.

**mtr(1), tcr(m,i,1)** for m = 0 to mpr(1) = temperature distribution on wall 1 etc.
**mtr(2), tcr(m,i,2)** for m = 0 to mpr(2)
**mtz(1), tcz(m,i,1)** for m = 0 to mpz(1)
**mtz(2), tcz(m,i,2)** for m = 0 to mpz(2)

**THERMAL CONDUCTIVITY AND SPECIFIC HEAT CAPACITY COEFFICIENTS FOR OIL**

When the lubricant is oil, the thermal conductivity and specific heat capacity are specified in this section.
**ck1, ck2, cc1, cc2** where
k = ck1 - ck2*T
cp = cc1 + cc2*T

**FLOW PARAMETERS**

General flow parameters are specified in this section.
**rc** = gas constant (i.e. p=den*rc*T).
**cp** = specific heat capacity when solving for air bearings.

**pr** = Prandtl number when solving for air bearings.
**prt** = turbulent Prandtl number bearings.
**grv** = gravity. This is generally ignored and so grv = 0. To model the earth's acceleration due to gravity grv = 9.81 m/s$^2$. This gravity parameter can also be used to model the effect of an accelerating fluid system i.e. that found in combat aircraft components.
**omref** = reference angular velocity. Used to evaluate the rotational Reynolds number written to screen and Couette velocity components for mean film properties cavitation model. Also, when using the exponential scheme to solve shaft temperature distribution omref is taken as shaft angular velocity. When solving in a rotating coordinate system omref is taken as the angular velocity of the coordinate system.
**pref** = reference point pressure.
**tref** = reference temperature for evaluating reference properties and initialising lubricant temperature.
**dtnu** = temperature difference used when evaluating Nusselt numbers.
**dtnorm** = temperature difference used when normalising energy equation.
**rref** = reference point radius.
**thref1** = reference point tangential coordinate.
**zref** = reference point axial coordinate.
**gvt** = tangential angle of gravity vector.
**gva** = axial angle of gravity vector.

**MULTIGRID PARAMETERS**
**ng** = number of multigrid levels in the axial and radial directions.
**ngt** = number of tangential multigrid levels.
**ni(1)** = number of axial grid nodes in lubricant.
**nj(1)** = number of radial grid nodes in lubricant.
**nk(1)** = number of tangential grid nodes in lubricant.
**na** = number of sweeps on finest grid.
**nb** = number of sweeps on intermediate grid.
**nc** = number of sweeps on coarsest grid.
**np** = number of times TDMA solver used for pressure correction equation.
**nntmp** = number times TDMA solver used for energy equation.
**ic** = compatibility for multigrid algorithm.
**inr** = Newton-Raphson for turbulent viscosity.

**UNDER-RELAXATION PARAMETER**
**urfu** = under-relaxation parameter for axial momentum equation.
**urfv** = under-relaxation parameter for radial momentum equation.
**urfw** = under-relaxation parameter for tangential momentum equation.
**urfp** = under-relaxation parameter for pressure.

**urft** = under-relaxation parameter for energy equation.
**urfd** = under-relaxation parameter for density.
**urfvis** = under-relaxation parameter for viscosity.
**urfvt** = under-relaxation parameter for turbulent viscosity.
**ex** = exponent for turbulent viscosity.
**c5** = Richardson number.
**alpha** = Gosman factor.
**beta** = multigrid under-relaxation parameter.

**CONVERGENCE CRITERION**
**maxit1** = maximum number of iterations for a time-step.
**maxit2** = maximum number iterations specified for whole run. For a steady state solution set maxit1 = maxit2.
**nwrite** = number of time-steps computed before data is written out, i.e. if nwrite = 10, data is written out to a file every 10th time-step. Several different data file names, to be written to, can be supplied in the input file.

**ROOT MEAN SQUARE TOLERANCE AND RESIDUAL TOLERANCE**
**tol** = rms change at which solution considered converged
**tolresid** = value of normalised residual at which solution considered converged (note, residuals printed to screen are expressed as percentages).
**nuspec** = 1 - normalise axial momentum equation residuals by normalising value (vnormu) specified in command file,
= 2 - normalise axial momentum residuals by denref*(wmax**2)*zmax*(rmax-rmin) where wmax is the maximum tangential velocity (this normalisation is probably most suitable for hydrodynamic bearings but other forms of normalisation can easily be added to the program).
**nvspec** = As for nuspec but applied to radial momentum equation.
**nwspec** = As for nuspec but applied to tangential momentum equation.
**npspec** = As for nuspec but applied to pressure correction equation. When npspec = 2 equation normalised by maximum tangential mass flow rate.
**ntspec** = As for nuspec but applied to energy equation. When ntspec = 2 equation normalised by maximum tangential mass flow rate times dtnorm (normalisation temperature difference specified in the command file).
**vnormu** = Normalising value for axial momentum equation (only used when nuspec = 1).
**vnormv** = Normalising value for radial momentum equation (only used when nvspec = 1).
**vnormw** = Normalising value for tangential momentum equation (only used when nwspec = 1)

**vnormp** = Normalising value for pressure correction equation (only used when npspec = 1).
**vnormt** = Normalising value for energy equation (only used when ntspec = 1).

A monitoring point should be defined. The values of the variables being solved for, at this point, are printed on the computer screen. The location of the monitoring point is defined by inputting grid point numbers.

**imon** = axial location of monitoring point.
**jmon** = radial location of monitoring point.
**kmon** = tangential location of monitoring point.
If you are not concerned about the monitoring point position then set imon = jmon = kmon = 1.

**GRID PARAMETERS**
**igty** = 0 - internally generated grid with grid refinement at walls based on cosine functions.
**igty** = 1 - internally generated grid with grid refinement at walls based on geometric expansions.
**igty** = 2 - grid read in externally from a data file.
**efz** = axial grid expansion factor (used when igty = 1).
**efr** = radial grid expansion factor (used when igty = 1).

**DATA FILE NAMES**
**ndfn1** = number of data file names. Several data file names
can be specified in the command file. This is useful when running time dependent problems, since the solution at different points in time can be written to a different data file.
**dfn(i)** = data file names (for i = 1 ...to ndfn1).
**gfn** = external grid data file name.

**WALL TEMPERATURE FILE SWITCH**
An external data file containing the temperatures at the grid points on walls 1 and 2 can be read in. This is useful if the temperature distribution on walls 1 and 2 is complex and not easily described by a polynomial.

**itempe** = 0 - no external temperature data file.
**itempe** = 1 - external temperature data file to be read in.

**WALL TEMPERATURE FILE**
**tfn** = external data file name containing the grid point temperatures for walls 1 and 2.

**INITIALIZATION**
**omi** = initial lubricant angular velocity for a new solution.
**pini** = initial lubricant pressure.
**ireset** = 0 - iteration counter not reset.
**ireset** = 1 - iteration counter reset.
**inpul** = new - there is no existing solution data file.
**inpul** = old - read in existing solution data file and continue solution. This option will be used if a solution has not converged, a new time-step of an unsteady solution is to be computed or a data file for another solution is to be used as an initial guess for a new solution.

*****SHAFT CONDUCTION INPUT PARAMETERS*****
**isols** = 0 - heat conduction in shaft not to be solved for.
**isols** = 1 - solve for heat conduction in shaft.

**MARKERS FOR SHAFT BOUNDARY CONDITION TYPES**
**nhtls** = 0 - isothermal boundary condition, 1 - heat transfer coefficient boundary condition
**nhtrs** = same as for nhtls.
**nhtts** = same as for nhtls.

**SHAFT HEAT TRANSFER COEFFICIENTS AND THERMAL CONDUCTIVITY**
**htls** = heat transfer coefficient on left hand end of shaft.
**htrs** = heat transfer coefficient on right hand end of shaft.
**htts** = heat transfer coefficient on cylindrical surface of shaft.
These heat transfer coefficients can be set to zero to produce adiabatic boundary conditions.
**tkshaft** = shaft thermal conductivity.

**SHAFT LENGTH FOLLOWED BY OTHER PARAMETERS**
**shaftl** = shaft length.
**nis** = number of axial grid nodes in shaft.
**njs** = number of radial grid nodes in shaft.
**efzs** = axial grid expansion factor for shaft (only a geometric can be used for the shaft).
**efrs** = radial grid expansion factor for shaft.

**imons** = axial location of shaft temperature monitoring point.

**jmons** = radial location of monitoring point.
**kmons** = tangential location of monitoring point.
**urfswaps** = under/over-relaxation parameter for temperature at the shaft-fluid interface (generally set to 1).
**urfqs** = under/over-relaxation parameter for heat flux at the shaft-fluid interface (generally < 1).
**nntmps** = number of times TDMA solver used for shaft temperature equation.
**dts** = time-step used for shaft when solving unsteady problem.

*****BEARING CONDUCTION INPUT PARAMETERS*****
**isolb** = 0 - heat conduction in bearing not to be solved for.
**isolb** = 1 - solve for heat conduction in bearing.

**MARKERS FOR BEARING BOUNDARY CONDITION TYPES**
**nhtlb** = 0 - isothermal boundary condition, 1 - heat transfer coefficient boundary condition.
**nhtrb** = as for nhtlb.
**nhttb** = as for nhtlb.

**BEARING HEAT TRANSFER COEFFICIENTS AND THERMAL CONDUCTIVITY**
**htlb** = heat transfer coefficient on left hand end of bearing.
**htrb** = heat transfer coefficient on right hand end of bearing.
**httb** = heat transfer coefficient on cylindrical surface of bearing.
**tkbearing** = bearing thermal conductivity.
**rmaxb** = bearing radius.

**njb** = number of radial grid nodes in bearing.
**erfb** = radial grid expansion factor (only a geometric grid can be generated for bearing)
**imonb** = axial location of bearing temperature monitoring point.
**jmonb** = radial location of monitoring point.
**kmonb** = tangential location of monitoring point.
**urfswapb** = under/over-relaxation parameter for temperature at the bearing-fluid interface (generally set to 1).
**urfqb** = under/over-relaxation parameter for heat flux at the bearing-fluid interface (generally < 1).
**nntmpb** = number times TDMA solver used for bearing temperature equation.
**dtb** = time-step used for bearing when solving unsteady problem.

### F.4 Boundary Conditions

The main boundary conditions for the *FLO_BEARING* program are mostly as discussed in Section E.4. The boundary condition types for the grooves or slots often found in hydrodynamic journal bearings are described in the program input section of this Appendix.

### F.5 Input Files

Sample command files are included on disc. The file names correspond to figure numbers in the main text. The Section E.5 post processing information generally applies. However no fv.f program is included.

# APPENDIX

# G

# FLO_ELEC PROGRAM USER INFORMATION

## G.1 *FLO_ELEC* Program

*FLO_ELEC* is a program for the prediction of three-dimensional flows in a Cartesian coordinate system. It is highly suitable for modelling flow in idealised electronic and room type geometries. This document provides a description of the program structure, subroutines, data input format and boundary conditions.

## G.2 Description of Program Structure and Subroutines

*Figure E.1* shows a simplified diagram of the program structure. The term RELAX is used to describe the process of setting up the coefficients of the discretized equations and applying the TDMA or GAUSS-SIEDEL solver (The solver choice must be specified locally in the subroutine SOLVE3. GAUSS-SIEDEL works best for the energy (T) equation when conjugate problems are being modelled. However, the COEFFT3 routine supplied here does not utilise the harmonic property means required for such problems) a specified number a times. The multigrid (or multilevel) algorithm and most of the algorithms used in this program are described in Chapters 2-7. Sections of the program that are redundant when not using the multigrid algorithm (i.e. making a single grid solution) are indicated using the superscript m.

**INITIALISE**

1. Read in problem definition.
2. Set up certain solution parameters, check for inconsistencies and that array dimensions are sufficient.

3. Call **GRID3** - calculates the fine grid mesh (or read in mesh from external file).
4. Call **INIT3** - initialises variables.
5. Call **READ3** - read in old data if running on an old solution.
6. Call **BOUNDW3** – set boundary conditions.
7. Call **GRIDC3**$^{m}$ - calculates all coarse meshes (every other main grid node omitted).
8. If incompressible, constant laminar viscosity, sets density and viscosity on all meshes.
9. Call **BLOCK3** - assigns indicator (nbl(i,j,k)) to each node to define whether solid (0), fluid (1) or derivative boundary (-1).
10. Call **BSPEC** – set nbl(i,j,k), for any pressure boundary conditions, at all grid levels .
11. Write to screen problem definition.
12. Call **COEFFWY3** – (if nwy=1) to solve Poisson equation for wall distance variable.
13. Call **SOLVE3** – TDMA or Gauss-Siedel solver for wall distance variable.
14. Section of code with RELAX and INTERPOLATE sections to yield a 'W' multigrid cycle. Call **CVISL3** - calculates laminar viscosity.
15. Call **CDEN3** - calculates density from gas law for air, or correlation for water.
16. Restricts previous values of time dependent variables onto coarse grids, routines described later.
17. Starts single grid solution cycle or multigrid cycle.

**RELAX 1** - (start of solution cycle) SIMPLE algorithm (SIMPLEC easily activated by entering COEFFU3, COEFFV3 and COEFFW3 subroutines) starting on finest grid.

1. Call **SMOTH** subroutines to smooth u, v ,w, upr (u at old time level), vpr and wpr for Speziale non-linear eddy viscosity model.
2. Call **NLINWS** assemble source term for non-linear Speziale model momentum equation contribution.
3. Call **COEFFW3** - sets up coefficients for z momentum equation.
4. Call **DCOEFFW3** – apply any w derivative boundary conditions.
5. Call **SOLVE3** – TDMA or Gauss-Siedel solver.
6. Call **SMOTH** - subroutines to smooth u, v ,w, upr, vpr and wpr for Speziale non-linear eddy viscosity model.
7. Call **NLINVS** - assemble source term for non-linear Speziale model momentum equation contribution.
8. Call **COEFFV3** - sets up coefficients for y momentum equation.
9. Call **DCOEFFV3** – Apply any v derivative boundary conditions.

10. Call **SOLVE3**
11. Call **SMOTH** - subroutines to smooth u, v ,w, upr, vpr and wpr for Speziale non-linear eddy viscosity model.
12. Call **NLINUS** - assemble source term for non-linear Speziale model momentum equation contribution.
13. Call **COEFFU3** - sets up coefficients for x momentum equation.
14. Call **DCOEFFU3** – Apply any u derivative boundary conditions.
15. Call **SOLVE3**
16. Call **COEFFP3** - sets up coefficients for pressure correction equation.
17. Call **SOLVE3** - for specified number of sweeps.
18. Call **PRC3** - corrects velocities and pressures.
19. Call **SMOTH** - subroutines to smooth u, v ,w, upr, vpr and wpr for Speziale non-linear eddy viscosity model.
20. Call **NLINKES** - assemble source term for non-linear Speziale model k equation contribution.
21. Call **COEFFK3** - sets up coefficients for k equation.
22. Call **SOLVE3** - for a specified number of sweeps.
23. Call **SMOTH** - subroutines to smooth u, v ,w, upr, vpr and wpr for Speziale non-linear eddy viscosity model.
24. Call **NLINKES** - assemble source term for non-linear Speziale model epsilon equation contribution.
25. Call **COEFFE3** - sets up coefficients for epsilon equation.
26. Call **SOLVE3** - for a specified number of sweeps.
27. For k-epsilon model call **CVTKE3** – evaluate turbulent viscosity.
28. For k-l model call **CVTKL3** – evaluate turbulent viscosity.
29. For l-vel model call **LVEL3** – evaluate turbulent viscosity.
30. For mixing length model call **MIXL3** – evaluate turbulent viscosity.
31. Call **COEFFT3** - sets up coefficients for temperature equation.
32. Call **SOLVE3** - for a specified number of sweeps.
33. Call **CVISL3**
34. Call **CDEN3**
35. Repeat for specified number of sweeps, then go to RESTRICT unless on coarsest grid, then go to INTERPOLATE.

**RESTRICT**

1. Call **RESTP3** - restricts variables on main nodes to next coarse grid, call for p, T, k, epsilon and turbulent viscosity$^m$.
2. Call **RESTU3** - restricts variables on u velocity nodes, call for x velocity$^m$.
3. Call **RESTV3** - restricts variables on v velocity nodes, call for y velocity$^m$.

4. Call **RESTW3** - restricts variables on w velocity nodes, call for z velocity$^m$.
5. Call **BOUNDW3**$^m$
6. Call **DBUVW3**$^m$ – Set u, v, w differential boundary conditions.
7. Call **CDEN3**$^m$
8. Call **CVISL3**$^m$
9. Work out defects/residuals for each variable.
10. Calculate right hand side of defect equation$^m$.
    i.e Call **COEFF** - fine grid.
        Call **RES3** - works out defects/residuals - fine grid.
        Call **REST** - restrict to coarse grid$^m$.
        Call **RHS** - calculates right hand side on coarse grid.
    do for u, v and w and pressure and temperature$^m$.
11. Call **OLD3** - saves restricted values of all principal variables$^m$.
12. Store residuals and rms values for finest grid
13. Go to **RELAX**$^{m.}$.

**INTERPOLATE**

1. Call **INTERP3** - calculates multigrid corrections for each variable, under-relaxes them, interpolates onto next finest grid and adds them onto variable$^m$.
2. Call **BOUNDW3**$^m$
3. Call **DBUVW3**$^m$
4. Call **CVISL3**$^m$
5. Call **CDEN3**$^m$

**RELAX 2**$^m$

1. As before, for a specified number of sweeps then go to **INTERPOLATE** unless on finest grid (then one multigrid cycle complete)$^m$.
2. Print out residuals and r.m.s and values at monitoring point.

**OUTPUT** stage at end of run or time-step.

1. Call **OUTPUT1** - writes data out (this first call is used when running unsteady cases to write out the data at the end of each time-step).
2. Call **CHANGE3** - makes all variables at the current time-step ( $t^{new}$ ) old ($t^{old}$) values ready to compute the next time-step.

### G.3 Program Input

In this section the program input is formally introduced. Generally only one input file is required to run the *FLO_ELEC* program. The program can be run on a UNIX machine by typing the following, flo_elec.e < flo_elec.com,

where flo_elec.com is the input file and flo_elec.e is the executable file. Generally the solution grid is generated in the main program (flo_elec.e). For more complex grids a data file containing grid coordinates can be read into the program. This data file name should be specified in the flo_elec.com file and can be generated using grid.f. The input file parameter layout, followed by a description of the parameters is given below. Program input should be in M.K.S units.

**ADAPTIVE TIME-STEP PARAMETERS**
nadapt, dtmin, dtmax, tmax, nadaptv

**FALSE DIFFUSION PARAMETER, CONVECTIVE SCHEME INDICATOR**
rupper, ncond

**TIME DEPENDENCE PARAMETERS**
itd, nts, dt,nstep

**CRUDE 'ALTERNATING DIRECTION IMPLICIT' CONTROLS (x,y,z)**
fqwe, fqns, fqfb

**SOLUTION TYPE**
lt, nke, nolin, nwy, incc, isnis, icp, iflu

**OVERALL DIMENSIONS**
xmax, ymax, zmax

**FLOW PARAMETERS**
rc, cp, pr, prt, grv
pref, tref
dummy, dummy, dummy

**MULTIGRID PARAMETERS**
ng, ngt, ni(1), nj(1), nk(1)
na, nb, nc, np, nntmp
ic, nwc

**UNDER-RELAXATION PARAMETER**
urfu, urfv, urfw, urfp, urft
urfvt, dummy, urfk, urfe
beta

**CONVERGENCE CRITERION**
maxit1, maxit2, nwrite

**ROOT MEAN SQUARE TOLERANCE AND RESIDUAL TOLERANCE**
tol
imon, jmon, kmon

**GRID PARAMETERS**
igty
efx,efy

**DATA FILE NAMES**
ndfn1
dfn
gfn

**BLOCK DEFINITION FILE**
bfn

**BOUNDARY CONDITION DEFINITION FILE**
bcfn

**ITERATION RESET & CASE STATUS**
ireset
inpu1

A description of the variables shown in the input file above will now be given.

**ADAPTIVE TIME-STEP PARAMETERS**
**nadapt** = 0 No adaptation
**nadapt** = 1 Crude gradient based
**nadapt** = 2 Altered order
**nadapt** = 3 Altered scheme
**nadapt** = 4 Altered step

**dtmin, dtmax** = Time-step clipping parameters
**tmax** = Maximum simulated time period
**nadaptv** = 1 Adapt time-step on u
**nadaptv** = 2 Adapt time-step on v
**nadaptv** = 3 Adapt time-step on w
**nadaptv** = 4 Adapt time-step on T
The program can be readily modified to allow time-steps to be set on combinations of the above.

**FALSE DIFFUSION PARAMETER, CONVECTIVE SCHEME INDICATOR**
The variable **ncond** is used to set the type of convective term treatment used.

**ncond** = 0 – First order upwind
**ncond** = 1 - HYBRID.
**ncond** = 2 - CONDIF
**ncond** = 3 - Standard CD
**ncond** = 4 - Positive coefficient CD

**rupper** = parameter to control false diffusion for CONDIF scheme. This parameter should be greater than or equal to one. The larger rupper the less the false diffusion (i.e the more accurate the solution) but the penalty is greater instability of the solution. rupper = 1 corresponds to first order upwind.

**TIME DEPENDENCE PARAMETERS**
**itd** = 0 - Steady solution
**itd** = 1 - Unsteady solution
**nts** = 0 - Not a new time-step (i.e converge current time-step further).
**nts** = 1 - New time-step
**dt** = Time-step
**nstep** = Number of time-steps

**CRUDE 'ALTERNATING DIRECTION IMPLICIT' CONTROLS (x,y,z) (fqwe, fqns, fqfb)**

**fq..** = 1.0 - Implicit time scheme
= 0.6 - Galerkin time scheme
= 0.5 - Crank-Nicolson
= 0.0 – 'Explicit'

**fqwe** - Time scheme used in x direction
**fqns** - Time scheme used in y direction
**fqfb** - Time scheme used in z direction

**SOLUTION TYPE**
**lt** = 0 - Laminar solution
**lt** = 1 - Turbulent solution
**nke** = 0 and **lt** = 0 – l-vel or mixing length model solutions (there are two subroutine Calls – mixl3 and lvel3, the desired zero equation model call should be un-commented. Different mixing length model damping functions can be activated by altering the local subroutine variable b2 (please see subroutine mixl3)

**nke** = 0 and **lt** = 1 – k-l model (the variable nwall switches wall/damping functions on and off. nwall = 1 gives wall functions, nwall = 0 gives a damping function)
**nke** = 0 and **lt** = 2 – k-epsilon model (the variable nwall switches wall/damping functions on and off. nwall = 1 gives wall functions, nwall = 0 gives a damping function)
**nolin** = 0 Gives linear k-l and k-epsilon models in the above
**nolin** = 1 Gives non-linear k-l and k-epsilon models in the above
**nwy** = 0 No wall distance Poisson equation evaluation
**nwy** = 1 Activate wall distance Poisson equation evaluation (necessary for l-vel, mixing length and k-l models)
**incc** = 0 - Incompressible fluid (By default, the Boussinesq approximation is set in the COEFFV3 source term. When modelling buoyant flows with incc = 1 the noted source term component must be commented out.)
**incc** = 1 - Compressible fluid
**isnis** = 0 - Isothermal flow
**isnis** = 1 - Nonisothermal flow
**icp** = 0 - Constant viscosity
**icp** = 1 - Variable viscosity
**icp** = 2 - Concentration equation solution
**iflu** = 0 - Air
**iflu** = 1 - Water

**OVERALL DIMENSIONS**
**xmax** – maximum system x dimension
**ymax** – maximum system y dimension
**zmax** – maximum system z dimension

**FLOW PARAMETERS**
General flow parameters are specified in this section.
**rc** = Gas constant (i.e. p= den*rc*T)
**cp** = Specific heat capacity
**pr** = Prandtl number
**prt** = Turbulent Prandtl number
**grv** = Gravity. This is generally ignored and so grv = 0. To model the earth's acceleration due to gravity grv = 9.81 m/s$^2$. This gravity parameter can also be used to model the effect of an accelerating fluid system i.e. that found in aircraft components
**pref** = Reference point pressure
**tref** = Reference temperature for evaluating reference properties and initialising fluid temperature

**MULTIGRID PARAMETERS**
**ng** = Number of multigrid levels in the x and y directions
**ngt** = Number of z multigrid levels
**ni(1)** = Number of (internal) x grid nodes
**nj(1)** = Number of y grid nodes
**nk(1)** = Number of z grid nodes
**na** = Number of sweeps on finest grid
**nb** = Number of sweeps on intermediate grid
**nc** = Number of sweeps on coarsest grid
**np** = Number of times TDMA/GAUSS-SIEDEL solver used for pressure correction equation
**nntmp** = Number times TDMA/GAUSS-SIEDEL solver used for energy equation
**ic** = Compatibility for multigrid algorithm
**nwc =** 0 – No 'W' cycle element
**nwc =** 1 – Activate 'W' cycle element

**UNDER-RELAXATION PARAMETER**
**urfu** = Under-relaxation parameter for x momentum equation
**urfv** = Under-relaxation parameter for y momentum equation
**urfw** = Under-relaxation parameter for z momentum equation
**urfp** = Under-relaxation parameter for pressure
**urft** = Under-relaxation parameter for energy equation
**urfvt** = Under-relaxation parameter for turbulent viscosity
**urfk** = Under-relaxation parameter for k
**urfe** = Under-relaxation parameter for epsilon
**beta** = Multigrid under-relaxation parameter

**CONVERGENCE CRITERION**
**maxit1** = Maximum number of iterations for a time-step
**maxit2** = Maximum number iterations specified for whole run. For a steady state solution set maxit1 = maxit2
**nwrite** = Number of time-steps computed before data is written out, i.e. if nwrite = 10, data is written out to a file every 10th time-step. Several different data file names, to be written to, can be supplied in the input file

**ROOT MEAN SQUARE TOLERANCE AND RESIDUAL TOLERANCE**
**tol** = rms change at which solution considered converged

A monitoring point should be defined. The values of the variables being solved for, at this point, are printed on the computer screen. The location of the monitoring point is defined by inputting grid point numbers.

**imon** = x location of monitoring point
**jmon** = y location of monitoring point
**kmon** = z location of monitoring point
If there is no concern regarding the monitoring point position set imon = jmon = kmon = 1.

**GRID PARAMETERS**
**igty** = 0 - Internally generated grid with grid refinement at walls based on cosine functions
**igty** = 1 - Internally generated grid with grid refinement at walls based on geometric expansions
**igty** = 2 - Grid read in externally from a data file
**efx** = x Grid expansion factor (used when igty = 1)
**efy** = y Grid expansion factor (used when igty = 1)
efz = 1 (set in program – can alter this)

**DATA FILE NAMES**
**ndfn1** = Number of data file names. Several data file names can be specified in the command file. This is useful when running time dependent problems, since the solution at different points in time can be written to a different data file
**dfn(i)** = Data file names (for i = 1 ...to ndfn1)
**gfn** = External grid data file name

**BLOCK DEFINITION FILE**
**nbl, xmin, xmax, ymin, ymax, zmin, zmax** - Reads maximum and minimum coordinates of solid blocks. When nbl = 0 block is solid. When nbl = 1 a fluid block is produced. This enables cut-outs to be made in solid blocks.

**BOUNDARY CONDITION FILE**
**xmin, xmax, ymin, ymax, zmin, zmax, velset, pset, tset, teset, tdset**
**xmin, xmax, ymin, ymax, zmin, zmax** - Coordinates used to define plane over which boundary condition applied.
**velset** - Velocity normal to plane
**pset** = 0 Implies velocity boundary condition, pset > 0 sets pressure boundary condition at 0 gauge pressure
**teset** - k at inlet plane
**tdset** - epsilon at inlet plane

**ITERATION RESET AND CASE STATUS**
**ireset** = 0 - Iteration counter not reset
**ireset** = 1 - Iteration counter reset
**inpul** = new - There is no existing solution data file
**inpul** = old - Read in existing solution data file and continue solution. This option will be used if a solution has not converged, a new time-step of an unsteady solution is to be computed or a data file for another solution is to be used as an initial guess for a new solution

**PRESSURE BOUNDARY CONDITIONS**
If pressure boundary conditions are defined, all non-normal velocity derivatives are set to zero. The normal velocity component is set to conserve mass.

## G.4 General Details

The program described can readily model the complex Cartesian based elements found in electronic assemblies. A modified version was used in the design of the latest NCR Cashpoint Dispenser. *Figure G.1* shows a flow solution that indicates the level of geometric complexity which can be easily be modelled. Post processing details are as given in Section E.5. Sample command files are given for the Chapter 10 Case Study. To run these simulations it is necessary to uncomment sections of code labelled cpu_mod. A similar labelling process is used in other programs and cases.

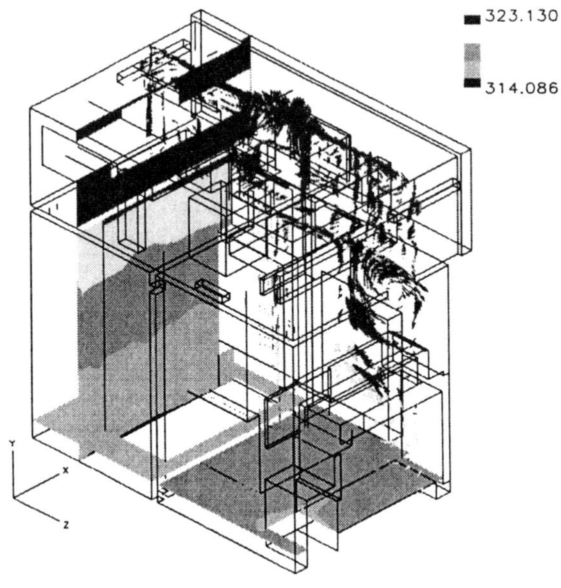


*Figure G.1* Flow solution for a complex system

# REFERENCES


Adair D. and Tucker P. G. Efficient modelling of rotating disks and cylinders using a Cartesian grid, *Appl. Math. Modelling,* Vol. 21, pp. 749-762 (1998)

Alsemgeest R., Shaw C. T., Richardson S. H. and Pierson S. Modeling the time-dependent flow through a throttle valve, *SAE Technical Paper 2000-01-0659, Proc. SAE 2000 World Congress, Detroit, Michigan, March 6-9* (2000)

Amano R. S. Laminar heat transfer in a channel with two right-angled bends, *ASME J. Heat Transfer*, Vol. 106, pp. 591-598 (1984)

Amon C. H. Spectral element-Fourier method for unsteady conjugate heat transfer in complex geometry flows, *J. of Thermophysics and Heat Transfer*, Vol. 9, No. 2, pp. 247-253 (1995)

Anderson J. D. *Computational fluid dynamics: The basics with applications*, McGraw-Hill, Inc. (1995)

Arad E. and Martinelli L. Application of a parallel multigrid algorithm for large eddy simulation, *2nd Symp. on Turbulence, Heat and Mass Transfer,* pp. 499-507 (1997)

Arakawa A. Computational design for long term numerical integration of the equations of fluid motion. Two-dimensional incompressible flow, *Part 1. Comp. Phys*. Vol. 1, pp. 119-130 (1966)

Armenio V., Piomelli U. and Fiorotto V. On the application of large eddy simulation to particle laden flows, *Proc. 1st Int. Symp. on Tubulence and Shear Flow (Eds Banerjee S. and Eaton J. K.), Santa Barbara,* pp. 139-144 (1999)

Ascher U. M. and Petzold L. R. *Computer methods for ordinary differential equations and differential algebraic equations*, SIAM (1998)

Banerjee R. K., Cho Y. I. and Back L. H. Numerical studies of non-Newtonian pulsed flow in a femoral branched vessel, *ASME Forum on Unsteady Flows*, FED-Vol. 157, pp. 229-238 (1993)

Barakos G. and Drikakis D. Numerical simulation of transonic buffet flows using various turbulence closures, *Proc. 1st Int. Symp. on Tubulence and Shear Flow (Eds Banerjee S. and Eaton J. K.), Santa Barbara,* pp. 995-1000 (1999)

Bardina J., Ferziger J. H. and Reynolds W. C. Improved subgrid models for large eddy simulation, *AIAA* Paper 89-1357 (1980)

Bartello P., Metais O. and Lesieur M. Coherent structures in rotating three dimensional turbulence, *J. Fluid Mech.,* Vol. 273, pp. 1-29 (1994)

Barton I. E. and Kirby R. Finite difference scheme for the solution of fluid flow problems on non-staggered grids, *Int. J. Numer. Meth. Fluids*, Vol. 33, pp. 939-959 (2000)

Barton I. E. Comparison of SIMPLE- and PISO-type algorithms for transient flows, *Int. J. Numer. Meth. Fluids*, Vol. 26, pp. 459-483 (1998a)

Barton I. E. Improved laminar predictions using a stabilised time-dependent SIMPLE scheme, *Int. J. Numer. Meth. Fluids*, Vol. 28, pp. 841-857 (1998b)

Barton I. E. Exponential-Lagrangian tracking schemes applied to Stokes law, *ASME J. Fluids Engrg*, Vol. 118, pp. 85-89, March (1996)

Beam R. M. and Warming R. F. An implicit finite difference algorithm for hyperbolic systems in conservation law form, *J. Comput. Phys.*, Vol. 22, pp. 87-110 (1976)

Beam R. M. and Warming R. F. Implicit numerical methods for the compressible Navier-Stokes and Euler equations, *1981-82 Lecture Series Programme, Computational Fluid Dynamics, von Karman Institute for Fluid Dynamics, Chaussee de Waterloo 72, B-1640 Rhode-Saint-Genese, Belgium, March 29 – April 2* (1982)

Bell C. B. and Surana K. S. A space-time coupled p-version least-squares finite element formulation for unsteady fluid dynamics problems, *Int. J. Numer. Meth. in Engrg*, Vol. 37, pp. 3545-3569 (1994)

Berger M. J. and Colella P. Local adaptive mesh refinement for shock hydrodynamics, *J. Comput. Phys.*, Vol. 82, pp. 64-89 (1989)

Blazek J., Kroll N., Radespiel R. and Rossow C. C. Upwind implicit residual smoothing method for multistage schemes, *AIAA 10$^{th}$ Computational Fluid Dynamics Conf., AIAA Paper 91-1533* (1991)

Bo T., Iacovides H. and Launder B. E. Convective discretization schemes for the turbulence transport equations in flow predictions through sharp u-bends, *Int. J. Numer. Meth. Heat Fluid Flow*, Vol. 5, pp. 33-48 (1995)

Book D. L., Boris J. P. and Hain K. Flux-corrected transport. II. Generalizations of the method, *J. Comput. Phys.*, Vol. 18, pp. 248-283 (1975)

Boris J. P. and Book D. L. Flux-corrected transport. I. SHASTA, a fluid transport algorithm that works, *J. Comput. Phys.*, Vol. 11, pp. 38-69 (1973)

Bosch G. and Rodi W. Simulation of vortex shedding past a square cylinder with different turbulence models, *Int. J. Numer. Meth. Fluids*, Vol. 28, pp. 601-616 (1998)

Bousquet N., Cazalbou J.-B. and Chassaing P. A proposal for taking into account the intermittency phenomenon in the calculation of wall-bounded turbulent flows, *Proc. 4th Int. Symp. on Engineering Turbulence Modelling and Measurements, Ajaccio, Corsica, France*, pp. 125-134 (1999)

Brackenridge J. B. Transverse oscillations of a liquid jet, *J. of the Acoustical Society of America*, Vol. 32, No. 10, pp. 1237-1242 (1960)

Brandt A. Multilevel adaptive computations in fluid dynamics, *AIAA J.*, Vol. 18, No. 10, pp. 1165-1172 (1980)

Brandt A. Multi-level adaptive solutions to boundary-value problems, *Mathematics of Computation*, Vol. 31, No. 138, 333-390 (1977)

Brebbia C. A. and Wrobel L. C. Steady and unsteady potential problems using the boundary element method, *Recent Advances in Numerical Methods in Fluids, Vol. 1, Pineridge Press Ltd, Swansea, U.K.*, Eds C. Taylor and K. Morgan, pp. 1-25 (1980)

Brenan K. E., Campbell S. L. and Petzold L. R. *Numerical solution of initial value problems in differential algebraic equations*, SIAM (1996)

Breuer M. and Rodi W. Large eddy simulation of turbulent flow through a straight square duct and a 180° bend, *Direct and Large Eddy Simulation I (Selected Papers from the 1st ERCOFTAC Workshop on Direct and Large-Eddy Simulation, P. R. Voke et al. (Eds.)*, pp. 273-285 (1994)

Briley W. R. and McDonald H. Solution of the three dimensional Navier-Stokes equations by an implicit technique, *Proc. of the 4th Int. Conf. on Numerical Methods in Fluid Dynamics, Lecture Notes in Physics, Vol. 35, Springer-Verlag, Berlin* (1975)

Briley W. R. and Walls H. A. A numerical study of time-dependent rotating flow in a cylindrical container at low and moderate Reynolds numbers, *Proc. 2nd Int. Conf. on Numerical Methods in Fluid Dynamics, University of California, Berkley (In Lecture Notes in Physics, Springer Verlag, Ed. Maurice Holt)*, pp. 377-384 (1970)

Burmeister J. and Horton G. Time parallel multigrid solution of the Navier-Stokes equations, *Int. Series of Numer. Mathematics*, Vol. 98, pp. 155-166 (1991)

Butler T. D. LINC method extensions, *Proc. 2nd Int. Conf. on Numerical Methods in Fluid Dynamics, University of California, Berkley (In Lecture Notes in Physics, Springer Verlag, Ed. Maurice Holt)*, pp. 435-440 (1970)

Cambon C., Benoit J-P, Shao L. and Jacquin L. Stability analysis and large eddy simulation of rotating turbulence with organized eddies, *J. Fluid Mech.*, Vol. 278, pp. 175-200 (1994)

Castrejon A. and Spalding D. B. An experimental and theoretical study of transient free-convection flow between horizontal concentric cylinders, *Int. J. Heat and Mass Transfer*, Vol. 31, No. 2, pp. 273-284 (1988)

Chapman M. FRAM – Nonlinear damping algorithms for the continuity equation, *J. Comput. Phys.*, Vol. 44, pp. 84-103 (1981)

Chen C. J. Nasseri-Neshat H. and Ho K. S. Finite analytic numerical solution of heat transfer in two dimensional cavity flow, *Numer. Heat Transfer*, Vol. 4, pp. 179-191 (1981)

Chergui J. and Daube O. Unsteady flow between a rotating and a stationary disc, *Proc. 6th Int. Conf. on Numerical Methods in Laminar and Turbulent Flow, Swansea, Pineridge Press*, Vol. 5, No. 1, p. 273.

Chesshire G. and Henshaw W, D. Composite overlapping meshes for the solution of partial differential equations, *J. Comput. Phys.*, Vol. 90, pp. 1-64 (1990)

Choi H. and Moin P. Effects of computational step on numerical solutions of turbulent flow, *J. Comput. Phys.*, Vol. 113, pp. 1-4 (1994)

Cochran W. G. The flow due to a rotating disc, *Proc. Cambridge Phil. Soc.* Vol. 30, pp.365-376 (1934)

Coelho P., Pereira J. C. F. and Carvalho M. G. Calculation of laminar recirculating flows using a local non-staggered grid refinement system, *Int. J. Numer. Meth. Fluids*, Vol. 12, pp. 535-557 (1991)

Crowe C. T., Sharma M. P. and Stock D. E. The particle-source-in cell (PSI-CELL) model for gas-droplet flows, *ASME J. Fluids Engrg*, pp. 325-332, June (1997)

Crowley W. P. Second order numerical advection, *J. Comput. Phys.*, Vol. 1, pp. 471-484 (1967)

Davidson L. Personal communication, *Dept. of Thermo and Fluid Dynamics, Horsalsvagen 7, Chalmers University of Technology, SE-412 96 Gothenburg, Sweden* (2000)

Davis R. L. and Dannenhoffer J. F. Three-dimensional adaptive grid-embedding Euler technique, *AIAA J.*, Vol. 32, No. 6 (1994)

Davis R. W. and Moore E. F. A numerical study of vortex shedding form rectangles, *J. Fluid Mech.* Vol. 116, pp. 475-506 (1982)

de Jongh F. M. and Morton P. G. The synchronous instability of a compressor rotor due to bearing journal differential heating, *ASME Int. Gas Turbine and Aeroengine Cong. and Exp.*, The Hague, Netherlands, Paper 94-GT-35 (1994)

De Matos A., Pinho F. A. A. and Silveira-Neto A. Large eddy simulation of turbulent flow over a two-dimensional cavity with temperature fluctuations. *Int. J. Heat and Mass Transfer*, Vol. 42, pp. 49-59 (1999)

De Socio L. M., Sparrow E. M. and Eckert E. R. G. The contrived transient-explicit method for solving steady-state flows: application to a rotating, recirculating flow, *Computers and Fluids*, Vol. 1, pp. 273-287 (1973)

De Vahl Davies G. and Jones I. P. Natural convection in a square cavity: A comparison exercise, *Int. J. Numer. Meth. Fluids*, Vol. 3, pp. 227-248 (1983)

Demirdzic I. and Peric M. Space conservation law in finite volume calculations of fluid flow, *Int. J. Numer. Meth. Fluids*, Vol. 8, pp. 1037-1050 (1988)

Derksen J. J., Kooman J. L. and van den Akker H. E. A. A parallel DNS implementation for confined swirling flows. *Proc. EUROSIM Int. Conf., Delft, The Netherlands*, pp. 237-244 (1996)

Dimitriadis K. P. and Leschziner M. A. Multilevel convergence acceleration for viscous and turbulent transonic flows computed with cell-vertex method, *Proc. of the $4^{th}$ Copper Mountain Conf. on Multigrid Methods, Copper Mountain, Colorado*, pp. 1-15 (1989)

Douglas J. and Gunn J. E. A general formulation of alternating direction implicit methods: parabolic and hyperbolic problems, *Numerische Mathematik*, Vol. 6, pp. 428-453 (1964)

Dukowicz J. and Ramshaw J. Tensor viscosity method for convection in numerical fluid dynamics, *J. Comput. Phys.*, Vol. 32, pp. 71-86 (1979)

Durbin P. A. Separated flow computations with the $k$-$\varepsilon$-$v^2$ model, *AIAA J.*, Vol. 33, No. 4, pp. 659-664 (1995)

Emerson D. R. *Introduction to parallel computers: Architecture and algorithms*, ERCOFTAC Series, High Performance Computing in Fluid Dynamics (Edited by P. Wesseling), Kluwer Academic Publishers, pp. 1-42 (1996)

Engelman M. S. and Sani R. L. Finite element simulation of incompressible fluid flows with a free/moving surface, *Recent Advances in Numerical Methods in Fluids*, Vol. 5, pp. 47-74 (1986)

Farthing P. R., Long C. A., Owen J. M. and Pincombe J. R. Rotating cavity with an axial throughflow of cooling air: heat transfer, *ASME J. Turbomach.*, Vol. 114, pp. 237-246 (1992a)

Farthing P. R., Long C. A., Owen J. M. and Pincombe J. R. Rotating cavity with an axial throughflow of cooling air: flow structure, *ASME J. Turbomach.*, Vol. 114, pp. 229-236 (1992b)

Ferziger J. H. and Peric M. *Computational methods for fluid dynamics*, Springer (1977)

Fletcher C. A. J. *Computational techniques for fluid dynamics: Fundamental and general techniques*, Springer, Vol. 1 (1997)

Flood C. R. Forecast evaluation, *Meteorological magazine*, Vol. 114, pp. 254-260 (1985)

Franke R. and Rodi W. Calculation of vortex shedding past a square cylinder with various turbulence models, *8th Int. Symp. of Turbulent Shear Flows, Munich*, pp. 189-204 (1993)

Fromm J. E. A numerical study of buoyancy driven flows in room enclosures, *Proc. 2nd Int. Conf. on Numerical Methods in Fluid Dynamics, University of California, Berkley (In Lecture Notes in Physics, Springer Verlag, Ed. Maurice Holt)*, pp. 120-126 (1970)

Fuchs L. A local mesh-refinement technique for incompressible flows, *Computers and Fluids*, Vol. 14, No. 1, pp. 69-81 (1986)

Gaitonde A. L. A dual-time method for two-dimensional unsteady incompressible flow calculations, *Int. J. Numer. Meth. in Engrg*, Vol. 41, pp. 1153-1166 (1998)

Gatski T. B., and Speziale C. G. On explicit algebraic stress models for complex turbulent flows, *J. Fluid Mech.*, Vol. 254, pp. 59-78 (1993)

Gebhart D., Jaluria Y. Mahajan R. L. and Sammalia B. *Buoyancy-induced flows and transport*, Hemisphere Publishing Corporation, New York, p. 227 (1988)

Germano M., Piomelli U., Moin P., Cabot W. H. A dynamic sub-grid scale eddy viscosity model, *Centre of Turbulence Research, Proc. of Summer Program, Stanford University* (1990)

Ghaddar N. K., Korczak K. Z, Mikic B. B and Patera A. T. Numerical investigation of incompressible flow in grooved channels. part. 1. Stability and self-sustained oscillations, *J. Fluid Mech.*, Vol. 163, pp. 99-127 (1986a)

Ghaddar N. K., Magen M., Mikic B. B. and Patera A. T. Numerical investigation of incompressible flow in grooved channels. Part 2. Resonance and oscillatory heat-transfer enhancement, *J. Fluid Mech.*, Vol. 168, pp. 541-567 (1986b)

Gharakhani A. and Ghoniem A. F. 3D vortex simulation of flow in an opposed-piston engine, *ESAIM: Proc. 3rd Int. Workshop on Vortex Flows and Related Numerical Methods*, Vol. 7, pp. 161-172 (1999) – http://www.emath.fr/proc/Vol.7/

Ghosal S. Mathematical and physical constraints on large eddy simulation of turbulence, *AIAA J.*, Vol. 37, No. 4, April, pp. 425-433 (1999)

Glanfield W. Personal communication (2000)

Glass J. and Rodi W. A higher order numerical scheme for scalar transport, *Comput. Meth. Appl. Mech. Engrg*, Vol. 31, pp. 337-358 (1982)

Gosman A. D. and Ioannides E. Aspects of computer simulation of liquid fuelled combustors, *J. Energy*, Vol. 7, pp. 482-490 (1983)

Gosman A. D. and Johns R. J. R. Development of a predictive tool for in-cylinder gas motion in engines, *Technical Paper Series, Society of Automotive Engineers, Presented at Congress and Exposition, Cobo Hall, Detroit, February 27 – March 3*, Paper No. 780315 (1978)

Gosman A. D. and Watkins A. P. A computer prediction method for turbulent flow and heat transfer in piston/cylinder assemblies, *Proc. of Symp. on Turbulent Shear Flows, University Park, Pennsylvania, (Eds F. Durst, V. W. Goldschmidt, B. E. Launder, F. W. Schmidt, J. H. Whitelaw) April 18-20*, pp. 5.23-5.30 (1977)

Gosman A. D., Koosinlin M. L., Lockwood F. C. and Spalding. D. B. Transfer of heat in rotating systems, *ASME Paper 76-GT-25* (1976)

Gresho P. M., Chan S. T., Lee R. L. and Upson C. D. A modified finite element method for solving the time-dependent, incompressible Navier-Stokes equations. Part 1: Theory, *Int. J. Numer. Meth. Fluids*, Vol. 4, pp. 557-598 (1984)

Gresho P. M. and Lee R. L. On the time-dependent solution of the incompressible Navier-Stokes equations in two and three dimensions, *Recent Advances in Numerical Methods in Fluids, Vol. 1, Pineridge Press Ltd, Swansea, U.K.*, Eds C. Taylor and K. Morgan, pp. 27-79 (1980)

Gupta R. S. and Kumar D. Variable time step methods for the dissolution of a gas bubble in a liquid, *Computers and Fluids*, Vol. 11, No. 4, pp. 341-349 (1983)

Harlow F. H. and Amsden A. A. A numerical fluid dynamics calculation method for all flow speeds, *J. Comput. Phys.*, Vol. 8, pp. 197-213 (1971)

Harlow F. H. and Welch J. E. Numerical calculation of time dependent viscous incompressible flow with free surface physics, *The Physics of Fluids* Vol. 8, No. 12, pp. 2182-2189 (1965)

Harlow F. H., Amsden A. A. and Hirt C. W. Numerical calculation of fluid flows at arbitrary Mach number, *Proc. 2nd Int. Conf. on Numerical Methods in Fluid Dynamics, University of California, Berkley (In Lecture Notes in Physics, Springer Verlag, Ed. Maurice Holt)*, pp. 447-452 (1970)

Hassan Y. A., Rice J. G. and Kim J. H. A stable mass flow weighted two dimensional skew upwind scheme, *Numer. Heat Transfer*, Vol. 6, pp. 395-407 (1983)

Hayase T. Monotonic convergence property for turbulent flow solution with central difference and QUICK schemes, *ASME J. Fluids Engrg*, June, Vol. 121, pp. 351-358 (1999)

Henkes R. A. W. M. *Natural-convection boundary layers*, PhD Thesis, Technical University Delft (1990)

Hide R. An experimental study of thermal convection in a rotating liquid, *Phil. Trans.* A, Vol. 250, p. 441 (1965)

Hignett P., White A. A., Carter R. D., Jackson W. D. N. and Small R. M. A comparison of laboratory measurements and numerical simulations of baroclinic wave flow in a rotating cylindrical annulus, *Quart. J. R. Met. Soc.*, Vol. 111, pp. 131-141 (1985)

Hignett P., White A. A., Carter R. D., Jackson W. D. N. and Small R. M. A comparison of laboratory measurements and numerical simulations of baroclinic wave flow in a rotating cylindrical annulus, *Quart. J. R. Met. Soc.* Vol. 111, pp. 131-145 (1985)

Hinze J. O. *Turbulence*, McGraw-Hill, New York (1975)

Hirsch Ch. and Ransbeeck P. Van. A general analysis of multidimensional convection schemes, Industrial computational fluid dynamics, *von Karman Institute for Fluid Dynamics, Lecture Series 1999-06*, pp. 61-87 (1999)

Hirt C. W. Heuristic stability theory for finite difference equations, *J. Comput. Phys.*, Vol. 2, pp. 339-355 (1968)

Hirt C. W., Amsden A. A. and Cook J. L. An arbitrary Lagrangian-Eulerian computing method for all flow speeds, *J. Comput. Phys.*, Vol. 14, pp. 227-253 (1974)

Huang P. G., Launder B. E and Leschziner M. A. Discretization of non-linear convection processes: a broad comparison of four schemes, *Comput. Meth. Appl. Mech. and Engrg*, Vol. 48, pp. 1-15 (1985)

Hujeirat A. and Rannacher R. A method for computing compressible, highly stratified flows in astrophysics based on operator splitting, *Int. J. Numer. Meth. Fluids*, Vol. 28, pp. 1-22 (1998)

Iacovides H. and Chew J. W. The computation of convective heat transfer in rotating cavities, *Int. J. Heat Fluid Flow*, Vol. 14, No. 2, pp. 146-154, (1993)

Iacovides H. and Theofanopoulos I. P. Turbulence modelling of axisymmetric flow inside rotating cavities, *Int. J. Heat Fluid Flow*, Vol. 12, No. 1, pp. 2-11 (1991)

Im H. G. Numerical studies of transient opposed-flow flames using adaptive time integration, *KSME Int. J.*, Vol. 14, No. 1, pp. 103-112 (2000)

Issa R. I. Solution of the implicitly discretised fluid flow equations by operator splitting, *J. Comput. Phys.*, Vol. 62, pp. 40-65 (1986)

James I. N., Jonas P. R. and Farnell L. A combined laboratory and numerical study of fully developed steady baroclinic waves in a cylindrical annulus. *Quart. J. R. Met. Soc.* Vol. 107, pp. 51-63 (1981)

Jameson A. Time dependent calculations using multigrid with applications to unsteady flows past airfoils and wings, *AIAA paper* 91-1596 (1991)

Jameson A. and Baker T. J. Solution of the Euler equations for complex configurations, *AIAA 6th Computational Fluid Dynamics Conf. AIAA, Paper 83-1929-CP* (1983)

Jameson A., Schmidt W. and Turkel E. Numerical solution of the Euler equations by finite volume methods using Runge Kutta time stepping schemes, *AIAA 5th Computational Fluid Dynamics Conf., AIAA Paper 81-1259* (1981)

Jenssen C. B. Industrial use of large eddy simulation, *Parallel Computational Fluid Dynamics – Recent Developments and Advances Using Parallel Computers, D. R. Emerson et al. (Eds.), Chapter 10,* pp. 587-592 (1998)

Jenssen C. B. and Weinerfelt P. A. Coarse grid correction scheme for implicit multiblock Euler calculations, *AIAA J.*, Vol. 33, No. 10, pp. 1816-1821 (1995)

Johansson S. H., Davidson L. and Olsson E. Numerical simulation of vortex shedding past triangular cylinders at high Reynolds number using a $k$-$\varepsilon$ turbulence model, *Int. J. Numer. Meth. Fluids,* Vol. 16, pp. 859-878 (1993)

Jones W. P. and Marquis A. J. Calculation of axisymmetric re-circulating flows with a second order turbulence model, *Proc. of the 5th Symp. on Turbulent Shear Flows, Connel University*, pp. 20.1-20.11 (1985)

Kais G., Long C. A. and Tucker P. G. Flow and heat transfer in rotating cavities with axial throughflow. *Proc. 10th Int. Heat Transfer Conf., Brighton,* Vol. 4, pp. 253-264 (1994)

Kato M. and Launder B. E. The modelling of turbulent flow around stationary and vibrating square cylinders, *9th Symp. on Turbulent Shear Flows, Kyoto, Japan*, pp. 10-4-1 – 10-4-6 (1993)

Khatir Z. *Discrete vortex modelling of near-wall flow structure in turbulent boundary layers*, PhD Thesis, Fluid Dynamics Research Centre, The University of Warwick (2000)

Kim J. and Moin P. Application of a fractional step method to incompressible Navier-Stokes equations, *J. Comput. Phys.*, Vol. 59, pp. 308-323 (1985)

Kimure I. and Hosoda T. 3-D unsteady flow structures around rectangular column in open channels by means of non-linear $k$-$\varepsilon$ model, *Proc. 1st Int. Symp. on Tubulence and Shear Flow (Eds Banerjee S. and Eaton J. K.), Santa Barbara,* pp. 1001-1006 (1999)

Kobayashi M. H. and Pereira C. F. Calculation of incompressible laminar flows on a non staggered non-orthogonal grid, *Numer. Heat Transfer*, Part B, Vol. 19, pp. 243-262 (1991)

Koosinlin M. L., Launder B. E. and Sharma B. I. Prediction of momentum, heat and mass transfer in swirling turbulent boundary layers, *Trans ASME, J. Heat Transfer,* May, pp. 204-209 (1974)

Kosovic B. Subgrid scale modelling for the large eddy simulation of high Reynolds number boundary layers, *J. Fluid Mech.,* Vol. 336, pp. 151-182 (1997)

Lacor C. Industrial computational fluid dynamics, *von Karman Institute for Fluid Dynamics, Lecture Series 1999-06*, Chapter 5, pp. 33-60 (1999)

Lacor C. *Programming aspects and algorithms for vector and parallel computers*, ERCOFTAC Series, High Performance Computing in Fluid Dynamics (Edited by P. Wesseling), Kluwer Academic Publishers, pp. 43-95 (1996)

Launder B. E and Sharma B. I. Application of the energy-dissipation model of turbulence to the calculation of flow near a spinning disc, *Letters in Heat and Mass Transfer*, Vol. 1, pp. 131-136 (1974)

Launder B. E, Reece G. J. and Rodi W. Progress in the development of a Reynolds-stress turbulence closure, *J. Fluid Mech.*, Vol. 68, Part 3, pp. 537-566 (1975)

Launder B. E. and Spalding D. B., The numerical computation of turbulent flows, *Comput. Meth. Appl. Mech. Engrg*, Vol. 3, pp. 269-289 (1974)

Leith C. E. Numerical simulation of the earth's atmosphere, *Meth. Comp. Phys.*, Vol. 4, pp. 1-28 (1965)

Lele S. K. Compact finite difference schemes with spectral like resolution, *J. Comput. Phys.*, Vol 103, pp. 16-42 (1992)

Leonard B. P. A stable and accurate convective modelling procedure based on quadratic upstream interpolation, *Comp. Meth. Appl. Mech. and Engng*, Vol. 19, pp. 59-99 (1979)

Leonard B. P. Locally modified QUICK scheme for highly convective 2-d and 3-d flows, *Proc. 6th Int. Conf. Numer. Meth. In Lam. and Turb. Flow, Swansea, Pineridge Press*, pp. 35-45 (1989)

Lerat A. Une classe de schemas aux differences implicites pour les systemes hyperboliques de lois de conservation, *Comptes Rendues Acad. Sciences Paris*, Vol. 288A, pp. 1033-1036 (1979)

Liu Jun. Computer Modelling of flows with a free surface, *PhD Thesis, University of London* (1986)

Long C. A. and Tucker P. G. Shroud heat transfer measurements from a rotating cavity with an axial throughflow of air, *ASME J. Turbomach.,* Vol. 116, pp. 525-534 (1994)

Long C. A., Morse A. P. and Tucker P. G. Measurement and computation of heat transfer in high-pressure compressor drum geometries with axial throughflow, *ASME J. Turbomach.*, Vol. 119. No. 1, pp. 51-60 (1997)

Long C.A. and Tucker P.G. Numerical computation of laminar flow in a heated rotating cavity with an axial through flow of air, *Int. J. Numer. Meth. Heat Fluid Flow 4*, pp.347-365 (1994)

Lonsdale G. Solution of a rotating Navier-Stokes problem by a non-linear multigrid algorithm, *J. Comput. Phys.*, Vol. 79, pp. 177-190 (1988)

Lund J.W., and Tonnesen J. An Approximate Analysis of the temperature conditions in a journal bearing. Part II: Application, *ASME J. Tribology,* Vol. 106, pp. 237-245 (1984)

Mallinson G. D. and de Vahl Davis G. The method of the false transient for the solution of coupled elliptic equations, *J. Comput. Phys.*, Vol. 12, pp. 435-461 (1973)

Mansour N. N. and Morel T. Computations for a curved shear layer and a backward-facing step using a $k$-$\varepsilon$ turbulence model, *Proc. AFOSR-Stanford-HTTM Conf. on Complex Turbulent Flows 1981, (Eds S. J. Kline, B. J. Cantwell and G. M. Lilley)* Vol. 2, pp. 842-862 (1983)

Martins L.-F. and Ghoniem A. F. Vortex simulation of the flow field in a planar piston chamber arrangement with an intake, *Int. J. Numer. Meth. Fluids*, Vol. 12, pp. 237-260 (1991)

Mason P. J. and Cullen N. S. On the magnitude of the subgrid-scale eddy coefficient in large eddy simultaneous of turbulent channel flow, *J. Fluid Mech.*, Vol. 162, pp.439-462 (1986)

Metais O., Flores C., Yanase S., Riley J. J. and Lesieur M. Rotating free shear flows. Part 2. Numerical Simulations, *J. Fluid Mech.*, Vol. 293, pp. 47-80 (1995)

Milojevic D. Lagrangian stochastic-deterministic (LSD) predictions of particle dispersion in turbulence, *Part. Part Syst. Charact.*, Vol. 7, pp. 181-190 (1990)

Mitchell A. R. and Griffiths D. F. *Finite difference methods for partial differential equations,* Wiley – Interscience, New York (1980)

Miyata H. and Nishimura S. Finite difference simulation of non-linear ship waves, *J. Fluid Mech.*, Vol. 157, pp. 327-357 (1985)

Morgan A. J. and Barton I. E., Application of Lagrangian tracking schemes to flow over a backward facing step, *To be published in Communications in Numer. Meth. in Engrg* (2000)

Morse A. P. Application of a low Reynolds number $k$-$\varepsilon$ turbulence model to high-speed rotating cavity flows, *ASME J. Turbomach.*, Vol. 113, p.98 (1991a)

Morse A. P. Assessment of laminar-turbulent transition in closed disc geometries, *ASME J. Turbomachinery*, Vol. 113, p.131 (1991b)

Morse A. P. Numerical prediction of turbulent flow in rotating cavities, *ASME Paper 87-GT-74, Presented at the ASME Gas Turbine Conf. and Exhibition, Anaheim, California,* (1987)

Morse A. P. Axisymmetric free shear flows with and without swirl, *PhD Thesis, University of London* (1980)

Morse A. P. and Ong C. L. Computation of heat transfer in rotating cavities using a two equation model of turbulence, *ASME J. Turbomach*, Vol. 114, p. 247 (1992)

Mullen J., Than C-F and Savilonis B. Modelling fire induced flow in a vertically vented enclosure, *ASME Forum on Unsteady Flows*, FED-Vol. 157, pp. 215-220 (1993)

Muramatsu T. and Ninokata H. Thermal striping temperature fluctuation analysis using the algebraic stress turbulence model in water and sodium, *JSME Int. J.*, Series 2, Vol. 35, No. 4 (1992)

Nasser A. G. F. A. and Leschziner M. A. Computation of transient recirculating flow using spline approximations and time-space characteristics, *Proc. 4th Conf. on Numerical Methods in Laminar and Turbulent Flows, Swansea,* pp. 480-491 (1985)

Nigen J. S. and Amon C. H. Effect of material composition and localised heat generation on time dependent conjugate heat transport, *Int. J. Heat Mass Transfer*, Vol. 38, No. 9, pp. 1565-1576 (1995)

Nigen J. S. and Amon C. H. Time-dependent conjugate heat transfer characteristics of self-sustained oscillatory flows in a grooved channel, *ASME J. Fluids Engrg*, Vol. 116, pp. 499-507 (1994)

Nishimura M., Tokuhiro A., Kimura N. and Kamide H. Computational study on quasi-planar jets mixing with low Reynolds number turbulent stress and flux equation models, *9th Int. Topical Meeting on Nuclear Reactor Thermal Hydraulics (NURETH-9) San Francisco, California, October 3-8*, CDROM (1999)

Orkwis P. D. and Vanden K. J. On the accuracy of numerical versus analytical Jacobians, *AIAA Paper 94-0176* (A94-20555) (1994)

Orszag S. A., Borue V., Flannery, W. S. and Tomboulides A. G. Recent successes, current problems and future prospects of CFD, *AIAA* Paper 97-0431, pp. 1-11 (1997)

Owen J. M., Pincombe J. R. and Rogers R. H. Source-sink flows inside a rotating cylindrical cavity, *J. Fluid Mech.*, Vol. 155, pp. 233-265 (1985)

Owen J. M. and Pincombe J. R. Vortex breakdown in a rotating cylindrical cavity, *J. Fluid Mech.,* Vol. 90, pp. 109-127 (1979)

Pan Z. and Tucker P. G. Transient and steady flow computations for an electro-mechanical system, *NAFEMS Int. J. CFD Case Studies,* Vol. 2, February, pp. 103-118 (2000)

Pascal H., Jakirlic S. and Hanjalic K. DNS and RANS-Modelling of in-cylinder turbulence subjected to axial compression, *Proc. of the 3rd Int. Symp. on Turbulence Heat and Mass Transfer, Nagoya, Japan – Turbulence, Heat and Mass Transfer 3, Eds Y. Nagano, K. Hanjalic and T. Tsuji, Publisher: Aichi Shuppan,* pp. 479-486 (2000)

Patankar S. V. *Numerical heat transfer and fluid flow,* Hemisphere, New York (1980)

Patankar S. V. and Baliga B. R. A new finite difference scheme for parabolic differential equations, *Numer. Heat Transfer,* Vol. 1, pp. 27-37 (1978)

Patankar S. V. and Spalding D. B. A calculation procedure for heat mass and momentum transfer in three-dimensional parabolic flows, *Int. J. Heat and Mass Transfer*, Vol. 15, 1787-1806 (1972).

Patel M. K. and Markatos N. C. An evaluation of eight discretization schemes for two dimensional convection-diffusion equations, *Int. J. Numer. Meth. Fluids*, Vol. 6, p. 129 (1986)

Patera A. T. A spectral element method for fluid dynamics: laminar flow in a channel expansion, *J. Comput. Phys.,* Vol. 54, pp. 468-488 (1984)

Patera A. T. and Mikic B. B. Exploiting hydrodynamic instabilities. Resonant heat transfer enhancement, *Int. J. Heat Mass Transfer,* Vol. 29, No. 8, pp. 1127-1138 (1986)

Peaceman D. W. and Rachford H. H. The numerical solution of parabolic and elliptic differential equations, *SIAM J.*, Vol. 3, pp. 28-41 (1955)

Piacsek S. A. and Williams G. P. Conservation properties of finite difference schemes, *J. Comp. Phys.* Vol. 6, pp. 392-393 (1970)

Pincombe J. R. Optical measurements of the flow inside a rotating cylinder, *DPhil Thesis, School of Engineering, University of Sussex, England* (1983)

Pinkus O., and Sternlicht B. *Theory of hydrodynamic lubrication*, McGraw-Hill Book Company Ltd., New York, Toronto and London (1961)

Piomelli U. and Junhui L. Large eddy simulation of rotating channel flows using a localized dynamic model, *Phys. Fluids*, Vol. 7, No. 4, pp. 839-848 (1988)

Pracht W. E. Implicit solution of creeping flows with application to continental drift, *Proc. 2nd Int. Conf. on Numerical Methods in Fluid Dynamics, University of California, Berkley (In Lecture Notes in Physics, Springer Verlag, Ed. Maurice Holt)*, pp. 452-457 (1970)

Press W. H., Flannery B. P., Teukolsky S. A. and Vetterling W. T. *Numerical recipes: The art of scientific computing*, University of Cambridge Press (1988)

Przulj V. and Younis B. A. Some aspects of the prediction of turbulent vortex shedding, *ASME Fluids Engineering Conf., Washington DC, FED*-Vol. 149, Separated Flows, pp. 75-81 (1993)

Pudykiewicz J. Simulation of the Chernobyl dispersion with a 3D hemispheric tracer model, *Tellus*, 41B, pp. 392-412 (1989)

Raithby G. D. and Schneider G. E. Erratum, *Numer. Heat Transfer*, Vol. 3, p. 513 (1980)

Raithby G. D. and Schneider G. E. Numerical solution of problems in incompressible fluid flow: treatment of velocity pressure coupling, *Numer. Heat Transfer*, Vol. 2, pp. 417-440 (1979)

Raw M. Robustness of coupled algebraic multigrid for the Navier-Stokes equations, *AIAA Paper 96-0297* (1996)

Raw M. A coupled algebraic multigrid method for the 3D Navier-Stokes equations *in: Hackbusch G. Wittum (Eds.), FAST Solvers for Flow Problems, Notes on Numerical Fluid Mechanics, Vol. 49, Vieweg, Braunschweig* pp. 205-215 (1995)

Rayner D. A numerical study into the heat transfer beneath the stator blade of an axial compressor, *D. Phil. Thesis, School of Engineering, University of Sussex* (1993)

Rayner D. Multigrid flow solutions in complex two dimensional geometries, *Int. J. Numer. Meth. Fluids*, Vol. 20, pp. 61-80 (1991)

Reindl D. T., Beckman W. A., Mitchell J. W. and Rutland C.J. Benchmarking transient natural convection in an enclosure, *ASME paper 91-HT-8*, pp. 1-7 (1991)

Rhie C. M. and Chow W. L. Numerical study of the turbulent flow past an airfoil with trailing edge separation, *AIAA J.*, Vol. 21, No. 11, November, pp. 1525-1532 (1983)

Richardson L. F. The approximate arithmetical solution by finite differences of physical problems involving differential equations with an application to the stresses in a masonry dam, *Trans. of the Royal Society London*, Series A, Vol. 210, pp. 307-357 (1910)

Roache P. J. A flux-based modified method of characteristics, *Int. J. Numer. Meth. Fluids*, Vol. 15, pp. 1259-1275 (1992)

Rockwell D. and Naudascher E. Self sustained oscillations of impinging free shear layers, *Ann. Rev. Fluid. Mech.*, Vol. 11, pp. 67-94 (1979)

Rodi W., Ferziger J. H., Breuer M. and Pourquie M. Status of large eddy simulation: results of a workshop, *Trans. of the ASME,* Vol. 119, pp. 248-262 (1997)

Roychowdhury D. G. *Numerical analysis of fluid flow and heat transfer around cylindrical bodies using non-orthogonal collocated grid,* PhD Thesis, Heat Transfer and Thermal Power Laboratory, Department of Mechanical Engineering, Indian Institute of Technology, Madras (2000)

Ruge J. and Stuben K. Algebraic Multigrid, *Arbeitspapiere der GMD 210* (1986)

Runchal A. K. CONDIF: A modified central-difference scheme for convective flows, *Int. J. Numer. Meth. in Engrg,* Vol. 24, pp. 1593-1608 (1987)

Saito H. and Scriven L. E. Study of coating flow by finite element method, *J. Comput. Phys.*, Vol. 42, pp. 53-76 (1981)

Satofuka N. Numerical solution of fluid dynamic equations using rational Runge Kutta methods, *Computing Methods in Applied Sciences and Engineering, VII, R. Glowinski and J-L Lions (Eds.), Elsevier Science Publishers B.V. (North-Holland),* pp. 249-259 (1986)

Schlichting H. *Boundary layer theory,* McGraw-Hill, New York (1979)

Schmidt W. and Jameson A. Recent developments in finite volume time dependent techniques for two and three dimensional transonic flows, *1981-82 Lecture Series Programme, Computational Fluid Dynamics, von Karman Institute for Fluid Dynamics, Chaussee de Waterloo 72, B-1640 Rhode-Saint-Genese, Belgium, March 29 – April 2* (1982)

Schonauer W., Glotz G. and Raith K. The solution of the laminar boundary layer equations by a variable order selfadaptive difference method, *Recent Advances in Numerical Methods in Fluids, Vol. 1, Pineridge Press Ltd, Swansea, U.K.,* Eds C. Taylor and K. Morgan, pp. 81-118 (1980)

Sedaghat A., Ackroyd J. A. D and Wood N. J. Turbulence modelling for supercritical flows including examples with passive shock control, *The Aeronautical J.*, pp. 113-125, February (1999)

Segerlind L. J. *Applied finite element analysis*, John Wiley and Sons (1984)

Seidl V., Peric M. and Schmidt S. Space and time parallel Navier-Stokes solver for 3D block adaptive Cartesian grids, *Proc. of the Parallel CFD'95 Conf., Pasadena*, pp. 577-584 (1995)

Sethian J. A. Fast Marching Methods. *SIAM Review*, Vol. 41, No. 2, pp. 199-235 (1999)

Shur M., Spalart P. R., Strelets M. and Travin A. Detached-eddy simulation of an aerofoil at high angle of attack, *Proc. 4th Int. Symp. on Engineering Turbulence Modelling and Measurements, Ajaccio, Corsica, France,* pp. 669-678 (1999)

Shur M., Spalart P., Stretlets, M. and Travin A. Navier-Stokes simulation of shedding turbulent flow past a circular cylinder and a cylinder with backward splitter plate, *Computational Fluid Dynamics '96, Proc. of the 3rd ECCOMAS Computational Fluid Dynamics Conf., Paris, France,* pp. 676-682 (1996)

Shyy W. and Chen M. Pressure based multigrid algorithm for flow at all speeds, *AIAA J.*, Vol. 30, No. 11, pp. 2660-2669 (1992)

Shyy W., Sun C., Chen M. and Chang K. C. Multigrid computation for turbulent recirculating flows in complex geometries, *Numer. Heat Transfer*, Vol. A23, pp. 79-98 (1993)

Skelboe S. The control of order and steplength for backward differentiation methods, *BIT*, Vol. 17, pp. 91-107 (1977)

Smagorinsky J. General circulation experiments with the primitive equations, *Ann. Rev. Fluid Mech.,* Vol. 16, pp. 99-165 (1963)

Smith R. M. and Hutton A. G. The numerical treatment of advection: a performance comparison of current methods, *Numer. Heat Transfer*, Vol. 5, pp. 439-449 (1982)

Spalart P. R. Strategies for turbulence modelling and simulations, *Proc. 4th Int. Symp. on Engineering Turbulence Modelling and Measurements, Ajaccio, Corsica, France,* pp. 3-17 (1999)

Spalart P. R. and Allmaras S. R. A one-equation turbulence model for aerodynamic flows, *La Rech. Aerospatiale,* Vol. 1, pp. 5-21 (1994)

Spalding D. B. Calculation of turbulent heat transfer in cluttered spaces, *Proc. 10th Int. Heat Transfer Conf.,* Brighton, UK (1994)

Spalding D. B. A novel finite difference formulation for differential expressions involving both first and second derivatives, *Int. J. Numer. Meth.. Engng*, Vol. 4, pp. 551-561, (1972)

Spalding D. B. A single formula for the law of the wall, *Trans ASME, Series* A, *J. Appl. Mech.*, Vol. 28, No. 3, pp. 444-458 (1961)

Speziale C. G. On non-linear *k-l* and *k*-ε models of turbulence, *J. Fluid Mech.*, Vol. 178, pp. 459-475 (1987)

Staniforth A. and Cote J. Semi-Lagrangian integration schemes for atmospheric models – a review, *Monthly Weather Review*, Vol. 119, pp. 2206-2223 (1991)

Steger J. L. and Kutler P. Implicit finite-difference procedures for the computation of vortex wakes, *AIAA J.*, Vol. 15, No. 4, pp. 581-590 April (1977)

Stone H.L. Iterative Solution of Implicit approximations of multidimensional partial differential equations, *SIAM J. Numer. Analysis,* Vol. 5, pp. 530-558 (1968)

Strietzel M. Parallel large eddy simulation of sheared and stratified turbulence, *Parallel Programming and Applications, P. Fritzson and L. Fimmo (Eds.),* Vol. 44, pp. 90-96 (1995)

Sun D.C., Brewe D.E., and Abel P.B. Simultaneous pressure measurement and high-speed photography study of cavitation in a dynamically loaded journal bearing, *ASME J. Tribology,* Vol. 115, pp. 88-95 (1993)

Tamamidis P. and Assanis D. N. Benchmarking high resolution schemes in two-dimensional unsteady flows, *ASME Forum on Unsteady Flows*, FED-Vol. 157, pp. 95-106.

Tanner R. I., Nickell R. E. and Bilger R. W. Finite element methods for the solution of some incompressible non-Newtonian fluid mechanics problems with free surfaces, *Comp. Meth. Appl. Mech. and Eng.,* Vol. 6, pp. 155-170 (1975)

Tasciyan T. A., Banerjee R. K. and Cho Y. I. A comparison on 2D pulsatile hemodynamic analyses within stenosed cartoid arterial bifurcation and its normal counterpart, *ASME Forum on Unsteady Flows*, FED-Vol. 157, pp. 221-228 (1993)

Tatineni M. and Zhong X. Numerical simulation of unsteady low-Reynolds-number separated flows over airfoils, *AIAA J.*, Vol. 37, No. 7, pp. 1295-1298 (2000)

Tatsumi K., Iwai H., Neo E. C., Inaoka K. and Suzuki K. Prediction of time-mean characteristics and periodical fluctuation of velocity and thermal fields of a backward-facing step, *Proc. 1st Int.*

*Symp. on Tubulence and Shear Flow (Eds Banerjee S. and Eaton J. K.), Santa Barbara,* pp. 139-144 (1999)

Thiriot K.-H. Uber laminare anlaufstromung einer flussigkeit uber einem rotierenden boden bei plotzlicher anderung des drehungszustandes, *Z angew Math Mech.*, Vol. 20, pp. 233-253 (1940)

Thompson M. C. and Ferziger J. H. A multigrid adaptive method for incompressible flows, *J. Comput. Phys., Vol. 82, pp. 94-121* (1989)

Timmermans L. J. P. and Nieuwstadt F. T. M. Parallelisation of a 3D code for direct and large eddy simulation of turbulent flow on the Cray T3D, *Proc. EUROSIM Int. Conf., Delft, The Netherlands,* pp. 297-304 (1996)

Tu C., Deville M., Dheur L. and Vanderschuren L. Finite element simulation of pulsatile flow through arterial stenosis, *J. Biomechanics*, Vol. 25, No. 10, pp. 1141-1152 (1992)

Tucker P. G. Prediction of turbulent oscillatory flows in complex systems, *Int. J. Numer. Meth. Fluids,* Vol. 33, pp. 869-895 (2000)

Tucker P. G. Assessment of geometric multilevel convergence and a wall distance method for flows with multiple internal boundaries, *Appl. Math. Modelling*, Vol. 22, pp. 293-311 (1998)

Tucker P. G. Numerical precision and dissipation errors in rotating flows, *Int. J. Numer. Meth. Heat Fluid Flow, Vol. 7, No. 7, pp. 647-658* (1997)

Tucker P. G. Numerical and experimental investigation of flow structure and heat transfer in a rotating cavity with an axial throughflow of cooling air. *D. Phil. Thesis, School of Engineering and Applied Sciences, University of Sussex* (1993)

Tucker P. G. Numerical and experimental work on rotating cavities, *Technical Report 90/TFMRC/TN91, Thermo-Fluid Mechanics Research Centre, School of Engineering and Applied Sciences, University of Sussex, Falmer, Brighton,* (1990)

Tucker P. G. and Long C. A. Numerical investigation into influence of geometry on flow in a rotating cavity with an axial throughflow, *Int. Comm. Heat Mass Transfer*, Vol. 23, No. 3, pp. 335-344 (1996)

Tucker P. G. and Long C. A. Second order finite volume jet stream computations, *Int. J. Numer. Meth. Heat Fluid Flow,* Vol. 6, No. 6, pp. 39-50 (1996)

Tucker P. G. and Pan Z. URANS Computations for a complex internal isothermal flow, *Accepted for publication in the J. Comput. Meth. Appl. Mech. and Engrg* (2000)

Tucker P. G. and Keogh P. S. On the dynamic thermal state in a hydrodynamic bearing with a whirling journal using CFD techniques, *ASME J. Tribology*. Vol. 118, pp. 356-363 (1996)

Van Doormaal J. P. and Raithby G. D. Enhancement of the SIMPLE method for predicting incompressible fluid flows, *Numer. Heat Transfer*, Vol. 7, pp. 147-157 (1984)

Vaughan C. M., Gilham S. and Chew J. W. Numerical solutions of rotating disc flows using a non-linear multigrid algorithm, *Proc. 6th Int. Conf. On Num. Meths in Laminar and Turbulent Flow, Swansea,* pp. 63-67 (1989)

von Karman Th. Uber laminare und turbulente reibung, *Z angew. Math. Mech,* Vol. 1, pp. 233-243 (1921)

Wilcox D. C. *Turbulence modelling for CFD*, DCW Industries (1998)

Wolfshtein M. The velocity and temperature distribution in one-dimensional flow with turbulence augmentation and pressure gradient, *Int. J. Heat Mass Transfer*, Vol. 12, pp. 301-318 (1969)

Xiyun L., Dalton C. and Zhang J. Application of large eddy simulation to an oscillating flow past a cylinder, *J. Fluids Engrg,* Vol. 119, pp. 519-525 (1997)

Yang K-S and Ferziger J. H. Large eddy simulation of turbulent obstacle flow using a dynamic subgrid scale model, *AIAA J.,* Vol. 31, No. 8, pp. 1406-1413 (1993)

Yao Y. F., Savill A. M., Sandham N. D. and Dawes W. N. Simulation of a turbulent trailing edge flow using unsteady RANS and DNS, *Proc. of the 3rd Int. Symp. on Turbulence Heat and Mass Transfer, Nagoya, Japan – Turbulence, Heat and Mass Transfer 3, Eds Y. Nagano, K. Hanjalic and T. Tsuji, Publisher: Aichi Shuppan,* pp. 463-470 (2000)

Yee H. C. and Harten A. Implicit TVD schemes for hyperbolic conservation laws in curvilinear coordinates, *AIAA J.*, Vol. 25, No. 2, pp. 266-274 (1987)

Zhou Q. and Leschziner M. A. Modelling particle dispersion in turbulent recirculating flow with an anisotropy resolving scheme, *ASME Fluids Engineering Division Summer Meeting, FEDSM'97*, June 22-26, pp. 1-8 (1997)

Zhu Z. W., Lacor C. and Hirsch Ch. A new residual smoothing method for multigrid, multi-stage schemes, *AIAA 11th Computational Fluid Dynamics Conf., AIAA Paper 93-3356* (1993)

# INDEX


**A**
Adams-Bashforth scheme, 25, 101, 118
Adams-Moulton scheme, 16, 25
Adams type scheme, 25
Adaptive grid embedding, 75
Adaptive meshing, 112
Adaptive time-stepping, 197, 212, 223, 234
Aero-engine, 183
ALE technique, 56
Algebraic stress models, 157, 233
Algebraic multilevel/grid methods, 87, 111, 267, 269
Aliasing, 57, 58, 63, 174
Altered order error estimate, 119, 197, 213
Altered scheme error estimate, 119, 197
Altered step error estimate, 119, 197, 213
Alternating Direction Implicit (ADI) scheme, 29, 80, 185
Anisotropy, 226, 227, 271
Annulus, 173, 193
Anticyclones, 173, 184, 185, 188
Approximate factorisation, 80
Artificial compressibility methods, 103
Attached flow, 218, 220
Attractor, 4
AVPI scheme, 98, 99, 116, 230, 234
AVPI/SIMPLEC algorithm, 185, 193
Axial throughflow, 184
Axis of rotation, 256

**B**
Back scatter, 155
Balancing Tensor Diffusivity (BTD), 60
Bearing, 201
Biconjugate solver, 81
Bifurcation, 4
Blocking off, 112
Block matrices, 105
Body fitted grids, 65
Boolean marker array, 113
Boundary conditions, 33, 186, 203, 213, 223, 230, 275
Bounding algorithms, 52
Boussinesq approximation, 162, 166, 167, 227
Branch, 5

**C**
Capacitance, 13
Cartesian tensor form, 157
Cell deletion, 114
Cell insertion, 114
Central differencing scheme, 50, 174, 178-180, 261

Centripetal acceleration, 99
CFX4, 105, 110, 111, 114, 118
Chimera method, 64
Collocated grid, 111
Compact schemes, 52, 56
Complex system cyclic flow, 211
Compressive heating, 163, 249
CONDIF scheme, 52, 174, 178-180, 185, 212, 230, 261, 264
Conditioned average, 159, 160
Conjugate gradient, 81, 112
Consistent formulation, 13
Consistent time-step variant of SIMPLE, 100
Continuity equation, 104, 105
Control volume faces, 252
Convection coefficients, 252
Convective term treatments, 50, 175
Convergence criteria, 85, 176, 186
Convergence radii, 112
Coordinate system angular velocity, 174, 175, 180, 181, 189-192, 204-210
Coriolis force, 99
Correction equations, 96
Correction storage (CS), 89
Couette velocity components, 277
Coupled problems, 171
Coupled solvers, 107, 111
Courant-Friedrichs-Lewy (CFL) condition, 32, 80, 115
Courant number, 32, 55, 223
Crank-Nicolson scheme, 16, 18, 101, 102, 156, 176, 185, 193, 223
Curvilinear grids, 66
Cyclic flow or steady state, 158, 178, 197, 211
Cyclones, 173, 184, 185, 188
Cylindrical polar coordinates, 157

**D**
Damping functions, 165
Deforming mesh, 114
Delta form, 20, 21, 106
Detached Eddy Simulation (DES), 156
Diagonal dominance, 50, 81, 83, 135
Diffusion coefficient, 262
Dilute flows, 169
Dimensionless wall distance, 163
Direct Numerical Simulation (DNS), 153, 154, 156
Dispersion, 58, 59
Dissipation, 58, 59
Domain decomposition in time, 92
Drag coefficient, 170
Drift, 174, 191
Driven box flow, 265
Dual time method, 103
DuFort-Frankel scheme, 16, 26
Dynamic cavitation modelling, 203, 276
Dynamic model, 155

**E**
Earth's acceleration due to gravity, 257
Eckert number, 185
Eddy lifetime, 171
Ekman layer, 176
Electronic system, 211, 268
Elliptic equations, 8
Ensemble average, 158
Estimated Streaming Terms (EST), 55
Euler error estimate, 213
Euler type equation, 261
Evolution processes, 90
Explicit forwards Euler, 185, 193
Exponential scheme, 16, 22

**F**

False diffusion, 48, 52, 60
False time-stepping, 83
Fast marching methods, 169
FIDAP, 71, 92, 104, 110, 114, 118
FIELDVIEW, 296
Finite element programs, 103
Finite volume, 111
First order forward difference Euler scheme, 203
First order upwind scheme, 50
Fixed mesh, 114
FLO_BEARING, 113, 299
FLO_CAV, 113, 272, 279
FLO_ELEC, 113, 265, 267, 296, 321
FLOTHERM, 110, 112, 114
FLOTRAN, 110, 112, 114
FLUENT, 110, 112, 114
Flux limiters, 52
Forcing function, 88
Fractional step methods, 101, 105, 156
FRAM, 57
Free disc flow, 144
Free surface modelling, 71
Free surface tracking technique, 72
Frictional heating, 163, 249
Full approximation, 88
Fully coupled solvers, 105
Fully explicit 1st order backwards Euler scheme, 176
Fully implicit 1st order backwards Euler scheme, 176

**G**

Galerkin scheme, 16, 22, 176, 185, 193
Galilean invariant, 47
Gaussian elimination, 80, 85
Gauss-Siedel, 81, 85, 112, 321
General grid structure, 297
Geometrical distortion, 272
Geophysical flows, 156
Global conservation of kinetic energy, 108
Grid aspect ratio, 271, 272, 273
Grid coarsening, 268
Grid distortion, 49, 204
Grid embedding, 267
Grid staggering approach, 105
Gosman factor, 84, 185
Grashof number, 196, 266

**H**

Hanging nodes, 112
Hexahedral mesh, 153
High pressure compressor, 183
Hopf bifurcation, 5
HOTBOX, 110, 112, 114
HYBRID scheme, 54, 174, 178-180, 185, 196, 203, 252, 267
Hydrodynamic journal bearings, 201
Hysteresis, 5

**I**

ICED-ALE method, 56
Impingement, 229
Implicit backwards Euler, 185, 193
Implicit coefficient Crank-Nicolson, 193
Implicit continuous-fluid Eulerian scheme (ICE), 104
Induced oscillations, 153
Inertial coordinate systems, 70
Instability, 4, 237
Integral length scale, 151
Intermittency, 158, 160

**J**

Jacobian, 19, 91, 92
Jacobi/Gauss-Siedel hybrid, 81

Jacobi solver, 81
Jet streams, 173, 179
Journal, 201

**K**
k-ε model, 165
k-l model, 164
Kolmogorov time scale, 151, 156
Kronecker delta, 162

**L**
Lagrangian frame, 226
Large Eddy Simulation (LES), 153, 155, 156
Law of the wall, 166
Lax-Wendroff scheme, 16, 25
Leap frog scheme, 16, 26
Length scales, 165
Limit cycle, 5, 188
Limiters, 52
Linearisation, 18, 93
Linear multilevel equation, 89
Local Extremum Diminishing (LED), 52
Logarithmic wall functions, 165
Long time period integrations, 156
Low frequency errors, 272
Lumped approach, 13
l-v model, 164

**M**
MACRL technique, 71
Magnetic bearing, 201
Marker and Cell (MAC) technique, 71, 73, 114
Mass diffusion term, 105
Mass operators, 13
Mass weighted skew upwinding, 111
Mean strain rate, 162
Method of Characteristics (MOC), 56
Method of lines, 11, 118
Metrics, 66
Mixed model, 155
Mixing length model, 152, 163, 212, 229
Modified equation, 58
Monotone Upstream Centred Scheme for Conservations Laws (MUSCL), 50, 51
Monotonicity conditions, 52
Moving boundaries, 49, 63
Moving grid transformations, 75
Multiblocking, 112
Multilevel/multigrid methods, 31, 75, 80, 85, 86, 90, 92, 105, 112, 115, 156, 169, 279
Multiple reference frame, 112
Multipoint method, 25

**N**
Navier-Stokes equations, 161
Newton method, 91, 92, 112
Newton-Raphson method, 81, 164
Non-linear eddy viscosity models, 153, 166, 212, 233
Non-physical oscillatory pressure fields, 107
Non-physical transient, 104
Non-reflecting outlet boundary conditions, 97
Normalised residual, 85
Normal wall distances, 168
Numerical advection correction, 111
Numerical diffusion, 52, 60
Numerical instabilities, 237
Nusselt number, 196

**O**
ODEs, 11
Oldroyd derivative, 166
Orbiting system, 202

**P**
Parabolic equations, 8
Parallel processing, 79, 92, 93
Particle dispersion, 33, 169, 224, 226
Penalty function, 103, 112
Periodic flow, 158
Phase averaging, 159
Phase error, 86
PHOENICS, 110, 112, 114, 118
PISO, 95, 98, 99, 112
Pitchfork bifurcation, 5, 186
Poiseuille velocity components, 277
Poisson equation, 167
Poisson pressure equation, 102, 104,
Post processing, 296
Prandtl number, 162
Predictor-corrector methods, 27, 118
Pressure correction equation, 97, 255
Pressure field, 94
Prolongation, 90, 258, 260
Pseudo-injection, 139, 223, 224
Pseudo time scale, 103

**Q**
Quadratic interpolation (QUICK) scheme, 50, 51, 54, 112, 152
Quadratic sources, 213, 223, 224
Quasi-Newton approach, 92
QUICKEST scheme, 55

**R**
Radial arm, 185
Rayleigh-Bernard problem, 184
Red black procedure, 81, 85, 92, 115
Reduced pressures, 107
Relaxation, 279
Residual equation, 88
Residual smoothing, 31, 88
Restriction, 90, 139, 258, 259
Reverse energy cascades, 151
Reynolds Averaged Navier-Stokes (RANS) equations, 151, 161
Reynolds averaging procedure, 160
Reynolds stress model, 111, 152, 162
Richardson corrected mixing length, 195
Richardson method, 26
Richardson number, 164
rms, 85, 86, 177, 186, 203
RNG, 111
Rotating coordinate system, 48
Runga-Kutta scheme, 27, 28, 34, 170, 226

**S**
Scalar transport, 224
Scalar variables, 163
Scale similarity model, 155
Scarborough criterion, 81, 83
Secant method, 91
Second moment closure, 157
Second order upwind scheme, 50
Segregated solver, 112
Semi-Lagrangian method, 56, 57
Separated flow, 218, 220
Sequential solvers, 105
Serial machines, 79
SHASTA, 57
Shear flows, 167
SIMPLE, 81, 84, 87, 88, 95, 96, 98-102, 104, 105, 112, 212, 230, 234
SIMPLE*, 98, 99
SIMPLE2, 98
SIMPLEC, 95, 96, 98, 112, 174, 185, 193, 203
SIMPLER, 98, 112

SIMPLEST, 112
Simultaneous equation solvers, 80, 104
Skew upwinding, 111
Sliding grids, 112, 207
Smagorinsky's constant, 155
Smoke flow visualisation, 224, 225
Smoothing, 31, 85, 86, 109, 272
Solution initialisation, 33, 176, 186, 213, 230, 275
Solution correction, 127
Source terms, 253, 255
Space conservation laws, 65, 67-70, 114
Spatial velocity variations, 216
Spectral methods, 62
Splitting method, 29, 80
Stability analysis, 62
Staggered grids or variables, 90, 107
Stagnation enthalpy, 249
Stagnation modification, 167, 230, 231
STAR-CD, 110, 111, 114
Stewartson layer, 176
Stiffness, 30, 32, 207, 208
Strange attractor, 4
Streamfunction vorticity method, 94
Streamline curvature, 229, 231
Streamline plots, 218
Streamline upwinding, 112
Strongly implicit procedure (SIP), 81
Sub-cycling, 109, 130, 131, 141, 235
Sub-grid scale model, 155
submatrix, 107
Substantial derivative, 46, 175, 190
Surface tension, 114
Swing, 126

**T**

TASC-flow, 110, 111, 114
Taylor series, 164
TDMA algorithm, 29, 81, 85, 88, 279, 299, 321
Temporal grid distortion, 204
Tetrahedral mesh, 153
Three-level schemes, 25
Time adaptive solutions, 74
Time-centred approach, 104, 105
Time adaptive scheme, 235
Time marching, 30
Toroidal vortex, 184
Total turbulence intensities, 215, 220
Total Variation Diminishing (TVD) concept, 52, 53
Transformations, 66
Transient flow, 158, 159, 229, 232
Turbulence averaging, 158
Turbulence energy cascade, 151
Turbulence intensity, 215, 221
Turbulent flow, 107, 268
Turbulent kinetic energy, 162
Turbulent transport equation, 162
Turbulent viscosity, 165
Two-level schemes, 17, 21, 101

**U**

Under-relaxation, 83, 84, 98, 109
Unstructured mesh, 153
URANS, 152, 157, 161, 193, 195, 212
URANS/VLES, 152, 157, 195, 196

**V**

Van Driest damping function, 163
Variable order, 118
Variable storage locations, 297
V cycle, 89
Vector machines, 79

VLES, 152, 195, 196
Viscous dissipation, 151
Volume Of Fluid (VOF) method, 114
von Karman constant, 155, 163
Vortex breakdown, 184, 185, 193-195, 218
Vortex methods, 33, 47
Vorticity equation, 94
Vorticity parameters, 168

**W**
Wall distance algorithms, 213
Wall functions, 165
Wavenumber, 180
Wiggles, 50

**X**
X cell technique, 112

**Z**
Zonal models, 152, 156, 166, 197, 212, 222, 229, 232

## ABOUT THE CD-ROM

The CD-ROM included with this book features computer programs that complement the material contained within the book.

The disk (CD-ROM) is distributed by Kluwer Academic Publishers with absolutely no support and no warranty from Kluwer Academic Publishers. Use or reproduction of the information on the disk for commercial gain is strictly prohibited. Kluwer Academic Publishers shall not be liable for damage in connection with, or arising out of, the furnishing, performance or use of the disk (CD-ROM).